W9-AQW-645

OK
9K11
.NS
H3
1970

THE VEGETATION
OF THE NEW JERSEY
PINE-BARRENS

FIGURE 1

Cedar swamp at Shamong, June 27, 1910, with an even stand of Chamaecyparis thyoides and a sphagnum basis out of which grow Rhus vernix, Acer rubrum, Clethra alnifolia, Myrica carolinensis, Vaccinium corymbosum. Note the fronds of Osmunda cinnamomea and the low herb, Trientalis americana. (Photograph by Henry Troth and J. W. Harshberger.)

THE VEGETATION
OF THE NEW JERSEY
PINE-BARRENS

AN ECOLOGIC INVESTIGATION

BY

JOHN W. HARSHBERGER, Ph.D.

DOVER PUBLICATIONS, INC.
NEW YORK

39389

Published in Canada by General Publishing Company, Ltd., 30 Lesmill Road, Don Mills, Toronto, Ontario.
Published in the United Kingdom by Constable and Company, Ltd., 10 Orange Street, London WC 2.

This Dover edition, first published in 1970, is an unabridged republication of the work originally published by Christopher Sower Company, Philadelphia, in 1916.
The foldout map attached to the inside back cover was reproduced in full color in the original edition.

Standard Book Number: 486-22537-2
Library of Congress Catalog Card Number: 78-106485

Manufactured in the United States of America
Dover Publications, Inc.
180 Varick Street
New York, N.Y. 10014

FOREWORD

The following pages will be devoted to a description of the pine-barren vegetation of the coastal plain of New Jersey from the phytogeographic and the ecologic aspects. The bulk of the monograph will present the observations and the research work which have been in progress for a period of at least twenty-five years, although the greater part of the material has been collected during the last ten years. Besides the original material, which comprises the larger part of the monograph, emphasis will be laid upon what has been published on the subject as it elucidates, or illumines, the problems which have been kept in view during the prosecution of this botanic study. The thought is to collate all the facts which are not purely statistic and systematic and which bear in any way upon the pine-barren region of New Jersey, or upon its interesting and characteristic vegetation. Full details of a general historic, statistic, floristic, and systematic character will be found in Witmer Stone's volume, issued in 1911 by the New Jersey State Museum, upon "The Plants of Southern New Jersey, with Especial Reference to the Flora of the Pine Barrens and the Geographic Distribution of the Species."* Many facts which the writer intended originally to incorporate in this monograph have appeared in the work of Stone, so that attention will be given chiefly to matters which concern the plant geographer, plant ecologist, and plant physiologist. As an almost complete bibliography of the flora is given by Stone, such a list will be omitted entirely, as will also facts which bear upon the distribution of plants in southern New Jersey as related to that of other parts of our country.

Although the material for this work has been gathered contemporaneously with that of Stone and our confrères in the Philadelphia Botanical Club, but entirely independent of them, it will be made to supplement that volume by including only that which has not been mentioned by Stone, or in a very casual and unemphatic way. The illustrations in the Annual Report of the New Jersey State Museum largely depict the species of plants which are characteristic of southern New Jersey and the pine-barrens, and a few that show actual conditions of vegetation.

---

* STONE, WITMER: The Plants of Southern New Jersey, with Especial Reference to the Flora of the Pine-Barrens and the Geographic Distribution of the Species. Annual Report of the New Jersey State Museum, 1910, pp. 25–828, with 129 plates; Trenton, 1911.

The illustrations of this monograph will show the natural vegetation of the region as it appeals to the pl... ecologist, landscape artist, and nature lover. The maps and other figures, therefore, will be of an entirely different character from those given in the bulky volume of the New Jersey State Museum.

The demand for such a work as this one following so closely upon that of the museum report has been emphasized by Harper* in his review of Stone's book, where he states that "with such a splendid floristic foundation to build on, the time is now ripe for some ecologically inclined botanist to make a detailed study of the vegetation of the same region, and thereby fill a long-felt want."

Such a study is pressing, because the original conditions of the country are fast changing. Forest fires have done great damage to the wild plants. The spread of areas of cultivation has destroyed many tracts of unusual botanic interest. The construction of dams, of railroad embankments and bridges across streams has wrought decided changes in the original pine-barren vegetation. The following account, therefore, will be a printed record, embellished with photographs and other illustrations, of the pine-barren vegetation in its original, wild condition, as far as that could be determined at the present time.

The author wishes to thank with deep appreciation the financial assistance given by the following persons, without whose aid the publication of the book would have been impossible.

American Philosophical Society.
Edward Bok.
Henry G. Bryant.
Herman Burgin.
Sarah C. DeHaven.
Frank R. Ford.
William H. Greene.
Joseph R. Grundy.
Charles C. Harrison.
Samuel Heilner.
A. J. Hemphill.
Alba B. Johnson.
Adolph W. Miller.

Henry D. Moore.
H. K. Mulford.
Clement B. Newbold.
Sarah Nicholson.
Harold Peirce.
James L. Pennypacker.
John T. Pennypacker.
Richard A. F. Penrose.
W. Hinckle Smith.
J. W. Sparks.
Thomas W. Synnott.
Joseph H. Taulane.
J. B. Van Sciver.

He desires to acknowledge the encouragement of Harold Peirce and James L. Pennypacker in the publication of the book, and Roland M. Harper and James L. Pennypacker, who gave the galley proofs a careful reading. Miss Elizabeth Caley, a graduate of the Pennsylvania School

* HARPER, ROLAND M.: Stone's Flora of Southern New Jersey. Torreya, 12: 216–225, September, 1912.

of Industrial Art, assisted in getting many of the drawings into final shape from pencil sketches furnished her by the author.

Through the interest of Alfred Gaskill, Chief of the Division of Forestry and Parks, Department of Conservation and Development of New Jersey, permission was given by the New Jersey Geological Survey to use its map of 1915 as a base map upon which the details of the pine-barren vegetation have been printed. This has been done as faithfully as the constant shifting of the relative positions of the wild and cultivated areas of the region has permitted, as also the changes produced by fires and the axes of lumbermen, especially in the removal of the valuable white cedar trees from the swamps of the southern part of the state. It is noteworthy that a white cedar swamp of a month ago may become a second growth deciduous swamp of the following year. Such changes are not always represented on a map based upon surveys made before such changes take place.

PHILADELPHIA, October 5, 1916.

TABLE OF CONTENTS

CHAPTER I

PAGES

PHYSIOGRAPHY OF THE REGION... 1–7

CHAPTER II

GEOGRAPHY OF THE PINE-BARREN REGION............................... 8–11
    Origin of the Designation "Pine-Barren," 8; Geographic Place Names, 9

CHAPTER III

INFLUENCE OF GEOGRAPHY AND VEGETATION ON THE PEOPLES AND THE INDUS-
    TRIES OF THE REGION... 12–30
    General Considerations, 12; Floral Districts, 15; Local Plant Names, 16;
    Industries of the Pine-Barrens, 17; Lumbering, 17; Turpentine Industry, 21;
    Cranberry Culture, 25; Sphagnum, 26; Peat, 26; Huckleberry Picking, 27;
    Drugs, 27; Greens and Flowers, 29

CHAPTER IV

RESEARCH INVESTIGATION OF PINE-BARREN SOILS......................... 31–47
    General, 31; Mechanic Analysis, 33; Mineralogic Analysis, 34; Chemic
    Analysis, 34; Experiments with Soils, 35; Capillary Rise of Water, 42; Re-
    tention of Water, 43; Wilting Coefficient, 45; Moisture Equivalent, 47

CHAPTER V

PHYTOGEOGRAPHIC FORMATIONS... 48–86
    Classification, 48; Pine-Barren Formation, 49; Architectural Forms of Pine
    Trees, 53; Facies of Pine-Barren Formation, 56; Method of Investigation,
    57; High Pine-Barrens, 60; Flat Pine-Barrens, 65; Low, or Wet Pine-Bar-
    rens, 75; Colors of Pine-Barren Vegetation (Winter Aspect, Spring Aspect,
    Summer Aspect, Autumn Aspect), 77; Coastal Pine-Barrens, 82; Contrast
    of German and New Jersey Formations, 85

CHAPTER VI

PINE-BARRENS TRANSITIONAL TO SEA-DUNE VEGETATION.................. 87–104
    Transition from Sea-Dunes to Pine Thicket at Sea Girt, 90; Bluff Forest
    along Great Egg Harbor Bay at Somers Point, 91; Tension Line Salt Marsh
    to Pine Forest at Somers Point, 94; Transition Salt Marsh to Pine Forest
    between Ocean Gate and Barnegat Pier, 95; Pine-Barrens at Northern
    Limit, 96; Transition Forest Western Pine-Barren Limit, 97; Notes on the
    Distribution of Transition Plants, 100; Successional Formations (Natural,
    Oak-Coppice Succession in Transition Region, Oak-Coppice Succession in
    Cape May Region—Mixed Pine-Oak Succession—Coastal Oak Forest Suc-
    cession—Oak-Bottoms), 100–104

CHAPTER VII

PINE-BARREN PLANTS AS WEEDS... 105–113
    Introduced Weeds, 105; Classification of Alien Plants, 106; Pine-Barrens of
    Long Island, 107; Tension Line between Pine Forest and Cedar Swamp, 109

ix

CHAPTER VIII   PAGES

CEDAR SWAMP FORMATION... 114–131
    General, 114; Synecology of Cedar Swamps, 120; Additional White-Cedar
    Swamp and Bog Plants, 126; Seasonal Aspects, 127; Fungous Diseases of
    the White Cedar, 129; Additional Fungi Found on White Cedar, 131

CHAPTER IX

TRANSITION PINE FOREST TO DECIDUOUS SWAMP......................... 132–142
    General, 132; Deciduous Swamp Formation, 133; Branch Swamps, 136;
    Deciduous Swamps of Transition Region, 136; Reed Marsh Formation, 138;
    Additional Swamp Plants, 138; Stream Bank Formation, 139

CHAPTER X

POND PLANT FORMATION... 143–150
    General Consideration, 143; Additional Plants, 144; Savanna Formation,
    145

CHAPTER XI

PLAIN FORMATION (COREMAL)... 151–171
    Geographic Location, 151; Synecology, 154; Measurement of Species, 159;
    Theories Concerning Coremal, 159; Experimental Treatment of Soils, 162;
    Contrast with Nantucket and European Conditions, 166; Hempstead Plains,
    Long Island, 170

CHAPTER XII

CULTIVATED PLANTS OF THE PINE-BARREN REGION....................... 172–174
    Forest Trees, 172; Fruit Trees, 172; Bush Fruits and Vines, 173; Small
    Fruits, 173; Root Crops, 173; Leaf Crops, 173; Fodder and Cereal Crops,
    173; Flower Crops, 173; Garden Crops, 174

CHAPTER XIII

GENERAL OBSERVATIONS ON PINE-BARREN VEGETATION.................... 175–184
    General, 175; Pine-Barren Vegetation by Quadrats, 176; Generic Coefficient,
    181; Families of Pine-Barren Plants, 182

CHAPTER XIV

BIOLOGIC TYPES OF PINE-BARREN PLANTS.............................. 185–192
    Raunkiaer's Classification, 185; Pine-Barren Formation (Trees, Shrubs,
    Undershrubs, Annual Herbs, Colorless Parasitic Plants, Ferns, Perennial
    Herbs, Evergreen Species, Deciduous-leaved Species), 186; Colors of Pine-
    Barren Flowers, 188; (do.) White-Cedar Swamp Formation, 190; (do.)
    Marshes, 192

CHAPTER XV

PHYTOPHENOLOGY OF PINE-BARREN VEGETATION.......................... 193–215
    Tables of Flowering and Fruiting Periods, 196–208

CHAPTER XVI

VEGETATIVE PROPAGATION AND STRUCTURE OF THE SHOOTS................ 216–244
    Hess Classification, 217; Detailed Root and Stem Studies, 219–244.

CHAPTER XVII

GENERAL REMARKS ON ROOT DISTRIBUTION.............................. 245–251

CHAPTER XVIII

PAGES

LEAF FORMS OF PINE-BARREN PLANTS.................................... 252–258
　　Classification, 253; General Remarks, 258

CHAPTER XIX

MICROSCOPIC LEAF STRUCTURE.. 259–285
　　General, 259; Classification, 260; Detailed Structure, 263; Synopsis, 283

CHAPTER XX

CONE AND SEED PRODUCTION OF THE PITCH-PINE.......................... 286–294
　　General Investigation, 286; Law of Cone and Seed Production, 292; Vivi-
　　pary in the Acorns of Quercus marylandica, 292

CHAPTER XXI

NOTES ON A FEW INSECT GALLS... 295–297

CHAPTER XXII

PINE-BARREN PLANTS FROM AN EVOLUTIONARY VIEWPOINT.................. 298–316
　　General, 298; Elementary Species, 299; Variation, 300; Mutations, 303;
　　Epharmony, 305; Plasticity of Species, 305; Response to Ecologic Factors
　　(Soil, Light, Wind, Water), 305; Convergent Epharmony, 308; Persistent
　　Juvenile Forms, 312; Hybridization, 313; Conclusion, 313

INDEX... 317–329

THE PINE-BARREN VEGETATION OF NEW JERSEY

CHAPTER I

PHYSIOGRAPHY OF THE REGION

The physiographic history of the coastal plain of southern New Jersey, which includes that portion of the state south and southeast of the "fall line," begins with the depression of land beneath the surface of the sea during the Lafayette epoch of the Miocene (Neocene) known as the Lafayette depression, or Miocene sinking. The Lafayette deposits were later elevated along the Atlantic and the Gulf Coasts, constituting in New Jersey the Beacon Hill formation and the older portion of the coastal plain land surface east and south of the "fall line." The materials deposited and later elevated above the sea consisted of gravel, sand, silt, and clay related variously to one another. This Post-Miocene period of uplift was associated with active erosion. The level plain which was developed as a result of the erosion of the Lafayette formation (the Beacon Hill of New Jersey) has been called the Pre-Pensauken peneplain in New Jersey.* This marginal plain extended along the Atlantic coast and was tenanted by a flora of fairly uniform character from north to south, for it is a noteworthy fact that out of a total of 386 New Jersey pine-barren plants, 183, or 48 per cent., represent an element common to the New Jersey and the southern pine-barrens, while 153 species are wide-ranging. This ancient Miocene coastal flora consists essentially of American generic and specific types.

With the beginning of the Pleistocene, or early Glacial Period, the coastal plains of New Jersey, Maryland, and other Atlantic states were depressed in part and differently beneath the ocean in the submergence generally known as the Columbia, or Pensauken by New Jersey geologists. The Beacon Hill (Lafayette) formation, however, remained constantly above water during the Pensauken submergence, constituting an

---

* In the correlation of these formations I have consulted CHAMBERLAIN, T. C., and SALISBURY, R. D.: Geology, III: 450; MCGEE, W. J.: The Lafayette Formation. Twelfth Annual Report, U. S. Geological Survey, 353–521, 1892.

island tenanted by the ancient vegetation of the Miocene coastal plain. The vegetation of this New Jersey Pensauken island had reached already a climax condition when it was isolated by the Pensauken submergence. Taylor* in a recent paper has elaborated on the theory which the writer first presented in detail on page 220 of his Phytogeographic Survey of North America, where it is stated that "during the Pensauken submergence of the New Jersey geologists New Jersey was depressed to such an extent as to drown the Delaware River at its lower end, allowing the sea to pass up its valley and over the peneplain which had been developed during the previous cycle of erosion, so that a broad sound was formed which connected Raritan and Delaware bays, *forming an island covered perhaps with pine-barren vegetation.* The mouth of the Delaware River during the Post-Pensauken uplift was transferred to Delaware Bay, followed by a cycle of erosion. It was during the Post-Pensauken uplift that the flora of the lower Delaware Valley and the coastal strip was probably developed, *so that the New Jersey pine-barrens* became surrounded by a fringe of vegetation developed along similar lines in the coastal strip and along the east and west banks of the Delaware River." This isolated island of pine-barren plants was removed still further from contact with the southern pine-barrens by the unequal depression of the coastal plain, so that with the exception of the island vegetation, the typic coastal plain plants were exterminated in the depressed portion of the plain in Delaware and Maryland during the Pensauken (the Wicomico of Maryland) submergence, for Shreve† has indicated the almost utter lack of pine-barren plants in Maryland. Thus, if we contrast the area occupied by the Beacon Hill formation and the area in central New Jersey covered by the pine-barrens, we find an unusual coincidence (Map, Fig. 2). This coincidence of the two areas is made clear when we remember that the Beacon Hill formation was an island during the Pensauken submergence and that it was covered with vegetation of the pine-barren type. The location of the present pine-barren area is due to this fact, to the present soil conditions, and to the fact that when the land surface was again elevated after the Pensauken submergence the deciduous types of vegetation invaded the recently elevated land in the coastal plains of Delaware, Maryland, and New Jersey, so that the pine-barren vegetation was surrounded on all sides by deciduous forests. The deciduous forests, on the other hand, were unable to supplant the pine-barren

---

* TAYLOR, NORMAN: The Origin and Present Distribution of the Pine-Barrens of New Jersey. Torreya, 12: 235, October, 1912; Flora of the Vicinity of New York. Memoirs of the New York Botanical Garden, v: 8–13, with suggestive maps, 1915.

† SHREVE, FORREST: Maryland Weather Service, III: 87 (1910).

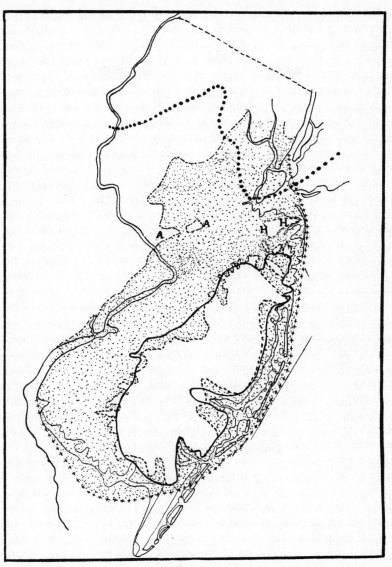

FIGURE 2

Map showing geologic history of pine-barren region

+ + + Outer limit of Pre-Pensauken Peneplain.
— . — . Inner limit (First Mountain) of Pensauken submergence.   Dotted area under
          water.
- - - -. Boundary of Beacon Hill area.   Not under water since the Miocene.
————   Present boundary of pine-barrens.   Note the overlapping of the Beacon
          Hill area and the pine-barren region.
. . . .   Great terminal moraine.   Southern limit of glacial ice.
A A. Rocky Hill.
H H. Atlantic Highlands.

vegetation, because the latter had reached a climax condition.   Both types of vegetation in the New Jersey coastal plain were mutually exclusive.   While the above outline was prepared before the appearance of Taylor's paper, yet in the elaboration of the theory by Taylor many important points are emphasized.   To quote: "For the phytogeographer the salient features of these changes are that the Beacon Hill has been uninterruptedly out of water since upper Miocene times, and that it has several times been partly and often entirely surrounded by water. These facts, together with the encroachment of the glacier, and its recession, with the probable deposition of a great deal of morainic material around Beacon Hill, make this formation the oldest in New Jersey, either on the coastal plain or in the glaciated regions northward, that could have been continuously covered with vegetation.   This, it seems to me, is why the Beacon Hill formation is the controlling factor in the origin and present distribution of the pine-barrens.   The area of the pine-barrens (see Fig. 2) is not exactly coextensive with Beacon Hill, but the differences are so slight that recent and local erosion of the formation would account for the failure of the two regions to superimpose, as it were."

"In other words, the New Jersey pine-barrens exist exclusively on this Beacon Hill formation, an area isolated by geological processes, and maintaining a relict or climax flora, the antiquity of which greatly antedates any of the rest of our vegetation hereabouts, so far as permanency of position and phytogeographical isolation are concerned.   This undoubtedly accounts for the composition of the flora, and it is interesting to note that zoölogists have found this same apparent isolation, the same endemism noted above."   All of the facts presented above seem incontestably to point to a geologic explanation of the origin and present distribution of the pine-barren vegetation.*   An interesting light is thrown upon the origin of the pine-barren vegetation on an island land mass by contrasting the vegetation of the Island of Nantucket, where a low heath vegetation, representing many of the species in the undergrowth of the New Jersey pine-barrens, is being invaded by the pitch-pine, so that some areas on Nantucket have become like the New Jersey pine-barrens in physiognomy and in floral constitution, illustrating the conversion of a treeless, wind-swept heath into a pine-heath, or Kiefernheide.†

* The writer thinks it unfortunate that Taylor should make only a partial quotation of the subject matter found on pages 218 and 219 of the Phytogeography of North America, ignoring the views presented at the top of page 220 of the same book.   The whole context should have been considered.

† HARSHBERGER, J. W.: The Vegetation of Nantucket.   Bulletin Geographical Society of Philadelphia, XII : 24–33, Apr., 1914.   Also consult the section of this monograph on the vegetation of the New Jersey plains, where additional evidence is presented.

The glaciation of the northern part of North America had very little influence on the pine-barren region, for we find no evidence of the deposition of glacial material, or of glacial terraces, but the surrounding region was not so fortunate, for the Post-Pensauken uplift inaugurated another period of erosion which was followed by a depression which allowed the deposit of materials that constitute the Cape May (Talbot of Maryland) subdivision of the Columbia. This effected only a small area of the state in early post-Glacial times.

If we consider that during the uplift in post-Glacial times the coastal plain extended northeastward, tying together Long Island, Block Island, Martha's Vineyard, Nantucket, Cape Cod, and other New England islands as far as Newfoundland, we have a ready explanation of the presence of many coastal plain plants in all of these islands, and in Nova Scotia and in Newfoundland. There are 116 species of the New Jersey coastal plain that range from Florida to Massachusetts.* Fernald† has shown that there is a southern coastal element in the flora of Newfoundland, especially in the sandy areas, or in the bogs on the carboniferous sandstones. These plants are of Nova Scotia, Sable Island, southern New England, Cape Cod, Long Island, New Jersey, and even farther south, such as: Panicum implicatum, Scirpus subterminalis, Carex silicea, C. hormathodes, C. sterilis, C. trisperma var. Billingsii, Eriocaulon septangulare, Juncus pelocarpus, Habenaria blephariglottis, Spergularia rubra, Arenaria peploides var. robusta, Pyrus arbutifolia var. atropurpurea, Elatine americana, Hudsonia ericoides, Myriophyllum tenellum, Vaccinium macrocarpon, Gaylussacia dumosa var. Bigelowiana, Utricularia clandestina, Solidago uniligulata, Aster radula, A. nemoralis, Potamogeton Oakesianus, Rhynchospora fusca, Schizaea pusilla, and Corema Conradii. This class of southwestern Carolinian types includes 60 species, or 7⅔ per cent. of the Newfoundland flora.

The distribution of Corema Conradii and Schizaea pusilla has long puzzled botanic geographers, but if we consider that in post-Glacial times the coastal plain continued far to the northeastward, in all probability these two plants, as well as others, had a continuous distribution from south to north. In general, the areas covered by the ice of the Glacial Period have risen since the ice melted, but the New England coast has been depressed, unequally separating the coastal plain into a number of islands. Now the distribution of Corema Conradii tallies with this formation of islands by depression of the coast, for it occurs on

---

* Stone, Witmer: Loc. cit.

† Fernald, M. L.: A Botanical Expedition to Newfoundland and Southern Labrador. Rhodora, 13: 135, July, 1911.

the plains of the pine-barren "island" of New Jersey, on Long Island (on authority of Dr. Torrey), on Nantucket (abundant in 1915), at Plymouth, Mass. (Tuckermann), Truro, Cape Cod (Watson), Bath, Maine (Gambell), Isle au Haut, Maine (Young), Kimball's Island, Maine, Mt. Desert, Maine, in Nova Scotia and Newfoundland.*

Presenting these facts in sequence we have: (1) The wide dispersal of coastal plain plants on the Lafayette deposits during the Post-Miocene; (2) the isolation of the pine-barren vegetation on Pensauken Island during the Pensauken submergence which destroyed the coastal plain vegetation of Delaware and the eastern shore of Maryland in late Glacial time; (3) the introduction of the deciduous type of vegetation into the middle district of New Jersey, into Delaware and the eastern shore of Maryland during the Pensauken uplift of Interglacial time; (4) the spread of coastal plain plants northeastward to Nova Scotia and Newfoundland during the post-Glacial uplift; (5) the isolation and separation of coastal plain species, such as Corema Conradii and Schizaea pusilla, by the unequal depression of the New England coast with the formation of islands upon which many of these species have been preserved, while in the depressed areas they have suffered extinction. Summarized in the form of a table, we have:

POST-MIOCENE TIME — Wide dispersal of coastal plain plants.

EARLY GLACIAL TIME { The isolation of coastal plain plants on Pensauken Island.

INTER-GLACIAL TIME { The introduction of deciduous vegetation into the coastal plain of New Jersey.

POST-GLACIAL TIME { A. The spread of coastal plain plants northward with the extension of the land areas.
B. The isolation of many such plants on coastal islands with later submergence.

The coastal plain of New Jersey, as it stands to-day, does not in its highest point reach an altitude of 121 meters (400 feet) and it descends to sea-level about the borders of the state. It is a relatively flat plain with slight relief, but of a rolling character, with hills here and there, and shallow valleys occupied by the master streams of the country. The ocean-directed streams are longer and more frequently branched than the shorter Delaware-directed streams. The pine-barren region is traversed almost entirely by the rivulets, creeks and rivers connected with the Atlantic Ocean drainage system. Such are the Manasquan,

* REDFIELD, JOHN H.: Corema Conradii and its Localities. Bulletin Torrey Botanical Club, XI: 97–100; see also Plate 90, Figs. 9 and 10, of Fernald's paper.

Toms, Wading, Mullica and Great Egg Harbor rivers. (folded map, Plate I.) The physiography of the country, as reflected in the areas of lowland and highland controlled by the streams, has important influence on the location of the important plant formations of the region. The salient features of this relationship will be emphasized in subsequent pages.

CHAPTER II

GEOGRAPHY OF THE PINE-BARREN REGION

The northern edge of the pine-barren region of coastal New Jersey is in approximately 40° 15′ N. latitude. The southern edge is found at 39° 10′ N. latitude. The extreme western border is at Maurice River, at latitude 75° 5′ W., while the eastern margin is approximately defined by the extensive salt marshes along the bays and channels shut off from the sea by barrier beaches (folded map, Plate I). In a northeast-southwest direction the region is about 121 kilometers (75 miles) long and from west to east in its broadest portion approximately 57 kilometers (35 miles). The coastal plain of this region is rolling with ridges, valleys and slopes, but notwithstanding that fact, when one views the country from an elevation he discovers its essentially flat character. So level is it in places that the greater height of the cedar trees in the swamps impresses the uninitiated as a ridge or hill covered with coniferous trees. The country is still in many places an unbroken wilderness. The drainage area of the Wading River is one of the wildest, most desolate portions of the eastern United States, and the writer has traveled a distance of 16 kilometers (10 miles) without seeing a house or a human being. Throughout the soil is a silicious sand underlaid with red colored gravel deposits. The fall of the eastward-flowing streams is quite uniform. Above the head of the tide the average fall of the Manasquan is 1.6 meters (5½ feet) per mile; Toms River, 1.9 meters (6½ feet) per mile; Cedar Creek falls 2.1 meters (7 feet) for 11.2 kilometers (7 miles); Mullica River, 1.5 meters (5 feet) for 26.7 kilometers (16 miles), and Great Egg Harbor River, 1.5 meters (5 feet) for 40.2 kilometers (25 miles).* The water of these streams is a pale coffee color and is known as cedar water. Its color is due to organic matter leached out of the soil and the peaty deposits of the swamps.

Origin of the Designation "Pine-Barrens"

With the general physiography and the general distribution of the vegetation of coastal New Jersey in mind where the forest of pine trees oc-

* VERMEULE, C. CLARKSON: Physical Description of New Jersey. Final Report 1, Geological Survey of New Jersey, 1888: 175.

8

cupies the central portion of the state and the deciduous vegetation the richer alluvial soils of the streams which flow into the Delaware River, as well as the richer interstream areas, we have an explanation of the designation "barrens" applied to the region of the pitch-pine forest. When the country was discovered first, before clearings and roads were made in the primeval forest, the early settlements were made along the larger streams, notably the Delaware, where the soil was rich and productive. After all the available land along the larger rivers had been patented, the early settler utilized the unsettled wilderness to the eastward. As the population became denser it encroached more and more upon the region covered with the forest of pitch-pines, and as the soils of this region were thinner and less productive, the region was looked upon as barren. Soon the name "barrens" was used by the early immigrants, and this name was coupled with pine, giving the well-known epithet of "pine-barrens." The word barren is used elsewhere as a descriptive geographic term. There are serpentine barrens in southeast Pennsylvania, where the unproductive outcrops of serpentine rock are distinguished by the presence of plants peculiar to that type of soil. Barren Hill, near Valley Forge, Pennsylvania, was occupied by the Continental soldiers on May 18, 1778, in order to oppose the advance of the British. The barren grounds of Canada represent the flat arctic country known as the tundra. All of these names are significant of the relative poorness of the land. Similar areas of pineland exist south of Virginia along the southern Atlantic and Gulf coastal plains and these have been named pine-barrens. But the plant geographer soon distinguishes several kinds of pine-barrens, which may be classed as follows:

PINE-BARRENS*

Pitch-Pine (Pinus rigida), Cape Cod, Long Island and New Jersey.

Loblolly-Pine (Pinus taeda), Virginia to Florida and Texas.

Long-leaf Pine (Pinus palustris), southeastern Virginia to Florida and Texas.

Slash-Pine (Pinus caribaea), Florida to South Carolina, westward to Mississippi, Bahamas, and Cuba.

Yellow Pine (Pinus echinata), Arkansas and northeastern Texas.

GEOGRAPHIC PLACE NAMES

Much has been made of the study of place names as indicative of the early settlement of countries by men of different nationalities and races.

* For the characterization of the different kinds of pine-barrens consult HARSHBERGER, J. W.: A Phytogeographic Survey of North America, 1911, vol. XIII, Die Vegetation der Erde.

When all trace of these peoples has been lost in a neighborhood, the names given to rivers, lakes, hills, and towns still attest to the former occupancy of the country by the people who bestowed the name as a geographic landmark. In this country considerable study has been made of the Indian names of places,* but the significance of the names as given by the early settlers has been touched barely by American philologists. Thus in England the name of the City of Chester is indicative of the location of a Roman camp, *castra*. In Pennsylvania, the location of the early Welsh towns is shown by such place names as Bryn Mawr, Cynwyd, Gwynedd, Radnor, Ithan (a creek), and the like.

Selections of the characteristic place names in the pine-barren region of New Jersey which are indicative of some local peculiarity and with special reference to the flora are here given. The white-cedar tree (Chamaecyparis thyoides) has been used in designating a considerable number of geographic features of the region, as: Cedar Bridge, Cedar Brook, Cedar Grove, Cedar Lake, and Cedar Run. Next to the cedar the white-oak has been used, as White Oak Bottom, White Oak Branch. Then a few well-known trees and shrubs have been chosen for place names, as: Chinquapin Branch, Gum Spring, Huckleberry Hill, Magnolia, Maple Root Branch, Pine Creek and Red-Oak Grove, while general impressions of the vegetation are expressed in such names as: Grassy Pond, Great Swamp, Green Branch, Wood Swamp. Names which have been applied during logging operations are Log Swamp Branch, Pole Branch, Rail Branch, Slab Causeway Branch and Tarkiln Branch. Indian names, although relatively little used, are found to some extent on the maps of the region. Such are Indian Jack, Manahawkin, Manopaqua Branch, Manumuskin, Mecheseatauxin Creek, Muskingum Brook, Nesochoque Branch, Papoose Branch, Shamong (Chemung = big horn), Tulpehocken Creek, Tuckahoe, Westecunk River. The following place names commemorate well-known animals. Those of exotic introduction are noted by italics: Bass River, Bears Head, Bear Swamp, Beaver Branch, Beaverville, *Bull Branch*, *Bulltown*, *Chicken Bone*, Fox Chase, *Goose Pond*, *Mare Run*, Panther Branch, Turkey Foot, Wild Cat and Wolf Run.

Names of places that express physiographic conditions are: Deep Run, Forked River, Great Swamp, Goodwater Run, Sandy Ridge, Wading River. A number of geographic features are indicated by well-known articles of human use, as: Apple Pie Hill, Bread and Cheese Run, Breeches Branch, Cabin Branch, Calico, Featherbed Brook, One Hun-

---

* BEAUCHAMP, WILLIAM M.: Aboriginal Place Names of New York. New York State Museum. Bull. 108, 1907.

dred Dollar Bridge, Tub Mill, and some have a personal flavor, viz., Comical Corner, Double Trouble, Friendship Bog, Good Luck, Hospitality Branch, Jake Branch, Jemima Mountain, Mary Ann Furnace, Mount Misery, Pleasant Mills, Stop-the-Jade Run.   The following few places have been designated by terms expressive of distance or dimensions: Five-Acre Pond, Four-Mile Run, Old Halfway, Ten-Mile Hollow.   The origin of such names as Aserdaten, Atco, Atsion, Batsto, is unknown to the writer, but in the miscellaneous assortment the following are noteworthy: Burnt Bridge Branch, Head of Snag, Hurricane Brook, Landing Creek, Long Cripple, Money Island, New Freedom, Oriental, Pole Bridge Branch, Stormy Hill and Tabernacle,   Some of the geographic names in the pine-barrens are names of persons compounded with forge, mill, town, ville, as: Bennett Mill, Buddtown, Hance Bridge, Hesstown, Monroe Forge, Smithville and Webb Mill.

How much more applicable and appropriate the names mentioned above are than the following examples of exotic ones that have been transplanted to the pine-barren region: Chatsworth, Chesilhurst, Lakehurst, Penbryn.   Railroad officials, county officers and the post office authorities should be very careful not to replace an old, well-established and expressive name which has a local flavor for a "high-falutin" term of no local significance.*

* GIFFORD, JOHN: Forests of New Jersey, 241, footnote.

CHAPTER III

INFLUENCE OF GEOGRAPHY AND VEGETATION ON THE PEOPLES AND THE INDUSTRIES OF THE REGION

GENERAL CONSIDERATIONS

The geography of any country modifies very profoundly the settlement, degree of civilization and industries of that country. Nowhere is that principle illustrated more clearly than in the pine-barren region of New Jersey. Until the opening of the first railroad through the region, travel was by horseback, wagon, or coach. The roads were cut through the wilderness and were few and far between. The inhabitants of the pine forest were isolated from their neighbors and by long isolation they have developed racial peculiarities of dress, speech and thought which are today characteristic. At present from an elevated hill we see an unbroken extent of dark-green pine forest, as far as the limit of vision, stretching away in long gentle swells. Much of the region is an unbroken wilderness save for the damage that has been wrought by forest fires and abandoned clearings scattered through the forest. Writing in 1885, Vermeule* says: "From Manchester southward to the Mullica River is one of the wildest, most desolate portions of the State. If we except the clearings on the shore road and along the marl border, not more than 2 per cent of the area is under cultivation. Here and there are narrow roads, barely wide enough for a single vehicle to pass clear of the trees, which thread their lonely way from clearing to clearing. They are relics of a time when the manufacture of iron from bog ore found in the swamps was an important industry of the region. Here and there one comes upon abandoned forge sites, or still more suggestive, abandoned villages, the relics of unsuccessful glass manufacture in the wilderness. An indescribable silence prevails. The soughing of the wind through the pines oppresses, while the crowing of a cock or the barking of a dog indicates the approach to a clearing and human habitation."

If the above quotation describes the condition of the pine-barrens in 1885, what must have been the character of this wilderness when the country was settled? So inaccessible and difficult was the travel across

* VERMEULE, C. CLARKSON: Final Report of the State Geologist, 1885, vol. 1: 177.

the continuous pine forest and so little was to be gained by settlement that the cultivable land was utilized very slowly by the early white settlers. Even the Indians of South or West Jersey rarely penetrated these pine-barrens except for hunting or fishing, for at the arrival of the white men we find the tribes of the Lenni-Lenape stock inhabiting the Delaware River Valley along both the east and the west banks. The most southern tribes, in what are now Salem and Cumberland counties, New Jersey, belonged to the Turkey Totem, while those opposite Philadelphia and extending to Trenton belonged to the Tortoise Totem. No tribes lived on the Atlantic coast of New Jersey, according to early authorities, and the shell heaps along the shore were formed by the Indians who now and then crossed the pine-barrens to enjoy the sea-fishing and the gathering of the shell-fish. The topography of the country influenced the location of the early settlements which were made along the Atlantic coast within easy access of the ocean by the coastal rivers and bays by light draft boats, and along the Delaware River, all points of which were easily accessible by vessels of greatest draft at that early day of sailing vessels. The principal settlements made in West Jersey before the country became a royal colony in 1703, and while in possession of Holland and Sweden (1609–1664); of Berkeley and Carteret Proprietary (1664–1673); of Province (east and west division, 1673–1703) were, in the order of the date of their foundation: Shrewsbury, 1664; Barnegat, 1668; Nevesink, 1670; Burlington, 1674; Manasquan, 1684; Mount Holly, 1685; Cape May Town (Portsmouth, Town Bank), 1686; Rancocas, 1686; Egg Harbor (Tuckerton), 1690; Bridgeton (Cohansey Bridge), 1690; Somerset Plantation (Somers Point), 1695; Fairfield, 1698; Cohansey (Cæsarea River), 1698; Haddonfield, 1701.

This vast, irregular-shaped tract of land, the "Pines," stood as a barrier between the early settlements on the Delaware and those along the sea-coast, where fishing and whaling were important industries.* Roads designed to connect these early settlements were early constructed across the pine-barren wilderness, where abounded red deer, bears, wolves, panthers, wild-cats, foxes, rabbits, opossums, pole-cats, hedgehogs, wild turkeys, pheasants, grouse and quails. Wolves were so abundant that a bounty of ten shillings was offered in 1682 for every head of that animal killed in West Jersey.†

Earlier than these roads several Indian trails‡ led through the pine-

---

* HARSHBERGER, J. W.: The Hardy Fishermen of New Jersey. Old Penn, XI: 773, March 25, 1913, with 7 figures.

† LEE, FRANCIS BAZLEY: New Jersey as a Colony and as a State, I: 286.

‡ Proceedings of the Surveyors' Association of West Jersey: 408.

barrens.   One started from Somers Point and extended along the east side of Great Egg Harbor River, so as to strike north of the heads of the several branches of Babcock Creek; and over the lowlands of that branch as it approached the tributaries of Little Egg Harbor, called the "Locks," by Blue Anchor and crossing the head of the Great Egg Harbor River at Long-a-coming (Berlin), passed a short distance south of Haddonfield, striking the Delaware at Coopers Point.   The second Indian trail came from the mouth of Little Egg Harbor River in a westerly direction and joined the first-mentioned trail near the head of Landing Creek, one of the tributaries of the last-named stream (folded map). The third began near Mullica's plantation near Batsto and went in a westerly direction between the streams to the first-named trail at the old Beebe place, about 1.6 kilometers (1 mile) south of Winslow.   The fourth trail, later called the Old Cape Road, started in Cape May County.   It crossed the head of Tuckahoe River in a northerly direction and to the west of the branches of the Great Egg Harbor River to the upper waters of Hospitality Creek at Cole's Mills, thence to Innskeep Ford, and joined the first-named trail at Blue Anchor, which was a central point.   These Indian trails were used by the white settlers between the Delaware River and the coast until they were converted into roads suitable for wagon and coach travel, while other highways were surveyed and built through the pine-barrens, which in the eyes of the proprietors of West Jersey were of comparatively little value.   By 1700 the pine-barrens had been crossed by surveyors and much of the timber had been purchased from the proprietors.*

The roads which ran through the pine forest so as to reach the timber grants, as well as the settlements on the east coast, were generally sandy, and during the dry season the tires of the wheels pulled heavily through the loose surface.   One of the most important of the later roads connected with the shore road at Barnegat.   It ran northwestward through what are now Cedar Grove, Woodmansie, Ong's Hat, South Pemberton, Birmingham, Mount Holly, and reached the Delaware at Burlington.

These highways opened up the country to the earlier English settlers whose names, as derived from ancient land grants and from their descendants, with whom the writer has come in contact, were Bartlett, Brown, Crane, Cranmer, Gaskill, Haven, Jennings, Moore, Seaman, Shaw, Southwick, Truax, and Wainwright, while at Barnegat we find the family names Arnold, Birdsall, Bird, Collins, Inman, Loper and Mott.†

* LEE, FRANCIS BAZLEY: New Jersey as a Colony and as a State, 1: 283.

† BLACKMAN, LEAH: History of Little Egg Harbor Township, Burlington County, from its First Settlement to the Present Time, 1868.

Considerable space has been devoted to this historic sketch of the early settlement, because we have reason to believe that that settlement was influenced by the character of the country and in turn profoundly modified the original boundaries of the pine-barren region. As the land along the shore road was converted into farms at a very early date, perhaps prior to March, 1713–14, when, owing to the excessive destruction of the forest, an act was passed prohibiting the common practices of stealing timber, cedar, pine staves and poles and of boring for the extraction of turpentine, the natural succession of the forest growth must have been greatly altered. And these changes occurred not only in the coastal strip of country back of the salt marshes, but in Cape May County as well.

Floral Districts

Stone,* in his book previously cited, recognizes five distinct floral districts in southern New Jersey: (1) The West Jersey, or Middle District, which covers, according to his colored map, not only the Delaware Valley region south of Trenton, but also all the country below the fall-line and north of the pine-barrens which terminate at Long Branch†; (2) the Pine-Barrens; (3) the Coastal Strip; (4) the Cape May District south of the Great Cedar Swamp; (5) the Maritime District. Stone's classification of districts is somewhat empiric. Having considered the history of the settlement of South Jersey, we are in a position to understand that before the year 1725 serious inroads had been made upon the forests, especially along the coast and on the peninsula of Cape May, where a number of flourishing towns had been started. These towns were largely inhabited by whalers and fishermen, who found ready access to the sea from the mainland. Farms were cleared in the pine forests, which probably extended down to the salt marshes. As the whaling and other industries of the sea declined in the Cape May peninsula and along the New Jersey sea-coast, many of these farms were abandoned and the open fields were invaded by deciduous trees, quite distinct from the pine trees which originally covered the country. Such settlement undoubtedly disturbed the original conditions, for the writer, in a former publication,‡ has referred to the influence of cattle on the forest at Wildwood. Cattle for many years ran wild on this island, which was granted by Charles II of England to his brother James, Duke of York, on March 12,

---

* STONE, WITMER: The Plants of Southern New Jersey, 57.

† More correctly, Deal Lake.

‡ HARSHBERGER, J. W.: An Ecological Study of the New Jersey Strand Flora. Proceedings Academy of Natural Sciences, 1900: 650.

1664. These cattle must have roamed for two hundred years over the island until they were shot only a short time ago. Many of the peculiar growths at Wildwood (now completely destroyed) were due undoubtedly to the trampling down of the young growth and the browsing of animals upon the tender buds and shoots, so that this irregular method of pruning and training caused the trees to assume curious forms and produced natural graft unions. The Cape May District and the Coastal Strip which Stone assumes are natural districts are, if our observations are correct, of artificial production. They exist today as well-recognized phytogeographic areas, but judging from the fact that in several places today the pine forest touches the salt marshes, they were covered by a pine forest integrally connected with the pine-barrens inland. If this surmise is correct, then theoretically we can reduce Stone's floral districts to four by excluding the Coastal Strip. If all of the peculiar plants known to exist in Cape May County, as well as the components of the deciduous forest, have been introduced with that forest type since the settlement of the country, then the Cape May District as distinct from the Pine-Barren District ceases to exist, reducing Stone's districts to three. The Cape May vegetation gives a strong impression of a mixed character, which would follow the occupancy of abandoned fields by pioneer vegetation of the deciduous type. As a result of this analysis the floral districts of South Jersey are three: (1) The Middle District; (2) the Pine-Barrens; (3) the Maritime District, so that the red of Stone's map would on theoretic grounds be extended on the east to the green color of the salt marshes, and on the south to the tip of the Cape May peninsula. Considering the encroachment of cultivated areas upon the pine-barrens, the region in all probability extended farther west toward the Delaware River.

The influence of the geographic environment on the development of the character and civilization of the inhabitants of a region is determined best by a study of their outlook on nature and by a consideration of their industries. We are concerned in this monograph with these problems, as far as the vegetation of the region has developed the powers of observation and habits of the people, or has influenced daily life and the development of the industries for which the region is noted.

LOCAL PLANT NAMES

Local names survive longest, as a rule, in the process of speech evolution, and bear evidence of the cultural stage and inventive genius of a people in introducing new terms for new phases of nature and in describing new animals, plants, and geographic places (see ante). The following is of interest:* The fishermen along Little Egg Harbor River call the eel-grass (Zostera marina) tiresome-weed, for, as they pull against the current in going to and from the ocean, the grass-wrack entangles their oars and retards the movement of their boats. The people in some districts in the pine-barrens call the orchid, Cypripedium acaule, "whippoorwill-shoe," which is better than that of lady's-slipper. The common pitcher-plant, Sarracenia purpurea (side-saddle-flower, huntsman's-cup), is called locally in Cape May County "dumb-watches." The flower grows from between the rosette of pitchered leaves, and after a time the incurved petals fall, exposing the top of the broad, expanded stigma. The five sepals remain and resemble, to the imagination of the New Jerseyman, a watch-case, while the convex surface of the stigma is likened to the face of a watch. This vegetable timepiece, with no hands to point the hours of the day, without the constant tick, tick, tick, is dumb. The whole plant is denominated very significantly and ingeniously "dumb-watch," or simply "watch." Other examples of this outlook upon nature might be cited, but sufficient has been given to indicate that a person interested in folklore would find an interesting field for investigation among the simple men and women of the pine-barrens.

INDUSTRIES OF THE PINE-BARRENS†

With the exception of the mining of bog-iron ore associated with the construction of ore furnaces and the utilization of the finer clay deposits and glass sands, the industries of the region from the earliest times have been dependent on the products of the forest and of the bogs so characteristic of the country.

LUMBERING.—The saw-mills driven by water power heralded the advent of permanent settlement. The first saw-mills were erected in the period 1700–1725 (Figs. 3 and 4). An act was passed in 1743 laying duties upon logs, timber, planks, vessel supplies, staves and headings, except firewood, exported to any of his Majesty's colonies upon the continent of America.

The pitch-pine, Pinus rigida, is the tree which gives character to the pine-barrens of southern New Jersey. It grew originally from 12 to 32 meters (40 to 75 feet) tall in the primeval forest, but at the present day, owing to frequent forest fires, it is rare to find trees as tall as 15 meters

* HARSHBERGER, J. W.: Local Plant Names in New Jersey. Garden and Forest, v: 395, 1892.

† See a suggestive paper along entirely different lines, WHITBECK, R. H.: Geographical Influences in the Development of New Jersey. Journal of Geography vi: 177–182, 1907–08.

(50 feet).   The tree as usually found ranges in height from 6 to 12 meters (20 to 40 feet).   The trunk can be manufactured into coarse lumber, and in the smaller sizes is made into charcoal or cut into firewood.   It has never been a favorite tree, owing to the poor quality of the timber, but yet it supplied a want in the early days before the pineries of the southern states were exploited, especially as it was possible to cut large trees that yielded straight and fair-sized boards, joists, and scantling for the construction of log houses (Fig. 5).   One or two hundred years ago, when the best pitch-pine lumber could be obtained, it was employed principally for the sills and beams of buildings.   Structurally the wood is soft, light, coarse-grained, and durable, and although little used for construction at present, the larger sizes can be used for railroad ties.   The ease with which the pitch-pine can be raised from seed, and its rapid growth on sterile soils, make it of especial value in silvicultural operations, where it is desired to cover barren areas with a forest cover, as in the Island of Nantucket and the dune sands on Cape Cod.

FIGURE 3

Old colonial saw-mill erected in 1750 and in continuous use since on Darby Creek, near Manoa, Pa., taken May 2, 1914.   Similar saw-mills were in operation in New Jersey and were scattered through the pine-barrens, being driven by water power.

The early saw-mills (Figs. 3 and 4) were driven by water power derived from the streams which were dammed at favorable localities accessible to the forest, where the trees were cut by the axe.   As compared with modern logging operations in the west and south, the methods of cutting and handling the logs in the forest over one hundred years ago were of the simplest description.   Horse and ox teams sufficed to transport the logs to the saw-mill, and the largest sizes were handled easily by one or two men.   As the country is level and flat, no insuperable difficulties were presented in getting the logs out of the woods, especially if the lumbering was done in the winter, when the frozen ground was covered with snow.   Sleds drawn by ox teams could easily remove the logs to the saw-mills driven by an overshot or an undershot waterwheel. Some of these old saw-mills are still used and are of great interest to the historian of the lumber industry in America.   The early builders of mills in New Jersey discovered that pitch-pine was well suited for the buckets and spokes of the waterwheels, which worked partially under water, because as a wood it resisted decay. The heartwood was used for boat building and for ship pumps. In Pennsylvania pitch-pine logs are used as mine props, wharf and bridge piles. Pitch-pine floors* wear well, look well, and last a long time. A pitch-pine floor in Pike County, Pennsylvania, laid with boards 0.6 meter (2 feet) wide and 3.8 centimeters (1½ inches) thick, did service one hundred and sixty years and was still in such a good condition that the boards were used in flooring a new house.†

FIGURE 4

Interior of old colonial saw-mill erected in 1750 on Darby Creek, Pa. The band saw with up-and-down movement is driven by wheels and other parts constructed of wood. The log-carrier with partly sawn log is shown in the foreground. May 2, 1914.

The wood has been used successfully for door and window frames, ceiling and other interior finish, and although knots are frequently abundant, yet, if properly finished, they give a decorative appearance to the wood. It is used to advantage for wood-pulp, excelsior, barrel headings, crates, nail-kegs, beds for wagons, and the inside boards of cupboards, ice-chests, cabinets, tables, and desks. One of the largest uses for pitch-pine is in the manufacture of boxes and crates.‡ In 1909 the box-makers of Massachusetts used 600,000 feet, and Maryland, 615,000 feet, of pitch-pine. As cordwood for fuel it has sup-

FIGURE 5

Old colonial log school house at Speedwell, Woodland Township, Burlington County, showing style of old pine-barren architecture and construction. (Courtesy of Herbert N. Morse.)

---

* HALL, WILLIAM L., and MAXWELL, HU.: Uses of the Commercial Woods of the United States. II. Pines, U. S. Forest Service Bulletin, 99: 34, 1911.

† DEFEBAUGH, J. E.: History of the Lumber Industry in America, II: 563.

‡ HALL and MAXWELL: Loc. cit.

plied brick-kilns, potteries, bakeries, steam engines, and houses for many years.

Fat pine-knots, which were obtained from old rotten logs by knocking away the decayed wood, or which were found in the forest in an excellent state of preservation after the softer, less resinous wood had disappeared, were used as a substitute for candles and lamps in the early days. These knots, which were gathered as fuel for winter's use, were split into thin pieces for illuminating purposes. These were bound into bundles with hickory, or yellow-birch, withes and were used as torches in night travel through the woods. "The dwelling-rooms of the Swedish settlers were lighted partly with tallow-candles and partly with so-called *perstickor*, or *spingstickor*. These were splint-sticks of pine used very extensively in this country and in Finland and Sweden, being mentioned in the Kalevala. They were thin and flat and were made by splitting the pine knots into proper size. They were fastened into a stick-holder of iron with a clasp for holding the pine splint and a sharpened end to stick into the crevices between the logs or into the wall. The other type was an upright shaft of wood, a yard or more in height, which could be moved from place to place. The splint was attached with the free end pointing downward, which was lighted. As rapidly as these torches were burned down they were replaced by fresh ones. Two or three burning sticks lighted a room for ordinary purposes, but they produced much black smoke."* The spearing of fish at night was facilitated by the use of pitch-pine torches. Charcoal was manufactured also from the inferior grades of pine logs and it was used in blacksmith shops and iron forges quite extensively. Lamp-black was made from pitch-pine.

The cedar swamps yielded and still furnish a fine lumber obtained from the white-cedar tree (Fig. 1). The durable qualities of white-cedar, especially in contact with the soil, were early appreciated. Although the wood is soft and rather weak, yet on account of its non-resinous character and its straight grain, it is an ideal material for boat-building, for cooperage (vats, tanks, churns, piggins, firkins, tubs) and various kinds of woodenware. Gottlieb Mittelberger, a German wagon-builder, considered the white-cedar as one of the finest materials for the construction of church organ pipes. Shingles made of white-cedar last many years. There is a specimen of a split shingle in the wood collection at the University of Pennsylvania, taken by the writer from the side of a house on Long Island built probably for at least two hundred years. The wood has been used satisfactorily for interior finish (floors, doors,

* JOHNSON, AMANDUS: The Swedish Settlements of the Delaware, 1638–1664, vol. 1: 351; RETZIUS: Finland, p. 73 ff.

joists, frames, rafters), fence posts, telegraph poles, and railroad ties. Coincident with the lumbering operations in the pine forest, the cedar swamps were culled of their largest and best trees.   Even today, after two hundred years of such exploitation, fair-sized cedar logs are removed from the forest.   According to Hall and Maxwell,* the cutting of southern white-cedar began probably three hundred years ago and was in full blast in New Jersey two centuries ago.   John Lawson, nearly two hundred years since, mentioned its use in the Carolinas for "yards, topmast booms, and bowsprits for boats, and shingles and pails."   The drain on the swamp forests for white-cedar lumber was so great that Benjamin Franklin published an essay in Poor Richard's Almanack (1749) in which he advocated forestry methods, especially the planting of red-cedar to supply the country when the white-cedar and other woods should fail.   Peter Kalm foretold the inadequacy of the whitecedar swamps to meet the demands for its excellent timber.   But the supply is still far from exhausted.

The indestructibility of the cedar wood is shown by the almost perfect state of preservation of logs that have been buried naturally in the cedar bogs.   The peaty soil is soft and spongy, and in some of the older bogs the swamp bottom for many feet is a tangled mass of fallen logs between which is found peaty material of a more or less decayed character, consisting of the remains of leaves, twigs and moss.   In some bogs these buried logs have been raised by mining operations, the position of the logs having been determined by probing with an iron rod.   The location of the logs by the iron rod is followed by the digging of a trench so that the log can be floated to the surface.   Many valuable pieces of timber have been secured by these mining operations.

The water percolating through the cedar swamps emerges with a light brown or tea color, and has a faintly pungent cedar taste.   As this water has antiseptic properties, it was long ago valued by sea captains, who filled their casks with cedar-water before going on a long voyage.   Some one had discovered that cedar-water kept sweet and potable longer than ordinary river-water.

TURPENTINE INDUSTRY.—The oleoresin which exudes from pine trees is known as crude turpentine, and it is obtained by incisions made in the tree, which is said to be boxed.   The early box system of gathering turpentine was imperfect and wasteful.   Boxes about 10 centimeters (4 inches) deep and 20 centimeters (8 inches) wide, according to the size of the tree, were cut in the living trees about 30 centimeters (1 foot) from

* HALL, WILLIAM L., and MAXWELL, HU.: Uses of Commercial Woods of the United States.   I.   U. S. Forest Service, Bulletin 95: 13, 1911.

the ground.   The trunk was scarified by slanting cuts above the box by an axe especially constructed for the purpose (Fig. 6).   The chipping began in March and continued well into the summer.   The oleoresin, which exuded from the stem, followed the converging, slanting cuts into the box.   A laborer could cut one hundred boxes in a day (Fig. 7).   The gathering, or the dipping, of the crude turpentine began three or four weeks after the chipping.   The first gathering was known as the virgin dip, and subsequent collections were made through the season.   At the end of the flow there was an accumulation of dried turpentine called "scrape," because it was scraped off into receptacles carried through the forest for the purpose.   The crude oleoresin was then taken to the still (Fig. 8). The old method of cutting deep boxes (Fig. 7) proved very destructive to the tree and reduced the term of operation to three or four years.   Recently a more economic method of collecting the crude oleoresin has been adopted through the experiments conducted in the forests of the south by the U. S. Forest Service.*   It is known as the cup-and-gutter method, where small galvanized iron gutters are attached in a slanting direction at the base of the scarified surfaces and the exudation runs down into earthen cups attached by a nail at a point where the gutters converge.   The amount of crude resin collected is greater than by the old system.   It is cleaner and the crude turpentine is more expeditiously handled.   The old method, however, was the one used in New Jersey when Gabriel Thomas†

FIGURE 6

Primitive method of boxing pine trees to obtain crude turpentine as practised about 1700 in New Jersey. (Photograph (No. 40,162) taken near Ocilla, Georgia. Courtesy of U. S. Forest Service.)

* Consult A New Method of Turpentine Orcharding.   U. S. Bureau of Forestry, Circular No. 24; HERTY, CHARLES H.: A New Method of Turpentine Orcharding.   U. S. Bureau of Forestry Bulletin 40, 1903; HERTY, CHARLES H.: Practical Results of the Cup and Gutter System of Turpentining.   U. S. Bureau of Forestry Circular No. 34; DAVIS, CHARLES: Making the most from Pine Orchards. The Country Gentleman, Aug. 31, 1912.

† THOMAS, GABRIEL: An Historical and Geographical Account of the Province and County of Pennsylvania and West Jersey in America, 32.

wrote of Robert Stiles (d. 1713), who lived on a farm on Pensauken Creek in Chester Township, Burlington County, just over the line from Gloucester, that "the Trade in *Gloucester County* consists chiefly in *Pitch, Tar* and *Rosin;* the latter of which is made by Robert Styles, an excellent *Artist* in that sort of Work, for he delivers it as clear as any *Gum Arabick.*"

Crude turpentine* is distilled now in copper stills. Formerly the use of iron stills was universal and water was used in the distillation (Fig. 8). A fifteen-barrel still (barrel weighed 127 kilos— 280 pounds) was charged early in the morning. Gentle heat was applied first until the mass was in a liquid state, when a coarse wire skimmer was used to remove the chips, the bark, the leaves, and such other foreign substances as rose to the surface. Meanwhile the temperature rose until 316° F. was reached. All the accidental water (that contained in the crude oleoresin as it came from the forest) having been distilled off, a small stream of water was directed into the still, so that the heat did not rise above 316° F., the boiling-point of oil of turpentine. The oil of turpentine and water distilled over, and the mixture was caught in a wooden tub. The distiller

FIGURE 7

Primitive method of boxing pine trees to obtain crude turpentine, as practised about 1700 in New Jersey. (Photograph (No. 40,168) taken near Ocilla, Georgia. (Courtesy of U. S. Forest Service.)

tested the quality of the flow from time to time and the distillation was continued until the proportion of water coming over was nine of water to one of turpentine. At this stage the heat was withdrawn, the cap of the still was taken off, and the hot rosin was drawn off by a valvular cock at the side of the still near the bottom. The rosin was

* Consult HAWLEY, L. F.: Wood Turpentines. U. S. Forest Service Bulletin 105, Washington, 1913.

passed through a strainer before it reached the barrel to rid it of foreign substances.

In the early days, when turpentine was made in New Jersey, besides the turpentine and rosin, which were the main products, pitch was also obtained.   Rosin when subjected to dry distillation yields a number of products: resin-oil (85 per cent.); resin-spirit; water; a powerful anæsthetic gas with a residue of pitch* amounting to 1½ cwt. from a ton of rosin.

Shoemakers' wax was made by melting together the best pitch and tallow in a vessel over a fire.   The resulting mass was rolled into balls and was used to wax the linen thread used by cobblers—the so-called "wax ends" being formed.

FIGURE 8

Primitive form of turpentine still as used in New Jersey about 1700. (Photograph furnished by U. S. Forest Service (No. 15,018), taken at Clinton, Sampson County, North Carolina, by Worth.)

Tar (Pix liquida) was used by country people in the early history of this country as axle grease, and the wagon without the tar bucket and its tar paddle swinging from the rear axle was an oddity.   Tar was manufactured by the backwoodsmen of New Jersey by using stumps, roots, and other waste materials of the pitch-pine lumbermen.   The important features in tar distillation was the pile of wood which had to be slowly and imperfectly burned from the top downward, so that the tar would be driven successively out of the wood next underneath the burning point, running down and out at the bottom, where it was collected.   The stacks of pine refuse consisted of circular mounds of earth about 6 meters (20 feet) in diameter, hard-packed and clay-coated, surrounded by a ditch used to collect the tar as it was formed.   As the wood was piled upon the mound, the diameter of the pile gradually increased to about 8 meters (25 feet) at the top and some 3 to 4 meters (10 to 12 feet) high.   The top was then covered with pine needles, and the sides of the heap were protected against a free current of air and the top ignited.   The tar was barreled as fast as it ran away. Such a stock burned for a week or ten days and yielded 100 barrels or

* For properties of rosin-pitch consult HOPKINS, ALBERT A.: The Scientific American Cyclopedia of Receipts, Notes and Queries, twenty-second edition, 1903: 606.

more of tar, which was a thick, viscid, semifluid mass, heavier than water, and of a deep, blackish brown color, having an unpleasant empyreumatic odor, bitter taste and of a complex chemic composition.

Another slightly different method employed in the manufacture of tar was used during the Civil War at Tuckerton, New Jersey, where it was a profitable industry. A large basin was constructed from which a small trough, or channel, led into another receptacle, and the whole was carefully cemented with a composition of sand, loam, and clay. The pine knots were piled in the large basin in a cone-shaped stack and covered with soil, sod, etc., similar to the preparation of a charcoal pit, and then lighted at the top. The tar oozing out passed into the bottom of the basin, thence through the channel into the receiver, from which it was collected.*

CRANBERRY CULTURE.—An important industry of southern New Jersey which illustrates what the natives of the region have done in the domestication and cultivation of a native wild plant is the cultivation of the cranberry, Vaccinium macrocarpon,† for the fruit of which a large demand has developed in the United States, for the total product in 1903 was 344,986

FIGURE 9

Cranberry bog and drainage ditch at Browns Mills. (Photograph by Henry Troth.)

hectoliters (977,516 bushels). The soils for the successful culture of the cranberry must be of a peaty character, with the water level within a few inches of the surface of the soil. The supply of water should be sufficient and the plantation so provided with dykes as to allow of flooding the area with water to the depth of 46 centimeters (18 inches) to 61 centimeters (2 feet) from November to May, in localities where it is necessary to protect the plant from insects and from late spring frosts. In the preparation of the bogs, many of which exist in southern New Jersey (Fig. 9), all bushes and trees must be removed, together with the top layer of soil to the depth of from five to ten centimeters (2 to 4 inches). After turfing, the surface must be graded, dams

* HILLMAN, SARAH C.: The Pine Lands in their Flora and Fauna. Proceedings Delaware County Institute of Sciences, III: 114-123, April, 1908.

† HARSHBERGER, J. W.: Peat Bogs and Peat. Old Penn Weekly, April 10, 1909, page 8.

and sluice gates must be constructed, and ditches dug to facilitate drainage.   The surface must be sanded to a depth of 8 to 10 centimeters (3 to 4 inches), and the young plants must be set out as cuttings by dibbling the soil in rows about the first of June.   The only cultivation necessary is to keep down weeds and grasses, which are not troublesome where the bogs are flooded periodically.   Harvesting the fruit, formerly done by hand, is now performed in some places by a cranberry scoop-rake, the sorting being done in sorting racks in which the berries are separated into the different marketable sizes.   The sorting sheds are low, rambling affairs, located in temporary villages in the higher pine-land, or larger, more substantial buildings (Fig. 10).   As the diseases of the cranberry have increased of late, agents* of the United States Department of Agriculture have investigated them with a view to overcoming the fungi and insects which attack the plant.   Much good work has been accomplished.

FIGURE 10

The Birches. Cranberry sorting and storage house, with wagon loaded with cranberry barrels. Note martin houses on poles. Tabernacle Township, Burlington County, November 4, 1915. (Courtesy of Herbert N. Morse.)

USES OF SPHAGNUM MOSS.—There are many bogs in southern New Jersey where bog moss, or sphagnum, is gathered for use in packing plants for shipment by wagon and railroad, and is baled for shipment to gardeners and nurserymen.   It is used in the cultivation of orchids, ferns and pitcher-plants, which require a constant supply of water about their roots for their best development.   Sphagnum is used in the operations of grafting and propagation of such plants as the rubber plant (Ficus elastica), which will produce new roots whenever a fistful of bog moss is tied around the stem, which, after the adventitious roots have appeared, is severed below the ball of moss.

PEAT.—The sphagnum bogs of New Jersey yielded peat, but of an inferior grade.   Owing to the structure of the moss, the bog is saturated with water.   The upper extremities of the moss stems continue their growth from year to year, while the lower portions die away and are converted gradually into peat by the pressure of the wet moss above

---

* SHEAR, C. L.: Cranberry Diseases.   Bulletin 110, U. S. Bureau of Plant Industry, 1907.

and by slow physic and chemic changes.* Such peat consists of the remains of bog mosses, as well as the leaves and stems of other bog-inhabiting plants. Buried in this peat, as previously noted, are perfectly preserved logs of white-cedar trees, which have been removed for their wood after the lapse of centuries.

HUCKLEBERRY PICKING.—Beneath the pine trees and in the swamps of New Jersey grow a number of species of huckleberries (Gaylussacia) and blueberries (Vaccinium). When the fruit of these bushes is ripe, the crop of berries is gathered, boxed, and crated for shipment to our large cities. This gives employment to many men, boys, women, and girls, now largely recruited from the foreigners who have immigrated to America. The blue huckleberry, Gaylussacia frondosa, is frequent in dry areas of the pine-barrens. It and the black huckleberry, Gaylussacia baccata, form the bulk of the huckleberry crop. The dwarf blueberries, Vaccinium pennsylvanicum, V. vacillans, yield fruit of great value when ripe in late June and early July. The swamp blueberry, Vaccinium corymbosum, frequent as a tall bush in swampy thickets in the pine-barrens, yields the largest and best-flavored berry, while the other tall blueberry, Vaccinium atrococcum, has a black fruit almost equally desirable. Experiments conducted by Coville indicate the practicability of cultivating the tall swamp blueberry by choosing soils with an acid reaction, so that in a few years the cultivation of this plant will be put on a commercial basis.†

DRUG PLANTS.—A list of the chief drug plants of the pine-barrens is given below, arranged according to the organs, or parts, of the plant used and the medicinal properties of each. The large drug houses still depend upon wild plants for part of their raw material, and the inhabitants of the districts in which these crude drug products are found make a livelihood collecting and curing the native plants for the large drug manufacturers.

ROOTS

Spurge (Euphorbia ipecacuanhae), diaphoretic, cathartic, emetic.
Wild indigo (Baptisia tinctoria), stimulant, emetic, cathartic.
Pitcher-plant (Sarracenia purpurea), unofficial.
Goat's rue (Cracca (Tephrosia) virginiana), unofficial.
Butterfly-weed (Asclepias tuberosa), unofficial.

* HARSHBERGER, J. W.: Peat Bogs and Peat. Old Penn Weekly, April 10, 1909, page 8; Bogs, their Nature and Origin. The Plant World, 12: 34, February, 1909; PARMELEE, C. W., and McCOURT, W. E.: A Report on the Peat Deposits of Northern New Jersey, Report N. J. State Geologist, 1905: 223–313.

† COVILLE, FREDERICK V.: Experiments in Blueberry Culture. U. S. Bureau of Plant Industry, Bulletin 193: 1–100. Plates 18, figures 31, 1911. Also Nat'l Geogr. Mag., June, 1916.

Rhizomes

Sweet-flag (Acorus calamus).
Colic root (Aletris farinosa), tonic, emetic, purgative.
Water-lily (Castalia (Nymphaea) odorata), demulcent, astringent.
False sarsaparilla (Aralia nudicaulis), stimulant, diaphoretic, alterative.
Regal fern (Osmunda regalis), unofficial.

Bark

Swamp magnolia (Magnolia virginiana), diaphoretic, tonic, febrifuge.
White oak (Quercus alba), astringent.
Black oak (Quercus velutina), astringent.
Bayberry (Myrica carolinensis), acrid stimulant, sialagogue, errhine.
Sassafras (Sassafras variifolium), stimulant, diaphoretic, alterative.
Red maple (Acer rubrum), unofficial.
Holly (Ilex opaca), unofficial.
Alder (Alnus rugosa), unofficial.
Wild cherry (Prunus serotina), stomachic and bitter tonic.

Leaves

Rattlesnake-plantain (Peramium pubescens), antivenine.*
Bearberry (Arctostaphylos uva-ursi), astringent, tonic, diuretic, nephritic.
Trailing arbutus (Epigaea repens), astringent, tonic.
Laurel (Kalmia latifolia), astringent and poisonous.
Holly (Ilex opaca), demulcent, tonic, emetic.
Pipsissewa (Chimaphila umbellata), astringent, tonic, diuretic, nephritic.
Wintergreen (Gaultheria procumbens), stimulant, astringent, diuretic, emmenagogue.
Sweet-fern (Comptonia asplenifolia), stimulant and astringent.
Poison-ivy (Rhus radicans), irritant, rubefacient.

Herbs

Sundew (Drosera rotundifolia), pectoral, rubefacient.
Golden-rod (Solidago odora), stimulant, carminative, diaphoretic.
Horse-mint (Monarda punctata), carminative, stimulant, nervine, emmenagogue.

Balsams and Oleoresins

Pitch-pine (Pinus rigida) (turpentine), stimulant, diuretic, diaphoretic, astringent; (tar) stimulant, irritant, insecticide.

---

* According to Captain John Carver, the Indians considered the juice of the leaves of the rattlesnake-plantain an absolute cure for rattlesnake poison. Applied externally and also swallowed, he says they were so sure of this fact that for a shilling they would let a rattlesnake bite them. This fact was contributed by Mr. J. L. Pennypacker, of Haddonfield.

During the two hundred and fifty years in which the pine-barren region of New Jersey has been known to white men, a considerable number of these drug plants have been gathered and utilized either as home remedies or else for shipment to Europe and to the larger American cities (notably Philadelphia), where they have been manufactured into useful drugs.* The revenue obtained by their sale has been a considerable asset to the herb gatherers of the New Jersey pine-barrens.

GATHERING OF GREENS AND FLOWERS.—The income derived from the sale of greens for decoration, especially at Christmas time, and of wild flowers for bouquets, has been considerable to the people of the New Jersey pine forest. The collection of these materials has been so careless and wasteful in past years that the supply of some of them is exhausted almost and all persons interested in the conservation of our vegetation and our natural resources should cry a halt. The following plants are those which have been gathered, according to the knowledge of the writer:†

GREENS

Ground-pine (Lycopodium obscurum).
Pitch-pine (Pinus rigida).
Greenbriar (Smilax laurifolia).
Laurel (Kalmia latifolia).
Mistletoe (Phoradendron flavescens).
Holly (Ilex opaca).
Inkberry (Ilex glabra).
Cat-tail (Typha latifolia).
Plume-grass (Erianthus saccharoides).

FLOWERS

Golden-club (Orontium aquaticum).
Turkey-beard (Xerophyllum asphodeloides).
Swamp-pink (Helonias bullata).
Turk's-cap-lily (Lilium superbum).
Moccasin-flower (Cypripedium acaule).
Water-lily (Castalia (Nymphaea) odorata).
Sweet-bay (Magnolia virginiana).
Wild lupine (Lupinus perennis).
Sweet pepperbush (Clethra alnifolia).
Pink azalea (Azalea nudiflora).
White azalea (Azalea viscosa).
Sand-myrtle (Dendrium buxifolium).

* Consult HENKEL, ALICE: Wild Medicinal Plants of the United States. Bureau of Plant Industry, Bulletin 89, Washington, 1906.

† See TREAT, MARY: Native Plants for Winter Decoration. Garden and Forest, VI: 141, March 29, 1893.

Laurel (Kalmia latifolia).
Arbutus (Epigaea repens).
Butterfly-weed (Asclepias tuberosa).
Buttonbush (Cephalanthus occidentalis).
White boneset (Eupatorium album).
Rough boneset (Eupatorium verbenaefolium).
Sweet goldenrod (Solidago odora).

The plants in the above list that have been most in demand until some of them are threatened with extinction are mistletoe (Phoradendron flavescens), water-lily (Nymphaea odorata), sweet-bay (Magnolia virginiana), holly (Ilex opaca), pink azalea (Azalea nudiflora), laurel (Kalmia latifolia), and arbutus (Epigaea repens). It has been no uncommon sight to see negro women on the streets near the large markets of Philadelphia selling large bouquets of wild flowers gathered in the pine-barrens of New Jersey, notably water-lilies, sweet magnolias, pink azaleas, laurel and arbutus, while at Christmas time boxes and crates of mistletoe, holly and laurel have been exposed for sale. All true lovers of our native vegetation look for some proper control of the ruthless destruction of our most beautiful trees, shrubs, and flowers for purely commercial purposes. The time has come to call a halt in the destruction of such natural resources.

CHAPTER IV

RESEARCH INVESTIGATION OF PINE-BARREN SOILS

Considerable attention has been given in recent years to an analysis of the soil as related to the growth of native vegetation. This study has developed several lines of investigation. The Bureau of Soils of the United States Department of Agriculture has emphasized the physic analysis of soils as of importance in determining the character of the soil, and it has emphasized also the importance of the toxic and useful organic constituents of the soil, as in part an explanation of soil sterility or of soil fertility. The second class of investigators, as Hilgard and Shantz, have considered natural vegetation to be a valuable indicator of the capabilities of the land for crop production, while the third group of students have laid stress upon the importance of all the factors which influence the character of the soil, whether they be chemic, physic, toxic, or biologic. The writer has expressed his views* on this important matter as follows:

"For many years the fertility of the soil was sought in the chemic substances which analysis proved to be essential to plants and which could be exhausted from the soil by the continual growth of a single crop upon it. To restore the fertility of the soil, it was necessary only to restore the ingredients necessary to keep a plant in a productive condition. Fertilizers were applied which were known to contain the most important materials of plant food and in an available form. Even to-day there are opposing camps of plant physiologists. One set holds to the principles, first clearly enunciated by Liebig, that the chemic condition of the soil is the most influential factor in the productivity of the garden or farm. The other group consider that the physic condition of the soil influences the tilth. This school teaches that all agricultural soils contain sufficient quantities of the essential mineral plant foods for many years to come. Recently a more advanced position has been taken by some students of the soil when they claim that the loss of fertility of many long cropped soils is due to the accumulation of toxic bodies, the accumulated excreta of plants that may have been grown without proper rotation. The true theory of soil fertility will probably be found to be

* HARSHBERGER, J. W.: The Soil, a living Thing. Science, N. S. xxxiii: 741, May 12, 1911.

one which will combine all of these theories with another one, which I believe must also be considered in reaching a satisfactory conclusion as to the relation existing between crops and the soil in which they grow.

"The theory is one which considers that the soil is a living thing apart from its chemic or physic structure, that in the reaction between the living soil and the growing plant is the true explanation of soil fertility. A fertile soil is a live one. An infertile soil is a dead one. Contrast the soil which is filled with organic matter (humus) and in which numberless fungous, bacterial and protozoan organisms are at work, with a mass of clay or sand without such organic material and associated living organisms. The one soil is fertile, because the organisms in the soil react favorably upon each other; the other soil is infertile, because the organisms present in this soil are antagonistic. Recent investigation has pointed the way along which future research on soils must proceed. That the soil is the seat of activities of much importance to growing plants is proved by the presence of the nitrifying and denitrifying bacteria, of the bacteria that produce the root nodules of the Leguminosae, of such organisms as *Clostridium pastorianum*, *Bacillus mycoides*, *B. ellenbachensis*, *Azotobacter chroöcoccum*, *A. Vinelandii* and the hyphæ of numerous saprophytic fungi, various putrefactive bacteria, which perform their rôle in making the soil the fit habitation of the higher flowering plants, producing the tilth or 'Bodengare' of the Germans. So too earthworms, insect larvæ, ants and burrowing animals assist in the task of aërating and mixing the surface layers of the soil. It is also evident that the production of toxic excretions by the roots of plants is undoubtedly a factor of importance in soil fertility. Following out a clue which the partial sterilization of the soil by chemicals or by steam gave, it was discovered that the bacteria which are useful in ammonia-making increased four-fold after such treatment, suggesting the presence in the soil of some agent which held them in check. After much painstaking study it was discovered* that the soil contained a living protozoön (*Pleurotricha*), which preyed upon the useful organisms, and that the heat and chemicals either destroyed these larger unicellular animals, or inhibited their activity. It can be said, therefore, that the problem of fertility of the soil is largely a biologic one, as well as dependent upon the physic, chemic and toxic condition of the surface layers."

The attempt has been made in the investigation of the pine-barren soils to make the study from an ecologic point of view. Nine stations were chosen in different parts of the region in order that the study of the pine-barren soils might be made as comprehensive as possible. As the

* HALL, A. D.: The Soil as a Battleground, Harper's Magazine, October, 1910, pp. 680–687.

sand of central New Jersey was derived originally from oceanic supplies, the first station (I) where soil samples were taken was the beach and the dunes at Belmar. The second station (II) was chosen at Como, where the pine-barren forest approaches the sea. This forest may be termed the coastal pine forest. Station III was located in the pine forest about 1.6 kilometers (1 mile) east of Lakehurst, and the other stations were located at Warren Grove (IV), at Shamong (Chatsworth) (V), at Atco (VI), at Sumner (VII), at Clementon (VIII), and in the deciduous forest at Clementon (IX). Some 28 samples were taken from the 9 different stations and some of the generalizations reached in this study are based on a study of these 28 samples representing a wide diversity of conditions of pine-barren soils.

MECHANIC ANALYSIS OF SOILS

Nineteen samples of soil were sent to the Bureau of Soils in Washington for mechanic analysis. The bureau very considerately undertook the task of analyzing these samples, and after a lapse of considerable time reported, under date of January 10, 1911. In the following table the analysis of 16 samples is given:

SERIAL NUMBER, U. S. BUREAU OF SOILS	NOTE BOOK NUMBER OF SAMPLE	LOCALITY	DEPTH	FINE GRAVEL 2 TO 1 MM.	COARSE SAND 1 TO 0.5 MM.	MEDIUM SAND 0.5 TO 0.25 MM.	FINE SAND 0.25 TO 0.1 MM.	VERY FINE SAND 0.1 TO 0.05 MM.	SILT 0.05 TO 0.005 MM.	CLAY 0.005 TO 0 MM.
				P. ct.	P. ct.	P. ct.	P. ct.	P. ct.	P. ct.	P. ct.
23821	1	Belmar	Beach Sand	8.5	59.2	28.0	4.1	0.1	0.1	0.0
23822	2	"	Surface Dune	1.2	47.4	39.3	9.7	0.7	1.0	0.4
23823	3	"	Dune 30 cm.	2.1	59.7	29.1	6.3	1.5	0.6	0.4
23824	4	Como	Surface Soil	3.1	20.5	26.5	34.9	8.0	5.1	1.3
23825	5	"	10 cm.	3.2	23.4	27.3	29.4	9.9	5.3	1.5
23826	6	"	24 cm.	9.8	28.2	27.2	21.6	7.7	4.3	1.0
23827	7	"	55 cm.	3.2	21.1	30.4	27.6	5.8	6.1	5.9
23828	8	Lakehurst	Surface Soil	5.3	22.3	18.6	43.1	4.1	4.1	1.6
23829	9	Lakehurst	15 cm.	4.9	20.0	21.1	42.6	7.4	2.5	1.2
23830	10	Lakehurst	50 cm.	11.8	18.6	19.4	35.4	5.6	4.4	4.5
23831	11	Upper Plains	Surface Soil	1.2	13.8	28.8	41.3	9.5	3.3	2.0
23832	12	Upper Plains	15 cm.	2.5	15.4	25.8	39.4	9.1	4.8	2.8
23833	13	Upper Plains	40 cm.	1.2	11.2	25.4	37.8	10.6	10.0	3.8
23837	17	Shamong	Surface Soil	2.9	28.3	23.4	35.3	6.5	1.8	1.5
23838	18	Shamong	25 cm.	3.2	18.9	23.0	43.0	9.1	1.4	1.3
23839	19	Shamong	50 cm.	3.0	19.4	22.7	40.2	8.0	2.6	3.6

It will be noted that fine gravel, where the particles range from 2 to 1 millimeters, is found in the coarse beach sand, the middle layer (24 centimeters deep) of the soil at Como (9.8 per cent.), at 50 centimeters deep in the soil at Lakehurst (11.8 per cent.), 15 centimeters deep in the soil from the Upper Plains (2.5 per cent.), and at 25 centimeters deep in the soil from the pine woods at Shamong (3.2 per cent.). The soils in general from the five stations given above range from coarse to fine sand (1 mm. to 0.1 mm.). Small percentages of silt not exceeding 6.1 per cent., with one exception, 10 per cent., in the soil 40 centimeters below the surface in the Upper Plains, are noted. The percentage of clay increases in the soils from the inland stations, being uniformly larger in the three samples from the Upper Plains. In general, therefore, it may be stated that the soils of New Jersey pine-barrens, from the surface to a depth of 55 centimeters, are sandy with slight percentages of silt and clay.

Mineralogic Analysis of Soils

A mineralogic examination of 19 samples of soils furnished by me to the U. S. Bureau of Soils is as follows: The sandy soils show approximately the same mineral composition, varying somewhat in the percentage of the same. They are generally poor in minerals other than quartz, *i. e.*, 3 to 5 per cent. and less. The feldspar content is very low and generally highly altered. The minerals commonly found, in the order of their occurrence, are: zircon, rutile, sillimanite, tourmaline, feldspar, and hornblende. Magnetite and ilmenite are present in some quantity. Augite, epidote, titanite, and chromite are present occasionally. The quartz grains, zircon and rutile crystals, indicate by their rounded appearance a great amount of attrition. There are no distinctive differences in the samples.

Chemic Analysis of Soils

Through the kindness of Dr. Daniel L. Wallace, of the Department of Chemistry of the University of Pennsylvania, the following partial analyses of surface, middle and subsoil layers are presented. The surface, middle and subsoil materials submitted to analysis were obtained by mixing soils from the different localities mentioned above and from all three levels:

	Nitrogen	Potash* (K_2O)	Phosphorus (P_2O_5)
Surface layers................	0.24 p. c.	0.53 p. c.	0.07 p. c.
Middle layers................	0.05 p. c.	0.24 p. c.	0.10 p. c.
Subsoil layers................	0.06 p. c.	0.35 p. c.	0.06 p. c.

* The original analysis for potash gave the percentages in terms of K_2CO_3, but to harmonize them with other analytic data they were converted into terms of K_2O. In terms of K_2CO_3 the percentages were 0.78, 0.35, 0.52.

It will be noted that the surface layers are richer in nitrogen. As to potash, the surface soils are richest, the middle layers less so, and the subsoils have a higher percentage than the middle layers. The middle layers are richer in phosphorus than either the surface or subsoil layer. Contrasting this with a high class soil such as Hilgard (Soils, 331) gives from Janesville, Wisconsin, with potash 0.59 per cent., and phosphoric acid (P_2O_5), 0.06, where the phosphoric acid is only just above the lower limit of sufficiency, we discover that the pine-barren soils of New Jersey, as far as phosphorus is concerned, are richer in this ingredient than the rich, friable loam of Wisconsin. The percentage of potash in the New Jersey soils is less than the percentage of that element in the Wisconsin soil. Hilgard, on page 352 of his book on Soils, gives for the pine woods of Mississippi potash (K_2O), 0.26 per cent., and for the pine flats, 0.06 per cent., and phosphoric acid (P_2O_5), 0.03 per cent. For high-class pinelands of Florida he gives potash, 0.19 per cent., and phosphoric acid, 0.11 per cent. In comparison with these Florida soils the potash content of the New Jersey is higher, but as regards phosphorus, there is a slight difference in favor of the Florida soils. The nitrogen content is, however, low as compared with the limits set by Hilgard (358) of 5 per cent. for humid regions and 1 per cent. for arid regions. Hence the New Jersey soils ought to respond to the use of nitrogen.

Experiments with Pine-Barren Soils

The results of a series of experiments upon pine-barren soils are given in the following table for 28 samples from 9 diverse localities in the region. In order to determine the water content of each sample, 500 c.c. of each kind of soil was taken and carefully weighed in the fresh condition. In each case the measure was filled heaping full, and the soil was shaken in order to compact it before the sample was emptied on to a piece of paper on the pan of the balance. After weighing the samples were spread upon newspapers and thoroughly air-dried before being carefully weighed again. The difference between the first weight and the second weight gave the weight of the water lost during the process of air-drying. The rate of percolation was estimated by taking a wide glass tube, 4.2 cm. wide and 49.8 cm. long, and marking upon it circular lines 5 cm. apart up to 30 cm. (Fig. 11). The lines were marked 0,5 cm., 10 cm., 15 cm., 20 cm., 25 cm., 30 cm. A piece of wet cheese-cloth was tied to the bottom of this glass cylinder, 2.3 cm. below the 30 cm. mark on the tube, which was filled with soil some distance above the zero (0) mark, 17 cm. below the top rim of the glass cylinder. The soil was shaken into place,

but not tamped.   The location of the zero point below the top of the
tube was purposely made, so that the sand above it would equalize the
flow of water before it reached the zero line of the tube.   The rate
of flow from the zero line to the line marked 30 cm. was noted by means
of a stop-watch reading to seconds (Fig. 11).   The time that it took for
the different columns of water to reach the different 5 cm. levels were
noted by the stop-watch.   As the 30 cm. mark was 2.3 cm. above the
cheese-cloth, the total time is not a sum of all the time intervals, but is
the time that it took for the water to pass from 30 cm. mark to the bottom of the tube, as indicated by the first drop falling to the catch-vessel

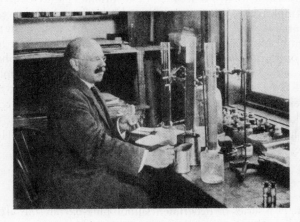

FIGURE 11

Apparatus used in conducting experiments with pine-barren soils.   Three glass cylinders
are shown with tripod supports, cheese-cloth to hold soil in place, and receptacles to
receive drainage waters, December 7, 1910.   The stop-watch is held in the left hand of the
author.

below, plus the time that it took to pass through the 30 cm. column of
soil.   As 500 c.c. of water were used for each percolation experiment, the
amount of percolation water was determined by measuring the amount
of water which passed through the column of sand into the receptacle
beneath after the flow of the water had ceased.   Deducting the water of
percolation from the original 500 c.c. used, the difference indicates the
amount of water in volume actually retained by the soil.   These figures
are given also in the table.

As these samples were taken at different seasons of the year, the water
content of the different soils, as indicated by the loss of weight of the air-
dried sample of soil, is of course different.   For example, it was 124

grams in the surface soil from the pine woods one mile east of Lakehurst, 128 grams from the sample taken 40 cm. below the surface in the deciduous woods one-quarter mile west of Clementon, and 209 grams in the sample taken 61 cm. below the surface in the same locality. The lowest was 5 grams from the beach sand gathered at Belmar. The two extremes, 5 grams for the beach sand and 128 grams and 209 grams for the subsoil upon which grew the deciduous forest, is of interest. For the pine woods, taking the mean for all of the samples of the surface layer, it stood 73 grams; for the root-containing layer it was 35 grams; for the deeper samples in the rootless subsoil it was 33.4 grams.

It is important to give a detailed description of the stations where the 28 soil samples were taken before considering the table of percolation experiments.

STATION I.—Belmar.
Sample 1.—Upper beach sand (surface).
Sample 2.—Dune sand (surface) between Ammophila arenaria and Solidago sempervirens.
Sample 3.—Dune sand, 30 cm. below the surface, with growth of Ammophila arenaria, Cassia chamaecrista, Solidago sempervirens, and Strophostyles helvola.

STATION II.—Pine woods along north shore of Como Lake.
Sample 4.—Surface, or humous, layer immediately beneath the covering of pine needles, which were removed.
Sample 5.—Sandy soil, 10 cm. below the surface.
Sample 6.—Sandy soil, 24 cm. below the surface and directly underneath the horizontal pine roots.
Sample 7.—Fine red-gravel layer, 55 cm. below the surface.

STATION III.—Pine woods, 1.6 kilometers (1 mile) east of Lakehurst. Position of Quadrat II.
Sample 8.—Center of Quadrat II. Surface soil.
Sample 9.—Soil 15 cm. below the surface.
Sample 10.—Soil 50 cm. below the surface.

STATION IV.—South edge of Upper Plains at Warren Grove, near George Cranmer's farm.
Sample 11.—Surface soil. Here grew Pinus rigida, Dendrium buxifolium, Gaylussacia resinosa, Ilex glabra, Pyxidanthera barbulata, Quercus ilicifolia, Tephrosia (Cracca) virginiana.
Sample 12.—Soil 15 cm. below the surface.
Sample 13.—Soil 40 cm. below the surface.

STATION V.—Pine-barrens at Shamong (Chatsworth).
Sample 17.—Surface, or humous, layer beneath the tall pine trees.
Sample 18.—Soil 25 cm. below the surface.
Sample 19.—Soil taken 50 cm. below the surface.

STATION VI.—Pine woods at Atco, November 8, 1910.
Sample 20.—Surface, or humous, soil.
Sample 21.—Soil 10 to 15 cm. below the surface in the root layer.
Sample 22.—Soil 50 to 60 cm.. below the surface.   This soil was wet
and cold, and was taken probably near the edge of a deciduous swamp.

STATION VII.—Pine forest at Sumner.   The humous layer was 20 cm.
deep, of dark-grayish sand.   Next came a white, sandy layer, 40
cm. thick, with reddish-yellow sand beneath.   The component
vegetation here consisted of Pinus rigida, Gaylussacia resinosa,
Gaultheria procumbens, Kalmia angustifolia, K. latifolia, Pteridium aquilinum, Quercus ilicifolia, Q. prinus.
Sample 23.—Humous layer down to 15 cm.
Sample 24.—Sandy soil, gray in color, 45 cm. below the surface.
Sample 25.—Subsoil reddish-yellow sand, 61 cm. below the surface.

STATION VIII.—A deciduous forest, 0.4 kilometer (one-quarter mile)
west of Clementon, with Liquidambar styraciflua, Liriodendron
tulipifera, Quercus alba, Q. velutina, Euonymus americanus, and
Viburnum dentatum.   The top soil was full of humus, about 30 cm.
thick, followed by a thin layer 25 cm. thick, succeeded by a coarser
sand saturated with ground-water.
Sample 26.—Surface soil below leaf-covering down to 10 cm.
Sample 27.—Soil 40 cm. below the surface.
Sample 28.—Soil 61 cm. below the surface, with water running into
hole as fast as the sand was dug out.

STATION IX.—Deciduous woods on a sloping hillside at Clementon.
The sandy ridge or hill with sand at least 6.4 meters (21 feet) thick
was covered with such trees as: Acer rubrum, Cornus florida, Diospyros virginiana, Fagus grandifolia, Ilex opaca, Liquidambar styraciflua, Pinus rigida, Quercus alba, Q. velutina, Sassafras variifolium, such shrubs as Kalmia latifolia, Smilax rotundifolia, Viburnum dentatum, and the herbs Epigaea repens, Maianthemum
bifolia and Mitchella repens.
Sample 29.—Top soil below forest litter down to 12 cm. June 9, 1911.
Sample 30.—Soil 23 cm. below the surface.
Sample 31.—Soil 38 cm. below the surface.

The rate of percolation, as shown by the table and by the graphic
curve, varied greatly in the different soil samples.   The curves for the
percolation through the surface soil samples show what diverse conditions exist in the humous layer in the pine-barren region.   The curves
1 and 2 for beach and dune sands show that the passage of water
through them was extremely rapid.   The curves of samples 4 and 8
conform nearly with those of the beach sand and show that the percolation of water through the surface soil in the pine woods at Como and
Lakehurst is rapid.   The curves for samples 17 and 18, although steeper
and more abrupt, indicate that the rate of percolation is slower, but still

rapid enough, so that the surface soils at Shamong and at Atco can be called pervious. Not so with the steep and abrupt curves of samples 23, 26, and 29 (Fig. 12), for the pine woods at Sumner, at the western edge of the pine-barrens, and for the deciduous woods at Clementon. Here the steepness of the curves indicates a slow rate of percolation. If we examine the curves which represent the rate of passage of water through the root-holding layers of the pine-barren soils, we discover that with the exception of the very rapid rate of sample 3, that of dune sand 30 cm. below the surface, curves 6, 9, 12, 18, 21, 30 approximate, showing that the conditions existing in the root-holding layer are more uniform throughout the region than are the conditions of the surface soil, as might be expected. The extremely abrupt and lengthy curves 24 and 27 respectively for the soil at Sumner, 55 cm. below the surface, and for the deciduous forest soil 40 cm. below the surface, 0.4 kilometer (one-quarter mile) west of Clementon, indicate that these soils are extremely impervious (Fig. 12). The series

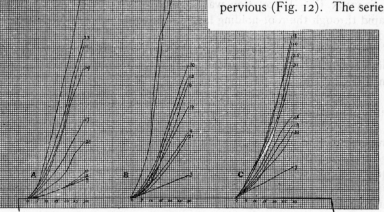

FIGURE 12

Curves showing rate of percolation of water through various soils. *A*, Surface soil; *B*, root-holding layer; *C*, rootless subsoil.

of curves for the rootless subsoil, with the exception of the extremely pervious subsoil 55 cm. below the surface in the pine woods at Como, numbered 7, are also fairly uniform. Curves 19, 22, 28 form one group, and curves 10, 13, 25, 31 form another group, the subsoil of the second group being less open and with a slower rate of percolation than the first group.

Now as the soil samples from the different stations were taken from the same hole, but at different depths, we must connect the different curves in a series by way of determining the rate of percolation under natural conditions of soil environment where the humous layer is super-imposed on the root-holding layer, and it in turn on the rootless subsoil. For Station II, curves 4, 6, 7 form a series; for Station III, curves 8, 9, 10 are a series; for Station IV, curves 11, 12, 13 form a series; for Station V, curves 17, 18, 19 should be grouped; for Station VI, that at Atco, curves 20, 21, 22 arrange themselves in a series; for Station VII, curves 23, 24, 25 can be grouped; for Station VIII group the curves 26, 27, 28, and for Station IX, curves 29, 30, 31 form a series (Fig. 12).   To sum up as to the rate of percolation of water through the soils at these nine stations:

STATION II.—Percolation rapid through the surface layer, slower through the root-holding layer, rapid through the subsoil.

STATION III.—Percolation rapid through the surface soil, not so rapid through the second soil layer, very slow through the subsoil.

STATION IV.—Percolation was slow through the surface soil, more rapid through the root-holding layer, and again slower through the sub-soil.

STATION V.—Percolation through the three layers of soil from the pine woods at Shamong was fairly uniform, being slightly retarded in the middle layer.

STATION VI.—The soil at Atco was relatively pervious, especially the surface layer.

STATION VII.—The soil of the pine woods at Sumner was relatively impervious, especially the middle layer.

STATION VIII.—The soil of the deciduous forest, 0.4 kilometer (one-quarter mile) west of Clementon, was as indicated by the curves 26, 27, very impervious, but the subsoil, indicated by curve 28, was more open and porous than the upper layers.

STATION IX.—The rate of percolation through the soil layers of the hillside deciduous forest was much slower than through the pine-barren soils, as indicated by the steepness of curves 29, 30, 31.

If we may be permitted to generalize on this series of experiments, it seems obvious that we have the dune sands at one extreme and the soils of the deciduous forest at the other extreme.   The dune sands are open and pervious to water.   The soils of the deciduous forest allow the water

TABLE SHOWING WEIGHTS OF SAMPLES, PERCOLATION, AND RETENTION OF WATER IN SOILS FROM A NUMBER OF PINE-BARREN LOCALITIES

Experimental Treatment of Soils. The lines 0 to 30 cm. refer to the intervals of the columns of soils, and the figures in the corresponding columns are seconds of time.

Column key:

- **I. Beach Sand, Belmar** — 1: Upper Beach, Surface Layer; 2: Dune Sand, Surface Layer; 3: Dune Sand, 30 cm. below Surface
- **II. Pine Forest, North Shore of Como Lake** — 4: Humus Layer beneath Layer of Pine Needles; 5: Sandy Soil, 10 cm. below Surface; 6: Sandy Soil, 24 cm. below Surface; 7: Red Gravel Soil, 55 cm. below Surface
- **III. Pine Forest, 1.6 Kilometers East of Lakehurst, Center of Quadrat No. 2** — 8: Surface Soil in Pine Forest; 9: Soil 15 cm. below Surface; 10: Soil 50 cm. below Surface
- **IV. South Edge of Upper Plains at Warren Grove** — 11: Surface Soil; 12: Soil 15 cm. below Surface; 13: Soil 40 cm. below Surface
- **V. Pine Forest at Shamong** — 17: Surface or Humus Layer; 18: Soil 25 cm. below Surface; 19: Soil 50 cm. below Surface
- **VI. Pine Forest at Atco** — 20: Surface or Humus Layer; 21: Soil 15 cm. below Surface; 22: Soil 50 cm. below Surface
- **VII. Pine Forest at Sumner** — 23: Surface or Humus Layer; 24: Soil 53 cm. below Surface; 25: Soil 61 cm. below Surface
- **VIII. Deciduous Forest 0.4 Kilometer West of Clementon** — 26: Surface Soil below Leaf Litter; 27: Soil 40 cm. below Surface; 28: Soil 61 cm. below Surface
- **IX. Deciduous Forest (Hillside) at Clementon** — 29: Surface Soil below Leaf Litter; 30: Soil 23 cm. below Surface; 31: Soil 38 cm. below Surface

Col.	500 c.c. Fresh Soil, Weight in grams	500 c.c. Air-Dried Soil, Weight in grams	Loss of Weight	0–5 cm.	5–10 cm.	10–15 cm.	15–20 cm.	20–25 cm.	25–30 cm.	Amount in c.c. of Percolated Water	Amount in c.c. of Retained Water
1	689	684	5	4	11	20	30	41		345	145
2	657	650	7	.7	16	30	43	57		315	185
3	667	639	28	8	16	25	35	45	56	330	170
4	464	417	47	.6	18	30	51	65		330	170
5	576	539	37	6	20	40	70	110	155	280	220
6	671	626	45	13	40	90	150	215	300	270	230
7	617	573	44	16	30	43	60	76		335	165
8	608	484	124	4	7	16	35		46	255	245
9	608	588	20	18	44	67	100	135	170	235	265
10	832	821	11	17	71	135	225	310	405	300	200
11	500	473	27	30	80	135	210	305	410	270	230
12	611	585	26	20	55	110	180	250	340	265	235
13	559	538	21	25	80	135	210	310	420	275	225
17	520	482	38	15	45	75	115	160	205	250	250
18	647	616	31	20	55	92	140	195	245	240	260
19	642	613	29	13	35	72	105	150	185	265	235
20	281	193	88	15	25	45	60	120	150	232	268
21	539	480	59	10	25	53	85	122	165	270	230
22	677	608	69	10	40	60	90	125	170	296	204
23	418	350	68	27	85	155	225	332	420	232	268
24	640	611	29	65	160	425	500	705	965	270	230
25	515	471	44	17	50	100	162	250	330	270	230
26	438	394	44	25	125	220	330	450	590	226	274
27	727	599	128	60	345	660	1020	1330		254	246
28	803	694	209	15	45	80	125	180	210	290	210
29	377	317	60	15	65	120	190	280	345	175	325
30	432	389	43	25	71	120	195	275	350	170	330
31	481	421	60	25	70	115	185	255	350	200	300

to pass less freely, in other words, they have a greater capacity for holding water. The pine-barren soils are sandy and pervious and may be said to occupy an intermediate position between the two extremes mentioned above. In all probability the deciduous forests exist on the less pervious soils, because they can exist solely in a soil which, by its water-holding capacity, can supply the water given off in transpiration from the enormous surface of the broad leaves. The pine, as a true xerophyte, occupies those soils which are more pervious and which show a more rapid rate of percolation, while the dune sands, being entirely without silt and clay, are covered by an arenicolous vegetation distinct from that of the pine forest. Geographically, the pine-barrens occupy an intermediate position between the strand vegetation on the east coast of New Jersey and the deciduous forests in the west.

Experiments with Capillary Rise of Water

A few experiments were conducted on the capillary rise of water in air-dried soil, but on account of the length of time that it took to complete the tests, they were discontinued. The experiments consisted in immersing the glass tube containing the soil in a bottle of water and noting the rate of capillary rise. The following figures are given, and the sample numbers are the same as given in the preceding pages. The numbers on the glass tubes are reversed, reading from bottom to top. The time is given in seconds for uniformity. The dry soil was shaken together into the tubes.

Sample 2:
 30–25 cm. = 11 seconds
 25–20 " = 85 "
 20–15 " = 1,637 "
Sample 3:
 30–25 cm. = 18 seconds
 25–20 " = 135 "
 20–15 " = 1,920 "
 15–10 " =23,880 "
Sample 4:
 30–25 cm. = 60 seconds
 25–20 " = 210 "
 20–15 " = 706 "
 15–10 " = 3,840 "

Sample 5:
 30–25 cm. = 25 seconds
 25–20 " = 62 "
 20–15 " = 270 "
 15–10 " = 1,050 "
Sample 6:
 30–25 cm. = 105 seconds
 25–20 " = 265 "
 20–15 " = 600 "
 15–10 " = 1,703 "
Sample 7:
 30–25 cm. =missed start
 25–20 " = 30 seconds
 20–15 " = 240 "
 15–10 " = 1,620 "

Large stones depress the rise of the water on the side of the glass tube where they occur. As the data for the capillary rise of the water in the column of sand are not complete, no graphic curves have been drawn.

However, as the time intervals for the rise of water between the successive 5 cm. are large, the curves would be sharp and steep for each of the soil samples. The rise of water from a saturated subsoil through surface layers of dry sand during periods of drought must be extremely slow, and that the lack of this water is felt by plants having superficial root systems is reflected in the xerophytic structure of these plants. The deep-rooted oaks and pines are more independent of the surface supplies of water, as their root systems usually penetrate into the wetter subsoil, while the superficially rooted plants, cut off from the deeper supplies of water, are dependent on the occasional light showers and are able to survive because of the structures which enable them to reduce their water loss.

RETENTION OF WATER BY PINE-BARREN SOILS

As the water which drains through a soil cannot be of any permanent value to the vegetation found on that soil, it is important to ascertain the capacity of a soil to retain the water which falls as rain. The retentive capacity of the different soil samples is given on the last horizontal line of the foregoing table. A graphic representation of that retentive capacity will enable us to interpret the phenomenon (Fig. 13). It will be noted from the accompanying graph that the dune sands (Station I) are at the extreme of minimum retentivity (145, 185, 170 c.c. respectively, out of 500 c.c. of water used), and that the deciduous forest soils at Clementon (Station IX) are at the other extreme of maximum retentivity of water (325, 330, 300 c.c.). The soils of the pine-barrens (Stations II, III, V, VI, VII) and Upper Plains (Station IV) stand midway (Fig. 13). With the exception of the coastal pine forest, the soil of which is more or less influenced by proximity to the dune sands, in none of the pine-barren soils does the amount of water retained by the soil out of 500 c.c. used fall below 200 c.c. The surface soil of the coastal pine-barrens retains only 170 c.c. of water, and the red gravel soil, 55 cm. below the surface, only 165 c.c. In many cases, chiefly for the surface soils of the pine-barrens, the number of cubic centimeters of water retained by the soil out of 500 c.c. used reads 245, 230, 250, 268, 268; for the middle soil layers the numbers are 230, 265, 235, 260, 230, 230; and for the subsoil they stand 200, 225, 235, 204, 230. The arithmetic mean of the figures expressing the amount of water retained by the dune sand is 166.66 c.c., or 33.33 per cent. of the 500 c.c. originally used in the experiments. The mean for the Upper Plains is 230, or 46 per cent.; the mean for the 16 pine-barren determinations is 229.37, almost that of the Upper Plains, or 45.87 per cent., while that of the deciduous forests is 280.83, or 56.16 per cent. It should be noted that the water-retaining capacity of the soils from the deciduous forest at Station VIII is not much different from that of the pine-barren soils. This may be accounted for by the close geographic position of the deciduous forest at Clementon near the western edge of the pine-barren region (Fig. 12).

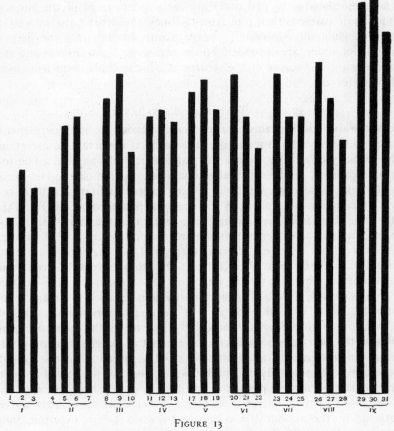

FIGURE 13
Graph representing the water-retentiveness of pine-barren soils.

From this study we conclude that the dune soils have the fastest rate of percolation and the least retentive character. The soils of the deciduous forest have the slowest rate of percolation and the greatest retentiveness, while the pine-barren soils have an intermediate rate of percolation and soils of fair retentiveness (Fig. 13).

Wilting Coefficient of Pine-Barren Soils

The higher plants obtain from the soil by means of their roots the water upon which their life and growth depend. It is taken up in the main by the root-hair cells near the growing tips of the roots. For each type of soil there is a definite limit of moisture content, below which the water cannot be absorbed by plants in sufficient quantity to supply their needs. This limit varies with the moisture-holding capacity of the soil, being high in clay and low in sand. The plants wilt when this limit is reached, and remain wilted until the soils get additional moisture. Hence this limiting moisture content has been called the "wilting coefficient." It is believed commonly that some plants are better able than others to extract the water from relatively dry soil. It would seem almost self-evident that those plants which thrive in regions where drought occurs must be able to lower the water content of the soil to a point far below that at which the roots of ordinary plants cease to take up moisture. Nevertheless the experiments of Briggs and Shantz* have shown that there is little difference in this respect, wilting in all cases taking place at practically the same limit of moisture content in a different soil.

The formula to determine the wilting coefficient of a soil when the water-holding capacity is known has been determined by Briggs and Shantz to be:

$$\text{Wilting coefficient} = \frac{\text{moisture-holding capacity} - 21}{2.90\ (1 \pm 0.021).}$$

The moisture-holding capacity is the percentage of water a soil can retain in opposition to the force of gravity when free drainage is provided. This percentage can be determined readily for the sample soils of the pine-barren region studied and described in the foregoing pages. In order to reduce the mathematic calculations to a minimum, the wilting coefficient will be determined for dune sands, Upper Plains soils, pine-barren soils, and deciduous forest soils by striking a mean of all the observations made on these four types of soils. The moisture-holding capacities derived from the arithmetic means of all the determinations made for the four types of soils are as follows:

Dune Sands	33.33	per cent.
Pine-Barren Soils	45.87	" "
Upper Plains Soils	46.00	" "
Deciduous Forest Soils	56.16	" "

* Briggs, L. J., and Shantz, H. L.: The Wilting Coefficient for Different Plants and its Indirect Determination, Bulletin 230, Bureau of Plant Industry, 1912.

From these figures by substitution in the above formula we obtain the wilting coefficient of the four types of soils. Two values are obtained for each coefficient if we use the plus or the minus quantity in the last parenthesis of the formula.

The wilting coefficients obtained in this way are:

Dune Sands	$\begin{cases} 4.1655 \text{ for } + \text{ value.} \\ 4.3569 \text{ " } - \text{ "} \end{cases}$
Pine-Barren Soils	$\begin{cases} 8.0641 \text{ for } + \text{ value.} \\ 8.4346 \text{ " } - \text{ "} \end{cases}$
Upper Plains Soils	$\begin{cases} 8.4635 \text{ for } + \text{ value.} \\ 8.8339 \text{ " } - \text{ "} \end{cases}$
Deciduous Forest Soils	$\begin{cases} 12.1628 \text{ for } + \text{ value.} \\ 12.3674 \text{ " } - \text{ "} \end{cases}$

According to the above computation, the pine-barren and the plains soils stand intermediate between the dune sands and the deciduous forest soils as to their wilting coefficients.

In order to determine whether the above formula had been applied correctly, a letter was addressed to Dr. H. L. Shantz. His reply, under date of December 28, 1912, makes clear some additional unpublished facts: "In applying the formula for the determination of the wilting coefficient from the moisture-holding capacity, or any of the formulæ given in Bulletin 230, it is best for ordinary purposes to ignore entirely the quantity within the parentheses, since it is only an expression of the uncertainty of the correlation from our results. The amount of variation which it will give in the values can be estimated as you have done, but this can be regarded by no means as the limits of error, when the wilting capacity is determined from the moisture-holding capacity. In handling these results I should simply deduct 21 from the moisture-holding capacity and divide the remainder by 2.9, and in case you wish to determine the amount of uncertainty with respect to a particular soil sample, the dividend would become in this case either 2.96 or 2.84, as the case might be. The most important thing in connection with the moisture-holding capacity is that the determinations be made under the conditions which were observed in our determinations, that is, in a column of soil one centimeter in height. In making this determination a great variation in values can be obtained, dependent upon the amount of shaking or jarring, and the determination is by no means a very accurate one."

Adopting, as we have done previously, Shantz's second alternative, the wilting coefficients of the four types of soils stand approximately as given above. If we use 2.9 as the divisor and carry our calculation to two decimal places, the numbers corresponding to the wilting coefficients stand:

Dune Sands...................................... 4.25
Pine-Barren Soils.............................. 8.57
Upper Plains Soils............................. 8.62
Deciduous Forest Soils........................12.12

It will be noted that there is a discrepancy between the values of the wilting coefficients derived by one or the other of these methods of determination, but the last figures may be taken as approximate expressions of the wilting coefficients of the different soils studied. Finally it ought to be stated that that portion of the soil moisture which is available for plant growth is represented by the difference between the actual water content and the wilting coefficient. The latter determination is consequently essential in any critical study of plant growth to soil moisture.

The Moisture Equivalent

Composite samples of pine-barren soils were sent to Drs. Shantz and Briggs, who examined the soils and reported as follows: "The moisture equivalent of the composite sample of the surface, or humous layers of pine-barren soils numbered 4, 8, 17, 20 is 11.3; for the middle soil layers numbered 5, 6, 9, 18, 21, it is 2.3, and for the subsoil layers numbered 7, 10, 19, 22, it is 4.4. The moisture equivalent of the soil is the percentage of water which it can retain in opposition to a centrifugal force 1000 times that of gravity. In making the determinations which are expressed as percentages of the dry weight of the soil used the soils are placed in perforated cups and moistened with an amount of water in excess of the quantity they can hold in opposition to the centrifugal force. After standing twenty-four hours the cups are placed in a centrifugal machine, which is operated at a constant speed so chosen as to exert a force 1000 times that of gravity upon the soil moisture. This method provides a means of determining and comparing the retentiveness of different soils for moisture when acted upon by a definite force, which is measured in absolute terms and is reproducible within narrow limits."

CHAPTER V

PHYTOGEOGRAPHIC FORMATIONS

Considerable space has been given in the foregoing pages to a consideration of the geography, physiography, settlement, and industries of the
New Jersey pine-barren region, with the object of assembling the facts
known about these subjects of general interest to those who have visited
the region. The preceding account has dealt with those phases of the
life of the people of south New Jersey which have been influenced by the
vegetation, and thus it has been an ethnobotanic survey as far as that
has been accomplished. The description of the pine-barren vegetation,
as a whole, will follow, and the subdivisions of this vegetation will be
dealt with under the head of the different and easily characterized plant
formations. These formations have been distinguished in a general
and unscientific way by the people who live in the pine-barrens with a
keen appreciation of the fundamental differences in the native vegetation, and this seems to be the case in other regions, as Graebner has
emphasized in "Die Heide Norddeutschlands," and as I have detailed
in my monograph on the Vegetation of South Florida.* It is left for
the trained botanist to interpret these general impressions and to trans-
late them into a scientific description based upon phytogeographic
principles.

When this is done, the phytogeographer recognizes nine natural plant
formations and four that are due to fire, repeated cutting of the forest,
or to the conversion of cedar swamps and bogs into areas suitable for
cranberry culture. Classified in the order of their importance they are:

Natural Plant Formations	Pine-Barren Formation.
	Cedar Swamp Formation.
	Deciduous Swamp Formation.
	Savanna Formation.
	Marsh Formation.
	Pond Formation.
	River Bank Formation.
	Bog Formation.
	Plains Formation.
Successional Plant Formations	Cranberry Bog Formation.
	Scrub-Oak Formation.
	Oak Coppice Formation.
	Mixed Pine-Oak Formation.

* Graebner, Paul: Die Heide Norddeutschlands. Die Vegetation der Erde, v: 14;
Leipzig, 1901; Harshberger, J. W.: The Vegetation of South Florida. Transactions
Wagner Free Institute Science, vii, Pt. 3: 146, 1914.

The relation of these different formations to each other, as to their possible derivation in an ascending series toward a climax condition, may be represented in a graphic way (Fig. 14), where the comparative age of the formation is shown by its distance above the lowest line and the probable succession by means of dotted lines and arrows.   As an illustration of the successions which the above diagram (Fig. 14) shows, the series beginning with the typic pine forest and ending with the mixed pine-oak formation may be chosen.   When a pine forest is destroyed by fire, or the trees removed in lumbering operations, it may be succeeded by a scrub-oak thicket.   This in turn is replaced by a coppice of taller

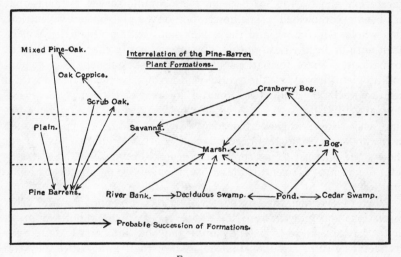

FIGURE 14
Successional arrangement of pine-barren formations.

trees, which, when admixed with pitch-pines, can be designated the Mixed Pine-Oak Formation.   The pines may ultimately suppress the oaks, and the vegetation returns to the original pine forest in which the series is bound sooner or later to culminate.   The full details of this succession and the interrelation of the different formations of the region will be emphasized as we proceed.

PINE-BARREN FORMATION

Historically, the occupancy of the recently elevated coastal plain during interglacial times by the pitch-pine and other associated species produced what we now call the pine-barrens.   Originally more widely distributed, this type of vegetation became restricted in area with the Pensauken submergence. The island, then formed, was probably covered with pine forest, and when a reëlevation of the land took place, the boundaries of the pine-barrens, which had been determined by the coast-line of Pensauken island, were maintained by the encroachment of a deciduous forest type of vegetation on the land surface elevated recently above the sea in post-Glacial times.

The pine which makes up the bulk of the New Jersey pine forest is Pinus rigida Mill, the pitch-pine, associated in places with the yellow

FIGURE 15

Open pine forest in Lebanon Reserve, with snow on the ground, December 29, 1908. The absence of a dense undergrowth is here noteworthy. (Courtesy of N. J. Forestry Commission.)

pine, P. echinata. If the ascent of a hill, or other elevation, is made, an unbroken extent of dark-green pine forest is seen stretching off into the distance as far as the eye can reach. A closer inspection, however, shows considerable diversity of vegetation as one passes through the open, sunlit pine forest from one point to another (Fig. 15). The average height of the pines in old stands from 3 to 4 decimeters (10 to 14 inches) in diameter was 20 to 22 meters (65 to 70 feet). Where fire has destroyed the original forest, the pine trees are much shorter, 7 to 14 meters (20 to 40 feet) being an average size for such trees, some of which, if badly injured by fire, show light-green sprouts growing from the trunk or base of the trees (Figs. 16 and 17).   The forest canopy consists of the dense crown of the standing pine trees, but only in exceptional cases do the crowns meet to shade the ground underneath.   The trees are spaced so that considerable sunlight reaches the forest floor, and hence the forest may be described as an open one (Fig. 18).   The pine trees are light

FIGURE 16

Regeneration of fire-denuded pitch-pines, Pinus rigida, near the Sym Place, July 29, 1913. Note the green sprouts on the central tree and the bushy young trees that have sprouted from the roots of old trees whose tops were killed by fire.

demanding, and they are dominant.   All other growth has been kept suppressed by the dominant pines.   The density of an old pine forest near Winslow is indicated in the following summary of measurement of twelve plots of one acre each.*   Likewise the quadrat method gives accurately the number of trees in the forest per 10 square meters, or any other unit taken arbitrarily as the unit of measurement.   Such an estimate will be shown graphically on a later page.

* Report of the N. J. State Geologist, 1899.   Report on Forests, p. 115.

FIGURE 17

Pitch-pine, Pinus rigida, in recently burned-over area, near the Sym Place, west of the Lower Plain, July 29, 1913. Trunks of the taller fire-denuded trees show the characteristic green sprouts formed by a regeneration of the branches. (Photo by George E. Nichols.)

FIGURE 18

Young open pine forest at Hanover, October 8, 1909. The low scrub oaks and other low shrubs are typic of many stretches of the high pine-barren formation. (Courtesy of N. J. Forest Commission.)

Plat Num-ber	Num-ber of Trees	Num-ber of Trees Over 6 Inches	Average Diam-eter, Breast High	Num-ber of Trees Over 10 Inches	Average Diam-eter, Breast High	Maxi-mum Diam-eter, Breast High	Average Height	Total Volume Over 6 Inches in Cubic Feet	Mer-chant-able Volume Over 10 Inches in Board Feet
1	67	66	13.9	34	14.4	22	69.1	2,882	10,170
2	70	65	14.0	51	16.2	18	68.8	2,917	9,925
3	78	72	13.0	56	14.8	18	66.9	2,812	8,940
4	86	74	12.7	56	13.9	20	69.7	2,548	8,828
6	92	78	12.0	55	13.4	18	68.1	2,557	7,739
7	65	60	13.0	46	14.3	21	67.5	2,228	7,521
8	71	69	12.4	54	13.5	19	66.3	2,388	7,488
9	68	63	12.5	49	13.1	19	68.1	2,127	7,208
10	104	78	11.2	51	12.6	17	68.1	2,048	6,134
12	45	43	13.1	35	14.1	20	68.1	1,613	5,631
14	85	62	11.0	35	13.1	19	66.8	1,708	4,618
17	74	57	10.4	27	12.8	16	66.8	1,275	3,369

ARCHITECTURAL FORM OF PINE TREES.—The shape of the forest-grown pine trees varies considerably, as the accompanying figures will show. The variation of the branching system of native forest trees has been studied very little in this country, although abroad considerable attention has been given to it. Miss Annie Oakes Huntington, by a series of descriptions and photographs, accomplishes this in her "Studies of Trees in Winter," published in 1902. Jepson, in his "Silva of Califor-nia," describes the architectural form of many Pacific Coast trees, as do Blakeslee and Jarvis in their "New England Trees in Winter" (1911). Dr. Ludwig Klein has undertaken, in Karsten and Schenck's "Die Vegetationsbilder," * to describe the Charakterbilder mittel europäischer Waldbäume, and has collected many beautiful photographs which accompany the text. Professor G. Haberlandt has characterized the form of tropic trees in Java in his "Eine botanische Tropenreise," published in 1893. "The Artistic Anatomy of Trees, their Structure and Treatment in Painting," by Rex Vicat Cole (1915), is a mine of information on this subject.

The pitch-pine, when growing in dense stands in the forest and under stress of competition for light, develops typically a central axis, with all the branches disposed laterally to it, with the lower branches killed by shading and in many cases broken off, leaving conspicuous stubs, which bristle in whorls beneath the green crown of the tree (Figs. 19, 20, and 21). The axis is not always perfectly straight, but may be inclined or bent, the bending being more or less ess-shaped (Figs. 22, 29, and 30). Other trees are conic (Figs. 19, 20, 21, and 22). The pitch-pine responds to the advantage of free space and branches more freely until it becomes round-headed (Fig. 28). Frequently the stem bifurcates. This bifurcation may be found at the base of the trees (Fig. 23), half way up (Fig. 24), or near the top (Fig. 25). Such trees are V-shaped or Y-shaped (Figs. 24 and 25). Another common form of bifurcated stem is the stag-

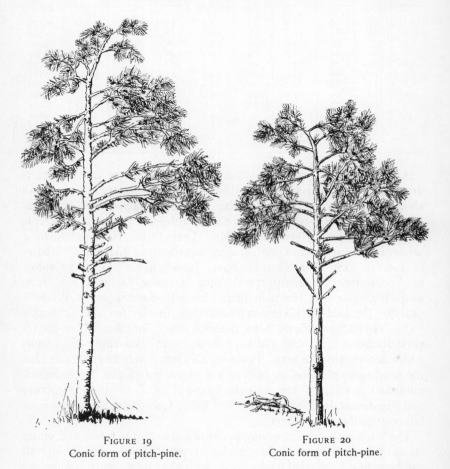

FIGURE 19
Conic form of pitch-pine.

FIGURE 20
Conic form of pitch-pine.

headed, where the main leader divides near the top into several branches, each of which branches again repeatedly. Occasionally a tree becomes candelabra-shaped (Figs. 26 and 27), when it represents a tree which has developed in an open space in the woods. Some of the forest patriarchs are of this form. At those localities where the pine-barrens approach the sea, as at Spring Lake, New Jersey, or the salt marshes, as at Barnegat Pier and Somers Point, the trees become reduced in size; the dead branches and old cones are present in some trees which become irregular and their architectural shapes are wind controlled. Here they become wind-swept and one-sided (Figs. 31 and 32). One of the usual forms is the bisected tree, where one side fails to develop, owing to the action of

FIGURE 22
Round-head form of pitch-pine.

FIGURE 21
Conic form of pitch-pine.

FIGURE 23
Bifurcated form of pitch-pine.

the wind on the exposed side of the tree (Figs. 33 and 34). Another form is the bush or shrub form (Garben-baum, or sheaf tree) which produces freely a number of clustered stems and more or less wind-blown (Fig. 35). This form may be called the bush shape. Another form may be called the candelabrum type, as the branches of the low shrubby form are disposed in a way to suggest a candelabra (Fig. 36). At the edge of the pine forest near the sea, low, prostrate, wind-swept trees are found, their top branches inclined, arched, or bowed in a direction contrary to that of the prevailing winds (Fig. 37). Such trees on a high mountain form the "Krummholz," or elfinwood of the ecologist. At the seashore these trees are called appropriately wind-swept dwarfs. One other architectural form is seen on the plains of New Jersey, where the pine trees are all low, but cone-bearing. Such low trees, showing nanism, may be denominated, because of their shape, cushion trees. Tabulated, the forms of pine-barren trees are as follows:

Conic (Figs. 19, 20, 21, 32)
Round-headed (Figs. 22, 28, and 38)
Bifurcated (Figs. 23, 24, 25, 26, 27, 34)
V-shaped (Fig. 23)
Y-shaped (Figs. 24 and 25)

Stag-headed (Fig. 26)
Bisected (Figs. 33 and 34)
Cushion-shaped (Fig. 135)
Candelabra-shaped (Fig. 36)
Bush- or sheaf-form (Fig. 35)
Wind-swept dwarf (Fig. 37)

FIGURE 24
Bifurcated and Y-shaped pitch-pine.

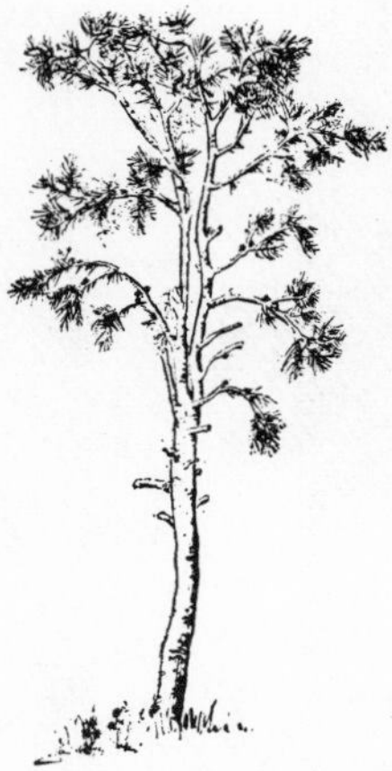

FIGURE 25
Bifurcated and Y-shaped pitch-pine.

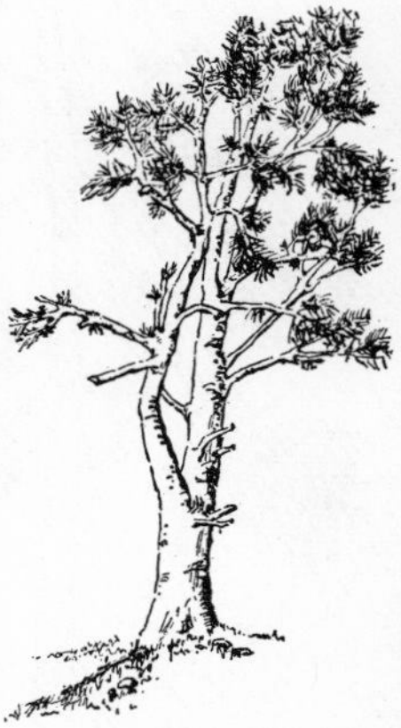

FIGURE 26
Bifurcated form of pitch-pine.

Finally, it might be stated that, although the typic pitch-pine trees have excurrent stems, yet in the forest we frequently find trees that are deliquescent, and an outline drawing of the branch system of such trees without foliage might be taken for that of a deciduous tree. In other trees the branches are drooping, but they never become typically weeping varieties, because the branches are thick, gnarled, and not pliant enough to assume the pendent habit.

FACIES OF PINE-BARREN FORMATION.—Several facies of the pine-barren formation can be distinguished. The character of these facies depends on the trees and shrubs which are associated in the pitch-pine forest. The species that are associated with the pitch-pine are determined to some extent by the soil conditions. Thus we can distinguish wet pine-barrens, flat pine-barrens, dry pine-barrens, and high pine-barrens. There are pine-barren facies with a bear-oak (Quercus ilicifolia) undergrowth, with a black-jack oak (Q. marylandica) undergrowth, and with a laurel (Kalmia latifolia) undergrowth. A common facies is where the heath-like plants (Dendrium, Vaccinium, etc.) are abundant, and there are facies where there is little undergrowth, but bare sand with prostrate trailing species, or reindeer lichen with associated mosses. There are pine-barrens transitional to the open formations of the sea dunes, to the cedar swamps of the interior, to the deciduous swamps of the sluggish streams. In fact, it can be said that a careful examination of the pine-barren region reveals almost

FIGURE 27
Bifurcated pitch-pine.

FIGURE 28
Round-headed pitch-pine.

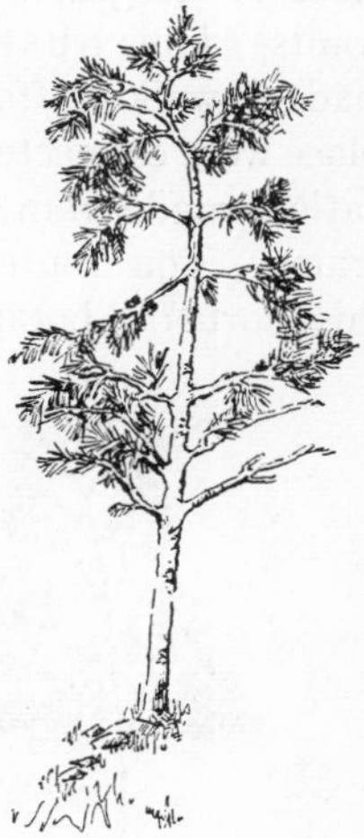

FIGURE 29
Excurrent pitch-pine tree.

as great a variety of types as there are physiographic and soil conditions, and as there are species which enter into competitive or complementary association in the facies. Before commencing the synecologic description of the pine-barren vegetation, it may be well to detail the method employed in the study of this vegetation, as such methods may be found helpful to other botanists working in the same field.

METHOD OF INVESTIGATION.—The observations of the writer were recorded in a series of field note-books. These note-books are of pocket size, 11.8 cm. (4¾ inches) wide by 18.7 cm. (7½ inches) long, with flexible backs, opening at the top. Each book, which cost five cents, contained approximately 60 pages of ruled paper. A map of the portion of the pine-barren region chosen for study was drawn in reduced size on the page of the note-books preceding the field notes for that district. Blue pencil indicated roads and railroads, the latter marked by cross lines. Red pencil showed rivers, streams, and swamps, while green indicated forest areas. The geographic names were written in black ink, and the localities where photographs and habitat observations were made were entered in lead pencil on the face of the map, while in the field. In journeying across the country by railroad, observations were made from the car windows and entered in the books with the names of the nearest stations or mile-posts as geographic reference points. Thus observations made in walking from place to place were connected by observations made from the moving trains. The routes taken in this sort of botanic survey

FIGURE 30
Pitch-pine with ess-shaped top.

FIGURE 31
One-sided pitch-pine tree.

work were indicated on a reference map. The following system was adopted to make the field notes as accurate as possible. Quite a number of plants, such as Pinus rigida, Quercus ilicifolia, Kalmia latifolia, K. angustifolia, Dendrium buxifolium, Xerophyllum asphodeloides, about the identity of which there could be no question, were entered in the note-books without verification. Frequently, when in doubt, the plant was identified in the field by consulting Britton's or Gray's Manual without preservation of the specimens. If unable to determine the name satisfactorily by the manual, a specimen was gathered for more careful study with herbarium material at hand. Hence, while not following the method of some botanists in insisting that each record should be accompanied by a dried specimen of the plant, yet the above system of verification can be used with a large degree of accuracy. The method advocated by Stone and others to have a herbarium sheet of the plant is desirable where purely syste-

FIGURE 32
Wind-swept pitch-pine tree.

FIGURE 33
Bisected pitch-pine tree.

FIGURE 34
Bisected pitch-pine tree.

FIGURE 35
Bush form of pitch-pine.

matic work is concerned, but the method used by the writer is the only feasible one where a synecologic study of the vegetation of a region of considerable size has been undertaken.

High Pine-Barren Facies.—In the heart of the New Jersey pine-barrens, as at Lakehurst, Whitings, and Bamber, are rolling hills not

FIGURE 36
Candelabra-shaped pitch-pine tree.

FIGURE 37
Wind-swept dwarf pitch-pine.

dissected by streams, which are separated from each other by large intervals of elevated country, sometimes with gravelly, sometimes with sandy, soil. The soil, being as a rule dry and well drained by seepage and surface run-off, is covered by a facies of the pine forest in which only the plants of the drier soils are associated with the pines (Fig. 39). The plants of moister situations are absent entirely. The prevailing pine, Pinus rigida, forms an even stand where the trees range in height from 7 to 12 meters (25 to 40 feet), as seen in Fig. 41. The canopy of this forest is fairly close, although there are intervals where the pine trees are spaced more widely (Figs. 39 and 40). From Quadrat 2, taken one mile east of Lakehurst, it has been ascertained that there are 15 pitch-pine trees to a quadrat of ten square meters, as compared with 14 in Quadrat 1 of the same size surveyed near the sea-coast

FIGURE 38
Pitch-pine tree, Pinus rigida, with dense witches'-broom, at Clementon, N. J. (Photograph by Genji Nakahara and J. W. Harshberger.)

at Como Lake, or as contrasted with Quadrat 3, taken at Sumner, N. J., with 13 trees.   Associated with the pitch-pine occur the short-leaf pine,

FIGURE 39
Level dry pine-barrens about two miles southeast of Lakehurst, N. J., 10 A. M., August 2, 1909.   (Photo by Roland M. Harper.)

FIGURE 40
Dry pine-barrens about a mile east of Toms River, N. J., 3.23 P. M., August 2, 1909.
(Stereoscopic photographs by Roland M. Harper.)

Pinus echinata, as at Lakehurst and Shamong, and an under story, or secondary layer, of medium sized trees, composed of Quercus marylandica (Figs. 42, 43, and 44), Q. stellata (Fig. 45), Q. velutina (Fig. 46), Sassafras variifolium sprinkled as tolerant species beneath and between the pine trees.   The third layer, or story, of this type of forest consists of the bear-oak, Quercus ilicifolia (Fig. 47), and the dwarf chestnut-oak, Quercus prinoides (Figs. 48 and 49).   These are low, rounded bushes, or small trees, with a number of branches rising from a powerful tap-root.   The bear-oak, Quercus ilicifolia (Fig. 47), is the more common of the low oaks in New Jersey.   It bears usually an abundant crop of acorns, and is very resistant to the destructive action of fire, as it readily sprouts from the charred stump after a destructive conflagration.   The dwarf chestnut-

FIGURE 41

Road through a scattered stand of pitch-pine, Pinus rigida, in Lebanon Reserve, December 29, 1908, with snow on the ground.   (Courtesy of N. J. Forest Commission.)

oak, Quercus prinoides, disputes the ground with the bear-oak, especially in the high pine-barrens.   It is usually of smaller size and more spreading in habit.   Occasionally these oaks close together to form a low thicket from 1 to 1.2 meters (3 to 4 feet) high beneath the pines.

As a fourth story, or layer, we find associated the sweet fern, Comptonia asplenifolia (Fig. 50), the huckleberries, Gaylussacia dumosa, G. resinosa, Ilex glabra, Kalmia angustifolia (Fig. 51), Rhus glabra (Lakehurst), while the fifth layer, or story, of low shrubs consists of the blueberries, Vaccinium pennsylvanicum, V. vacillans, and such grasses and

FIGURE 44

Plowed fire line with Prof. C. von Tubeuf, of Munich, Germany, on its inner, undisturbed edge, with Quercus marylandica, and the outer area cleared of undergrowth with a few dwarf oaks. October 11, 1913.

FIGURE 42

Low black-jack oak, Quercus marylandica, in pine forest at Shamong, June 27, 1910. (Photograph by Henry Troth and J. W. Harshberger.)

FIGURE 43

Pine-barrens at Shamong (Chatsworth), October 11, 1913. On the right of Prof. Adolf Engler, of Berlin, Germany, who is holding Sarracenia purpurea in one hand, is Quercus ilicifolia (nana), with white cedar in the background. Immediately back of Dr. Engler and to the left is Quercus marylandica.

FIGURE 45

Post-oak, Quercus stellata, beneath Pinus rigida, in forest at Shamong (Chatsworth), June 27, 1910. To the right is Quercus marylandica, and in the foreground Comptonia asplenifolia, Vaccinium vacillans and Hudsonia ericoides. (Photograph by Henry Troth and J. W. Harshberger.)

herbs as the beard-grass, Andropogon scoparius, Panicum depauperatum (Whitings), and goat's-rue, Tephrosia virginiana. The sixth and forest-floor layer consists of the bearberry, Arctostaphylos uva-ursi, the arbutus, Epigæa repens, the spurge, Euphorbia ipecacuanhae, the wintergreen, Gaultheria procumbens, the pyxie, Pyxidanthera barbulata, and the lichen, Cladonia rangiferina. Occasional bare sandy stretches are found between the pine trees that are covered with pine needles. In these areas

FIGURE 46

Black-oak, Quercus velutina Lam., beneath pitch-pine in flat pine-barrens, between Whitings and Bamber, March 25, 1910. (Photograph by Genji Nakahara and J. W. Harshberger.)

in the high-pine barrens we find spreading masses of the bearberry, Arctostaphylos uva-ursi, the sweet-fern, Comptonia asplenifolia, a few specimens of the moccasin-flower, Cypripedium acaule, and the huckleberry, Vaccinium vacillans. The dry leaf layer is in some places 5 centimeters (2 inches) thick. A hilltop may be ascended on the road from Lakehurst to Whitings, where the undergrowth thins out and where the plants are more scattered. The bare ground is covered with pine needles, and the oaks disappear almost entirely. The ground is covered with flat, radiating growths of the bearberry, Arctostaphylos uva-ursi (Fig. 52), associated with the huckleberry, Gaylussacia resinosa, and the grass, Panicum depauperatum. The trailing stems of the bearberry may reach a length of 1.8 to 2.4 meters (6 to 8 feet), are crowded densely, and the plant, with its dark-green, leathery, evergreen leaves, that turn a rich reddish-brown color in some localities, is always a conspicuous object of

FIGURE 47
Bear-oak, Quercus ilicifolia, as undergrowth in pine forest at Shamong (Chatsworth), June 27, 1910. (Photograph by Henry Troth and J. W. Harshberger.)

botanic interest (Fig. 52). This low shrub and the dwarf chestnut-oak are the most distinctive species of the high pine-barrens. The abundance of the forest litter in some places indicates the long absence of forest fires.

FLAT PINE-BARREN FACIES.—The distinction between the high pine-barrens and the flat pine-barrens is hard to draw absolutely, as one type blends imperceptibly with the other. But the arbitrary distinction, if not absolute, is a very convenient one. The flat pine-barren facies represents by far the most common type of forest in southern New Jersey. It may be taken as typic. The forest occupies country of little relief, as interstream areas where hardly any dissection has taken place. The soil drainage is by percolation during most of the year. It is usually very dry except during spells of heavy rainfall. During seasons of heavy rain the run-off is sluggish and the surface, as long as the rain is falling, may be covered with a sheet of water. Such a forest type in the southern states would be termed flat woods. This kind of pine forest was studied at Ancora, Barnegat, Browns Mills Junction, Lakehurst, between Lakehurst and Toms River, May's Landing, New Goshen, New Lisbon, Newtonville, Ocean Gate, Shamong (Chatsworth), head of the north arm of

FIGURE 48
Clump of dwarf oak, Quercus prinoides, at Shamong (Chatsworth), June 27, 1910.
(Photograph by Henry Troth and J. W. Harshberger.)

Shark River, Sumner, Toms River, Whitings, and Woodmansie. The following description is a general one, including the facts obtained by a study of this type at the above widely separated localities.

The height of the pitch-pine trees varies in different localities. No general height measurements were made in this study, but in the undisturbed forest, where fires have been kept out of the woods, it may range from 6 to 12 meters (20 to 40 feet). The height of 7.6 meters (25 feet) may be taken as an average one. The flat pine-barren facies has a richer ground flora of low shrubs and herbs than the high pine-barren facies.

The yellow-pine, Pinus echinata, mingles occasionally with the pitch-pine, Pinus rigida. The tolerant, or secondary trees of this forest (Fig. 53) are, in the order of their frequency: Quercus marylandica (dwarf to 15 feet), Q. stellata, Kalmia latifolia [Atsion, Ancora], Sassafras variifolium, Prunus serotina, Quercus alba [Lakehurst to Toms River], Q. velutina, Q. coccinea [Newtonville to Richland], Q. falcata [New Goshen], Q. prinus [Ancora], Q. phellos [Ocean Gate], Amelanchier intermedia [Browns Mills], Cornus florida [New Goshen]. The red-cedar, Juniperus virginiana, is occasional as scattered avivectent trees [Browns Mills]. Sourgum, Nyssa sylvatica [Ocean Gate], and holly, Ilex opaca [Mays Landing], are found. The two most important secondary oaks of this type of forest are Quercus marylandica and Q. stellata. The laurel becomes an important element in some localities.

FIGURE 49

Pinus rigida to left of Prof. A. Engler, surrounded with a low, bushy form of Quercus prinoides and Quercus marylandica to the right in the foreground. October 11, 1913.

The black-jack oak usually has a straight or slightly bent main stem, with short lateral branches that are bent downward, while in some trees the lower branches are long enough and drooping enough to sweep the ground.

The post-oak, Quercus stellata, has a more robust, erect habit (Fig. 45), covered in summer with leathery leaves with divergent, rounded lobes, brownish, downy underneath.

FIGURE 50

Sweet-fern, Comptonia asplenifolia, at Clementon, June 4, 1910.

Its twigs are stout, light orange to reddish-brown, strongly contrasting with the older growth, which is much darker, sometimes almost black. The buds are broadly ovate, often as broad as long, and hemispheric. The bud scales are bright reddish-brown, sparingly hairy. The black-oak, Quercus velutina, is found also in the pine forest, but is less abundant than the other two (Fig. 46). It is

FIGURE 51

Low pine-barren with Pinus rigida, black oak (Quercus velutina), sheep laurel (Kalmia angustifolia), wintergreen (Gaultheria procumbens), between Whitings and Bamber, March 25, 1910. (Photograph by Genji Nakahara and J. W. Harshberger.)

not present in many localities. The twigs of this oak are reddish, mottled with gray, very bitter to the taste, and coloring the saliva yellow. The buds are ovate to conic, large, ending in a sharp point. The bud scales are covered with a dense, pale, yellowish-gray to dirty white tomentum. The other oaks are less common, or infrequent, especially Quercus prinus, Q. falcata, Q. phellos, which are hardly to be considered as true pine-barren species.

The third layer of the forest, or the dwarf-oak story, is of great interest (Fig. 53). The plants of this layer constitute the bulk of the undergrowth in the flat pine-barrens. Here are found generally distributed throughout the region the bear-oak, Quercus ilicifolia (Fig. 53),

a shrub 0.9 to 1.5 meters (3 to 5 feet) tall, with slender, yellowish-green twigs provided with greenish-yellow to reddish down, which often disappears in winter on exposed parts, while it can be seen on protected parts of the shoot. The buds are ovate to conic, sharp or blunt-pointed, small, generally not over 3 mm. The acorns mature the second season and are produced in great abundance clustered about the stem. Dwarf wild cherry, Prunus serotina, is occasional in dry, sterile sand, where fires are frequent, as between Lakehurst and Toms River. The sassafras,

FIGURE 52

Radiate patches of bearberry, Arctostaphylos uva-ursi, on the ground with low Quercus marylandica, Gaylussacia resinosa, etc., at Shamong, June 27, 1910. (Photograph by Henry Troth and J. W. Harshberger.)

FIGURE 53

Even stand of Pinus rigida at Shamong (Chatsworth), June 27, 1910, with an under-growth of Quercus marylandica, Q. stellata, Q. ilicifolia, Q. prinoides, Gaylussacia frondosa, G. resinosa, Ilex glabra, Sassafras variifolium, Vaccinium pennsylvanicum, and Pteris aquilina. (Photograph by Henry Troth and J. W. Harshberger.)

Sassafras variifolium, is also a common constituent, as a small tree with flat-topped crown and upright branching habit. The huckleberry, Gaylussacia frondosa, reaching the height of 0.9 to 1.2 meters (3 to 4 feet), may be considered to be a constituent of this facies, especially where the pine woods begin to blend with the cedar swamps, as at Point Pleasant. The sumac, Rhus copallina, probably a recently introduced species, is found also as an element of the third story. The sweet pepper-bush, Clethra alnifolia, one of the most characteristic plants of the flat pine-barrens, enters into competition with the shrubby oaks in their demand for light and room. The inland waxberry, Myrica carolinensis, is associated with the inkberry, Ilex glabra, while Smilax glauca ascends the lower parts of the trunks of the pine trees and other plants of this layer, or stands erect in the forest. The laurel, Kalmia latifolia (Fig. 54), either remains as a constituent of the dwarf-oak story or it ascends to the height of the second-story trees, as at Atsion, Ancora, and Clementon. The hawthorn, Crataegus uniflora, is found in a few places and should be classed as an element of this layer. It has been found at Ancora, Atco, Cedar Brook, Quaker Bridge, and a number of other localities in the pine-barrens. The beach-plum, Prunus maritima, as a spreading shrub, is found at Browns Mills, Hammonton, and at three other places in the pine-barrens. The wild honeysuckle, Azalea nudiflora, grows in the pine-barrens at Hammonton and Speedwell, according to Stone.

FIGURE 54

Laurel, Kalmia latifolia, beneath Pinus rigida, at Clementon, June 4, 1910. (Photograph by Mrs. Helen B. Harshberger.)

Closely associated with the foregoing, and filling up the spaces between the larger shrubs and small trees, sometimes as a continuous and exclusive growth, we find a number of undershrubs and herbs that are important and characteristic plants of the pine-barrens. Thus there exists a fourth story, consisting, in the order of their abundance, of Gaylussacia resinosa, Kalmia angustifolia, Gaylussacia frondosa, Comptonia asplenifolia, Ilex glabra, Pyrus arbutifolia, Dendrium buxifolium [Lakehurst, Browns Mills] (Figs. 55 and 56), Gaylussacia dumosa, and dwarf Clethra alnifolia. The southern black blueberry, Vaccinium virgatum, occurs locally at New Lisbon, Shamong, and Lakehurst. The following herbaceous plants vary in their abundance—sometimes they are common, at other places they are rare or entirely absent: Pteris aquilina, Andropogon scoparius [plentiful, as at Mays Landing], Xerophyllum asphodeloides, Baptisia tinctoria, Lespedeza capitata [Woodmansie],

FIGURE 55

Sand-myrtle, Dendrium (Leiophyllum) buxifolium, in pine forest, Browns Mills, May 13, 1911. (Photograph by Mrs. Helen B. Harshberger.)

Lupinus perennis (Fig. 58), Tephrosia virginiana (Fig. 57), Aralia nudicaulis [Ancora], Asclepias obtusifolia [West Plains to Shamong], Gerardia pedicularia [Whitings], G. purpurea [Barnegat to Cedar Grove], Linaria canadensis, Melampyrum americanum, Aster concolor [Lakehurst[, Aster spectabilis [Shamong], Chrysopsis mariana, C. falcata, Liatris graminifolia pilosa [Shamong], Hieracium venosum [Woodmansie], Solidago bicolor, S. odora, and S. stricta [Shamong]. In addition the following species occur in this layer of the pine-barren forest: Panicum Addisonii [Whitings], P. Ashei [Albion], P. columbianum [Atco], P. Commonsianum [Bamber], P. meridionale [Atsion, Lakehurst], P. villossimum [Cedar Brook], Danthonia sericea [Folsom], Cyperus cylindricus [Woodmansie], Meibomia marylandica [Manahawken], M. obtusa [Woodbine],

FIGURE 56

Clump of Dendrium (Leiophyllum) buxifolium, in full flower, at Browns Mills, May 13, 1911. (Photograph by Miss Lydia P. Borden.)

FIGURE 57

Goat's-rue, Tephrosia (Cracca) virginiana, in flower. Pine forest at Shamong, June 27, 1910. (Photograph by Henry Troth and J. W. Harshberger.)

M. rigida [Albion], M. stricta [Mays Landing], Lespedeza angustifolia [Shamong], Linum floridanum [Winslow Junction], Monarda punctata [Hammonton], Dasystoma flava [Winslow], Gerardia Holmiana [Woodmansie], Nabalus virgatus [Bamber], N. serpentarius [Atsion], Eupatorium pubescens [Clementon], E. rotundifolium [Forked River], Solidago erecta [Albion], S. puberula [Whitings], Euthamia tenuifolia [Island Heights Junction], Sericocarpus linifolius [Richland], Aster dumosus [Forked River], A. gracilis [Bamber], A. novi-belgii [Island Heights Junction], A. patens [Cedar Brook], A. undulatus [Malaga].

The most abundant low shrub of the fifth story, or forest layer, is Vaccinium vacillans (Fig. 59), which rarely exceeds a height of 23 centimeters (9 inches). As an element of the same layer occurs another low blueberry, Vaccinium pennsylvanicum, with which in dry sand are associated Panicum depauperatum [Lakehurst to Whitings], Crotonopsis linearis [Atsion], Helianthemum canadense [Whitings], Hudsonia tomentosa [Toms River], Lechea racemulosa [Speedwell], Linum medium [Egg Harbor City], Opuntia vulgaris [Lakehurst to Toms River], Oenothera fruticosa [Shamong], Sabatia lanceolata [southeast of Lakehurst], and Sericocarpus asteroides.

The forest floor is covered with an accumulation of pine needles and oak leaves. When these have been blown away, or from other causes have been removed, patches of bare sand or gravel are found in the pine

FIGURE 58

Pine forest between Sumner and Albion, with Lupinus perennis foreground, Sassafras variifolium and Gaylussacia resinosa. (Photograph by Genji Nakahara and J. W. Harshberger.)

forest. The ground vegetation constitutes a sixth layer or story. Here are found prostrate radiating growths of the bearberry, Arctostaphylos uva-ursi, whose local presence in the forest suggests an approach to the high pine-barrens. Growing out of the leafy litter, and sometimes almost covered by it, are found Cladonia rangiferina, Cypripedium acaule, Galactia regularis, Chimaphila umbellata [Browns Mills], Epigaea repens, Gaultheria procumbens, Pyrola rotundifolia [Atco], Pyxidanthera barbulata (Fig. 61), and in the autumn various fleshy fungi.

Attention should be drawn to localities where the conditions of growth are somewhat different from the above general description. Such a locality is found along the railroad between Lakehurst and Toms River. Here, beneath the scattered pine trees on a sloping bank of exposed white sand, grew Arenaria caroliniana in cushions (Fig. 69), Euphorbia ipecacuanhae in prostrate rosettes with a great variety of colors and shapes of leaves, tufts of Hudsonia ericoides (Fig. 60), mingling with Hudsonia tomentosa, Chrysopsis falcata, scattered, flat cushions of Pyxidanthera barbulata (Fig. 61), upright fragrant golden-rod, Solidago odora, clumps of Tephrosia virginiana, and bunches of the beard-grass,

FIGURE 59

Road through typic pine forest where pitch-pines, Pinus rigida, are approaching maturity. Note the secondary oaks and heather thicket beneath the pines. (Courtesy N. J. Forest Commission.)

Andropogon scoparius. The bracken-fern, Pteris aquilina, also grew in the same sterile sand with Asclepias obtusifolia (=amplexicaulis).

It can be stated as an almost general rule that where the undergrowth of low shrubs becomes too thick, the bearberry, Arctostaphylos uvaursi, has no chance of growth or is crowded out. On flatter levels the low ericaceous shrubs grow close together, catching the leafy litter in which grows the wintergreen, Gaultheria procumbens. The floweringmoss, Pyxidanthera barbulata, usually grows in compact, prostrate cushions in dry sand, or in hard, gravelly soil, but where in the shade it is covered with forest litter in looser soil, it grows in open, not in a compact form. It was found on March 10, 1910, with rich, brown evergreen leaves in dense cushions, studded over with closed white buds. From sheets of it in the Herbarium of the University of Pennsylvania it has been determined that the pyxie was in full bloom at Ancora on April 12, 1879, at Browns Mills on April 12, 1908, at Sumner April 27, 1907, at Medford on May 8, 1888. When in flower, the cushion is studded with small white, star-like flowers with conspicuous yellow anthers. At Woodmansie on June 13, 1910, were found capsules of this plant dehiscing, with the seeds falling out. The whole period of flower and fruit development is accomplished in about three months—March 10th to June 13th.

LOW OR WET PINE-BARREN FACIES.—Occupying the lower levels of the region, as the flat land along a meandering stream, or the bottom of broad, shallow valleys between low, undulating hills, are found the low, wet, or moist pine-barrens. The soil is kept moist by the slow descent of the water-level from the hill-slopes, and also because, after a rain, the run-off is slow and portions of the level are subjected to a periodic overflow of the adjacent stream, or to the seepage of water through the sand from the same source.

FIGURE 60

Rounded clumps of pine-barren heather, Hudsonia ericoides, in full flower, growing in pine forest one mile south of Shamong (Chatsworth), May 27, 1916.

A number of additional species are found in the low pine-barrens that are not found in the dry pine-barrens.

The pitch-pine, Pinus rigida, is the dominant tree, but where the forest slopes down toward a cedar swamp, the white-cedar, Chamaecyparis thyoides, becomes associated with the pine and the secondary oaks become of less importance, if they do not disappear entirely. The sourgum, Nyssa sylvatica, is also an associate of the pine as a codominant tree. The sweet-gum, Liquidambar styraciflua, becomes an element of the pine forest at the outer edge of the pine-barren region, where the soil is richer, as on the road between Browns Mills and New Lisbon, as also the trembling-aspen, Populus tremuloides, and white-birch, Betula populifolia. As elements of the first story near the eastern border of the pine-barrens, as along Shark River, we find Quercus alba, Q. coccinea,

Q. prinus. The story of secondary trees in the wet pine-barrens consists of the red-maple, Acer rubrum, sweet-bay, Magnolia glauca, service-berry, Amelanchier intermedia [Browns Mills], and in drier soil, Quercus marylandica, Q. falcata [Allaire, Hammonton, Winslow Junction], Q. velutina, and Sassafras variifolium.

The third story consists, in the order of their abundance, of Azalea viscosa, Clethra alnifolia, Rhus vernix, Vaccinium corymbosum, Gaylus-sacia frondosa, Myrica carolinensis, Leucothoë racemosa, Kalmia lati-

FIGURE 61

Bed of flowering moss, Pyxidanthera barbulata, from New Jersey pine-barrens. (Photograph by Henry Troth.)

folia (Fig. 62), Vaccinium atrococcum, Ilex glabra, Cassandra calyculata, Quercus ilicifolia, while the lianes are Smilax glauca, S. rotundifolia, S. tamnifolia (Sumner, Cedar Brook, Vineland, etc.).

The fourth story, or layer, of the low pine-barrens consists of Kalmia angustifolia, Pteris aquilina, Osmunda cinnamomea, Dendrium buxifo-lium (in drier places), Pyrus arbutifolia var. melanocarpa, Andromeda mariana, Ilex glabra, Gaylussacia resinosa, Comptonia asplenifolia, Solidago stricta [Woodmansie, Atsion, Toms River], and Xerophyllum asphodeloides, which is usually abundant (Fig. 63). The cinnamon-fern, Osmunda cinnamomea, in its growth forms large tussocks. After the death of the fern the pitch-pine becomes established on these tussocks, and it thus invades the wet places of the deciduous swamps by being raised in the early stages of its development above the general water-level. The fifth layer, or story, consists of Aletris farinosa, Anemone nemorosa [Williamstown Junction], Bartonia tenella [Spring Lake, Woodbine],

FIGURE 62

Laurel, Kalmia latifolia, in pine forest at Clementon, June 4, 1910, when in full flower.

Chrosperma muscaetoxicum [Bear Swamp, Vineland], Cyperus dentatus [Atsion, Batsto, Lakehurst], Drosera filiformis [Atco, Forked River, Hammonton], Eleocharis tuberculosa [Atco, Mays Landing, Toms River], Gaultheria procumbens, Gratiola aurea [Winslow, Woodbine], Lobelia Canbyi [Bamber, Lakehurst], L. Nuttallii [Cedar Grove, Winslow], Lysimachia stricta, Panicum ensifolium [Forked River], Polygala cruciata [Woodmansie], P. lutea [Albion, Island Heights], Rubus cuneifolius, Scleria triglomerata [southeast of Lakehurst], Schwalbea Americana [Woodmansie].

FIGURE 63

Several plants of turkey's-beard, Xerophyllum asphodeloides, associated with cinnamon-fern, Osmunda cinnamomea, in wet pine-barrens at Clementon. (Photograph by Genji Nakahara and J. W. Harshberger.)

COLORS OF THE PINE-BARREN VEGETATION

The seasonal aspects of the vegetation of the New Jersey pine-barrens are perhaps not so striking as in the prairie-plains region of the west, or in the desert region of Arizona, because the dominant tree of the region is the pitch-pine, which is an evergreen. We miss also the familiar spring flowers of the deciduous forest areas of the Piedmont region and the Allegheny Mountains, which are gay with wood anemones, spring-beauties, blood-root, rue-anemones, yellow dog's-tooth violets, and hepaticas before the tender green leaves of the forest trees appear by a bursting of the dominant buds. Associated with the pitch-pine are other evergreen species, such as the laurel, Kalmia latifolia, so that the somber greens of the pine forest prevail throughout the entire year. And yet, when we come to study the pine-barren vegetation intimately, we can distinguish a spring or vernal aspect, a summer or aestival aspect, an autumnal or serotinal aspect, and a winter aspect. Although these different aspects are characterized by the plants which are in vegetative vigor and flower at different seasons of the year, yet we find that we can distinguish to some extent the seasonal changes in the vegetation by the change in color of the foliage and of the constituent species of the forest in general. The following are some of the color schemes as they have impressed the writer in his tramps through the pine-barrens at various seasons of the year.

WINTER ASPECT.—The prevailing overhead color is the somber green of the pine trees, which meet, as the eye follows the crown into the perspective distance, to form a continuous mass of dark-green color (Fig. 65). The serried columns of the pine trees are of a reddish-brown color, relieved by the oaks, which at various seasons of the year, owing to their deciduous habit, display different colors. With the exception of a few herbaceous species, the winter conditions of the forest prevail until the end of March. Quercus marylandica frequently retains its yellowish-brown leaves over winter, and, therefore, there are splashes of that color against the dark reddish-brown color of the pine-tree boles. In some localities the white-oaks, Quercus alba, which are only 1.5 meters (5 feet) tall, retain their ochraceous leaves until the following spring. The bear-oak, Quercus ilicifolia, is a common element of the pine woods, growing about 1 meter (3 feet) high. Its dull, reddish-brown to purplish-brown persistent leaves, dark brown, smooth bark, and slender, yellowish-green to reddish-brown twigs add materially to the winter coloration of the pine woods. Although the leaves of Ilex glabra are of a dark-green color, yet in contrast with the darker browns of the forest they appear of a bright green shade, as do those of the sheep-laurel, Kalmia angustifolia, and wintergreen, Gaultheria procumbens. Mrs. Mary Treat,* who lived at Vineland, New Jersey, studied the pine-barrens from the stand-

* HARSHBERGER, J. W.: The Botanists of Philadelphia and their Work, 1909: 298–302.

point of an artist and a naturalist. She has studied the vegetation at all seasons of the year, and many facts presented in this section have been suggested from her published descriptions, but reset in botanic phraseology. The plants of the heath family, Ericaceae, are conspicuous in the winter aspect of the pine-barrens. The diversified forms and colors of the denuded branches and twigs of the deciduous members of the family are noteworthy. The downy young shoots of the stagger-bush, Andromeda mariana, are reddish-brown, and below them are the scaly flower-buds. The blueberry, Vaccinium pennsylvanicum, is a dwarf, straggling bush with large scaly flower-buds, arising from green branches that are also characteristic of the low shrub, Vaccinium vacillans. The sweet-pepper bush, Clethra alnifolia, has a dull-gray color, which is increased by the persistent gray capsules in spikes. As previously mentioned, Kalmia angustifolia is of a fresh green color throughout the winter. The sand-myrtle, Dendrium buxifolium, of low, bushy form, bears small, shining leaves, which become purplish-green on some sun-exposed plants. The ground color is a grayish-brown, which contrasts with the colors of the low, or creeping, ericaceous plants of the forest. The white and pink tips of the trailing-arbutus, Epigaea repens, with its evergreen leaves, are seen. Chimaphila maculata is green with white veins, and the persistent capsules of Pyrola rotundifolia stand well above the shining green leaves. Small downy buds are arranged along the stems of the sweet-fern, Comptonia asplenifolia. The long, curved, grass-like leaves of the turkey's-beard, Xerophyllum asphodeloides, are of a light green color, enlivening in patches the dull color of the forest floor.

SPRING ASPECT.—The advent of spring is marked by the blossoming of a number of plants. One of the first is the flowering moss, Pyxidanthera barbulata (Fig. 61), which in dense, green mats is studded over with star-like, white flowers. The heath-like Hudsonia ericoides (Fig. 60) lightens up the gray, sandy places with its bright yellow flowers. The stagger-bush, Andromeda mariana, and several of the blueberries are in flower, and as these plants grow at places in dense masses, the white and pinkish-white colors appear continuous. The male flowers of the pitch-pine in May elongate, and the bright, glaucous green color of the elongating new shoots change the somber green color of the pines into a lively light-green color, greatly different from the summer, autumn, and winter coloration of the trees. The sand-myrtles later burst into bloom, and their masses of white flowers enliven the dark-green color of the pine forest (Figs. 55 and 56). The leaves of the deciduous trees have appeared by the beginning of June. The secondary oaks in their early summer growth have a light grayish-green color. This is due to the persistence of the tomentum, which falls off as the leaves reach maturity. The light chlorophyll greens of Pteris aquilina, Osmunda cinnamomea (Fig. 64), Vaccinium pennsylvanicum, V. vacillans, and Gaylussacia resinosa are noteworthy. The sweet-fern, Comptonia asplenifolia, is of a dark-green color by this time.

SUMMER ASPECT.—One of the finest displays of color in the pine-barrens in early June is when the laurel, Kalmia latifolia (Figs. 54 and 62), is in full flower, varying from rose-pink to pure white, while the low laurel, Kalmia angustifolia, has rich clusters of deep rose-red flowers. The turkey's-beard, Xerophyllum asphodeloides (Fig. 63), is showy, with its compact racemes of white flowers which arise from a mass of vivid green, wiry leaves. The fly-poison, Chrosperma (Amianthium) muscaetoxicum, is another conspicuous, white-flowered herbaceous perennial. The goat's-rue, Tephrosia virginiana, adds its flower tints to the passing display in the pine forest. The wild-indigo, Baptisia tinctoria, with its yellow flowers, and the lupine, Lupinus perennis, with racemes of blue flowers (Fig. 66), are among the most conspicuous flowering plants of the region. The star-grass, Aletris farinosa, with spikes of white flowers, and the butterfly-weed, Asclepias tuberosa, with orange-red flowers, form spots of color in the forest.

FIGURE 64

Developing circinate fronds of cinnamon-fern, Osmunda cinnamomea, in low pine forest near Browns Mills, N. J., May 15, 1915. (Photograph by S. S. Schumo.)

The gay-colored bracts of the strong-scented horsemint, Monarda, are displayed in late summer.

AUTUMN ASPECT.—The low-growing oaks are slow to lay aside their rich summer green, but in October they assume their autumn tints. The sassafras is brilliant in yellow and scarlet, while the sumach and many shrubs of the heath family have turned a deep crimson, shading to purple. The golden asters, Chrysopsis falcata and C. mariana, display their bright yellow disc and ray flowers. The asters (Aster) and golden-rods (Solidago) are ablaze with gold and azure. The blazing-star, Liatris graminifolia, is conspicuous with its spikes of rose-purple flowers. Where the bunch-grass, Andropogon scoparius, occurs, it covers extensive areas with its brownish-yellow foliage cured to hay. The red fruits of the wintergreen, Gaultheria procumbens, are noticeable at short range, as are also the white ones of the waxberry, Myrica carolinensis.

"The autumn color of the pine-barrens varies from season to season. Some years the swamp maples are a blaze of red. The shrubs, too, make brilliant masses of color, especially those of the heath family. The Vacciniums are purple, crimson, and scarlet; Andromedas gleam through various shades of red to a bronze-purple, while the varied shades of red in Leucothoë mingle with the yellow of Clethra and Azalea viscosa. The fruit of black alder shines brightly red among its greenish-yellow leaves

FIGURE 65

Road through pine forest in Lebanon Reserve, with snow on the ground, December 29, 1908. Note the second and third layers of vegetation beneath the dominant pines. (Courtesy of N. J. Forest Commission.)

and contrasts well with the red-brown foliage of Alnus serrulata near by. Very handsome, too, are the cranberries, trailing through the sphagnum, with purple and green leaves and scarlet and crimson fruit. Different species of Smilax are clambering everywhere, mantling dead trees and every other unsightly object, and all this wealth of color mingled with the green of the pines, cedars, and laurel."*

A visit to the pine forest in November, 1912, enabled the writer to make the following notes: The evergreen species of the forest, as pre-

* TREAT, MARY: Garden and Forest, VIII: 452 (1895).

viously mentioned, are Ilex glabra, Kalmia latifolia, K. angustifolia, Dendrium buxifolium, Arctostaphylos uva-ursi, and Gaultheria procumbens. These all grow in different relationship to each other and in different shades of green. Against this dark-green background one sees splashes of bright-red color which prove to be the bright red leaves of Smilax glauca and Gaylussacia resinosa and a few lower leaves of Quercus stellata, while in slightly wet places the bright red to reddish-purple leaves of Vaccinium atrococcum are seen. But the color tones of the oaks in late autumn are deep brown to yellow-brown of

FIGURE 66

Blue lupine, Lupinus perennis, in full flower at Browns Mills, May 13, 1911. (Photograph by Miss Lydia P. Borden.)

the persistent leaves.   In a piece of pine woods used as a picnic ground, where the ground has been cleared repeatedly, and where it had been tramped hard, the bright colors of the low leafy sprouts contrasted strongly with the lifeless browns of the same species in the nearby forest. The sprout leaves of Quercus prinoides were orange-yellow to yellow-brown, then red to claret-red; those of Q. stellata, and Q. velutina were bright red; those of Q. ilicifolia, red to reddish-claret, reddish-brown to orange-red.

FIGURE 67

Pine forest of pitch-pine, Pinus rigida, at Mays Landing, April 6, 1910. Lower branches of pines draped with green brier, Smilax rotundifolia. Litter of pine needles broken with green mosses and beard-grass, Andropogon scoparius. In the background holly, Ilex opaca, post-oak, Quercus stellata, bayberry, Myrica carolinensis, etc. (Photograph by Henry Troth and J. W. Harshberger.)

Coastal Pine-Barrens

The pine-barrens reach the Atlantic coast between Sea Girt on the south and Asbury Park (Deal Lake) on the north, at Sea Girt, Spring Lake, Belmar, Avon, Bradley Beach, Ocean Grove and Asbury Park.

Rapid inroads have been made in this coastal forest since about 1876 by the establishment of various summer resorts, so that the original condition of the vegetation must be inferred from the remnants. At several points, notably Sea Girt, Spring Lake, and Belmar, this forest approaches the sea in almost unaltered state. It is, therefore, important to place on permanent record the facts obtained by a study of these original conditions. This study was made more complete by noting the more conspicuous plants on each of the building lots in the borough of Belmar, and by noting the vegetation left in isolation at various points between Belmar and Asbury Park. At Avon, along the north shore of Shark River Bay, were noted Pinus rigida, Quercus alba, Q. coccinea, Q. marylandica (Fig. 68), Q. rubra, Q. stellata, Prunus serotina, Sassafras variifolium, the waxberry shrub, Myrica carolinensis, and the tall herb, Baptisia tinctoria. In addition to the above the black-oak, Quercus velutina, was noted at Bradley Beach, Deal Lake, which is approximately the northeastern limit of the pine-barrens in New Jersey. Here were noted, August 7, 1909, Acer rubrum, Liquidambar styraciflua, Nyssa sylvatica, Pinus rigida, Quercus alba, Q. phellos, Q. rubra, Q. velutina, and Sassafras variifolium.

The remnant of the pine-barren vegetation, as determined by a lot to lot census at Belmar, consists of the trees and plants enumerated in the following list:

HERBS

Chrysopsis mariana.
Gerardia purpurea.
Lespedeza capitata.

Melampyrum americanum.
Pteris aquilina.
Solidago odora.

LIANES

Ampelopsis quinquefolia.
Smilax glauca.

Smilax rotundifolia.
Vitis labrusca.

Rhus radicans.

LOW WOODY PERENNIALS

Epigaea repens.

Rubus cuneifolius.

Vaccinium macrocarpon.

SHRUBS

Amelanchier canadensis.
Azalea viscosa.
Clethra alnifolia.
Gaylussacia frondosa.
Gaylussacia resinosa.
Ilex glabra.
Myrica carolinensis.
Prunus maritima.
Pyrus arbutifolia.

Quercus ilicifolia.
Rosa lucida.
Rhus copallina.
Sambucus canadensis.
Sassafras variifolium.
Vaccinium corymbosum.
Vaccinium pennsylvanicum.
Vaccinium vacillans.
Viburnum dentatum.

FIGURE 68

Barren oak, Quercus marylandica, growing beneath pitch-pine near Belmar, N. J., in coastal pine-barrens. (Photograph by E. L. Mix and J. W. Harshberger.)

TREES

Acer rubrum.
Carya alba.
Diospyros virginiana.
Juniperus virginiana.
Liquidambar styraciflua.
Magnolia glauca.
Nyssa sylvatica.
Pinus rigida.

Prunus serotina.
Quercus alba.
Quercus coccinea.
Quercus marylandica.
Quercus phellos.
Quercus rubra.
Quercus stellata.
Quercus velutina.

CONTRAST OF GERMAN AND NEW JERSEY FORMATIONS

It is important in phytogeographic work to contrast the different types of vegetation of the world with one another. One of the most important results of the international phytogeographic excursions, begun auspiciously in Switzerland and in England and continued by American ecologists through the United States in the summer of 1913, has been the comparison of the different kinds of vegetation types found on the different continents. Professor C. Schröter, on the trip of the phytogeographers to the pine-barrens of New Jersey, suggested the similarity of the physiognomy of the pine-barrens and the pine-heaths (Kiefernheide) of Europe.

FIGURE 69

Sandwort, Arenaria caroliniana, associated with spurge, Euphorbia ipecacuanhae, sweetfern, Comptonia asplenifolia, blueberry, Vaccinium pennsylvanicum, etc., at Shamong (Chatsworth), June 27, 1910. (Photograph by Henry Troth and J. W. Harshberger.)

In following out this suggestion, a reading of Graebner's "Die Heide Norddeutschlands" has enabled the writer, as one of the leaders of the excursion in New Jersey, to make a more intelligent comparison of the two types of vegetation. The heathland of North Germany is developed in a region controlled by an oceanic climate. Where the continental climate begins to prevail pine trees invade the heathland and it is converted into a pine-heath, with the pines as dominant plants and the heath flora as subordinate. A sojourn on the Island of Nantucket, 45 kilometers (28 miles) out at sea, during two summers, suggested in a study of the vegetation there an hypothesis as to the origin of the New Jersey pine-barrens. Remove the pine trees from the forests of New Jersey and some of the pine-barren species found there and you would have the conditions on Nantucket controlled by an oceanic climate (Fig. 150). The Nantucket vegetation is a true heath with many of the southern coastal plain species missing.   Add the pitch-pine, which is now invading the central part of the Island of Nantucket, and you would have a typic pine-barren vegetation or pine-heath (Kiefernheide) (Fig. 151).

Similarly, as in Germany, with the control of a continental climate, the pine trees in the New Jersey region become dominant and the heath-plants, in the form of Arctostaphylos, Dendrium, Gaylussacia, Kalmia, Vaccinium, and dwarf forms of Quercus become subordinate to the pines and form the characteristic undergrowth which has been described in detail on former pages.   The different facies of pine-barren vegetation which have been described may be compared with the Waldheiden of the Germans, for we find that Graebner distinguishes two types: " Kiefern-heide" and "Laubwaldheiden," and the distinctive facies of these two types are classified by him as follows:

WALDHEIDEN

1. *Typus Kiefernheide.*
    Facies b.   Kiefernheide mit Vorherrschen von Juniperus com-munis.
    Facies c.   Kiefernheide mit Vorherrschen von Rubus-arten.
    *Facies d.   Kiefernheide mit Vorherrschen von Arctostaphylos.
    *Facies e.   Kiefernheide mit Vorherrschen von Gräsern.
    *Facies f.   Feuchtige moosige Kiefernheide.
    *Facies g.   Kiefernheide mit Vorherrschen von Vaccinium myrtillus und V. Vitis-idaea.

2. *Typus Laubwaldheiden.*
    Facies a.   Birkenheide.
    *Facies b.   Eichenheide.

Those facies of the forest-heath in Germany which are similar physi-ognomically to the ones in New Jersey are marked with an asterisk. Facies g in Germany, with the prevalence of two species of Vaccinium, is represented in New Jersey by a pine forest with an undergrowth of Gaylussacia resinosa, Vaccinium pennsylvanicum, V. vacillans, and Kalmia angustifolia.   The oak-heath (Eichenheide) is represented in New Jersey by the secondary formation which we have designated oak-coppice, which in some places is prominent after the pines have been re-moved.   Oak-heath in New Jersey succeeds pine-heath.   The bear-berry, Arctostaphylos uva-ursi, is one of the most abundant plants on the heaths of Nantucket, where it forms carpets an hectare (1 to 2 acres) in extent, representing the bearberry heath.   It also occurs as a common undergrowth in the high pine-barrens of New Jersey, which thus may be contrasted with Facies d of the German pine-heath. A more detailed study by instrumental methods by travel in this country and abroad would be extremely profitable in working out the details of vegetational types.

CHAPTER VI

PINE-BARRENS TRANSITIONAL TO SEA-DUNE VEGETATION

From the ocean end of Como Lake, inland from the built-up portion of Belmar south to Como Lake, the prevailing vegetation, as indicated by the sky-line, is coniferous. The ocean front of this forest consists of low pine trees, as at Belmar Park. South of Como Lake, on the Como side, the prevailing tree is Pinus rigida, and the composition of the forest may be studied now at close range.

Crossing the dune complex covered with Ammophila arenaria, Myrica carolinensis, Solidago sempervirens, Strophostyles helvola, Triodia cuprea, we come to the strip of forest facing the northeast, with dead, wind-swept cedars and clumps of spire-shaped, living ones (Fig. 70), with an association of wind-swept trees of Pinus rigida, Magnolia glauca, Nyssa sylvatica, and Sassafras variifolium, intermingled with such shrubs as Cephalanthus occidentalis, Ilex opaca, Myrica carolinensis, Rhus copallina, and Vaccinium corymbosum, together with the switch-grass, Panicum virgatum, and the boneset, Eupatorium album (common). The vines are Ampelopsis quinquefolia, Lonicera japonica (introduced and naturalized), Rhus radicans, Smilax rotundifolia and Vitis labrusca. The vegetation of the dune complex merges with that of the thicket formation. We now enter the pine-barren formation, which extends inland. The composition of this forest in Spring Lake (Como) is as follows: the forest floor, covered with pine needles, is dotted over with clumps of Ammophila arenaria, Baptisia tinctoria, with a sprinkling

FIGURE 70

Extreme outer edge of pine-barrens fronting the sea dunes at Spring Lake, August 24, 1910. Note wind-swept trees of Nyssa sylvatica, Juniperus virginiana, Pinus rigida, with which as lianes are associated Rhus radicans, Vitis labrusca, Ampelopsis quinquefolia. (Photograph by E. L. Mix and J. W. Harshberger.)

of Sericocarpus asteroides.   There are isolated bushes of Myrica carolinensis and Quercus ilicifolia.   As we walk inland we find that the
marram-grass, Ammophila arenaria, disappears, and out of the pine
needles grow Baptisia tinctoria, Cypripedium acaule, Melampyrum
americanum, Smilax glauca, and
low trees of Prunus serotina.   Here
grow also Andropogon scoparius,
Chrysopsis mariana, Gaylussacia
resinosa, and Quercus ilicifolia.

Back of the transitional pine
forest (Fig. 71), which immediately
faces the ocean on the Spring Lake
and Belmar sides of Como Lake,
we enter the pine-barrens proper
(Fig. 72), where Pinus rigida forms

FIGURE 71
Open coastal pine forest with little undergrowth near Manasquan.

FIGURE 72
Pine forest (coastal) at Belmar, August 24,
1910.   Clump of waxberry, Myrica carolinensis, in foreground, and bear-oak, Quercus
ilicifolia, at the right edge of the picture.
(Photograph by E. L. Mix and J. W. Harshberger.)

an open growth in association with
the oaks Quercus alba, Q. marylandica (Fig. 73), Q. phellos, Q.
stellata, Q. velutina, and the red
cedar, Juniperus virginiana, occasional and of secondary habit.
Other trees of less importance
are Populus grandidentata, Prunus
serotina, and Sassafras variifolium.   The shrubs of this forest
formation are Amelanchier inter-

media, Andromeda mariana, Clethra alnifolia, Comptonia asplenifolia, Gaylussacia resinosa, Ilex glabra, Kalmia angustifolia, Myrica carolinensis, Prunus maritima, Quercus ilicifolia, and Rhus copallina, together with the lianes, Rhus radicans, Smilax glauca, and S. rotundifolia. The herbaceous plants noted in this coastal forest were the grasses Ammophila arenaria, Andropogon fur-

FIGURE 73

Black-jack oak, Quercus marylandica, at edge of sand dune, Belmar, N. J. Note that the lower part of the tree to the left has a cluster of adventitious roots formed when the sand covered the base of the tree. These roots no longer function, as the sand has been removed by wind action. (Photograph by E. L. Mix and J. W. Harshberger.)

FIGURE 74

Coastal pine forest at Belmar. This forest is second growth on an old abandoned field, as it is almost entirely devoid of undergrowth, and shows old furrows, as indicated by the parallel shadows running back in the picture. (Photograph by Mrs. Helen B. Harshberger.)

catus, together with Ascyrum hypericoides (in clumps), Asclepias obtusifolia, Aster ericoides, Baptisia tinctoria, Chimaphila maculata, Chrysopsis mariana, Cypripedium acaule, Erigeron canadensis, the prostrate, sand-loving Euphorbia ipecacuanhae, Gaultheria procumbens, Gerardia purpurea, Hudsonia tomentosa, Melampyrum americanum, Polygonella articulata, Pteris aquilina, Sericocarpus asteroides, and Solidago odora (Fig. 74).

TRANSITION FROM SEA DUNES TO PINE THICKET AT SEA GIRT

Starting inside the narrow dune area, the thicket formation is approached by crossing a narrow tension strip of Ilex glabra, I. opaca, low red cedars, Juniperus virginiana, bayberries, Myrica carolinensis, Prunus serotina, Rhus copallina, and the lianes, Ampelopsis quinquefolia and R. radicans. Here are found such herbs as Chrysopsis mariana, Eupatorium album, and Solidago sempervirens.   These plants enter the thicket and mingle with the wind-swept cedars, which are draped with the vines Ampelopsis

FIGURE 75

Blending of bay beach vegetation with Panicum virgatum, golden-rod, Solidago sempervirens, and dense forest. On bluff and inside forest grow Ilex opaca, red cedar, Juniperus virginiana, Quercus stellata, etc. Somers Point, April 6, 1910. (Photograph by Henry Troth and J. W. Harshberger.)

quinquefolia, Celastrus scandens, Rhus radicans, Smilax rotundifolia, and Vitis labrusca. The pitch-pine, Pinus rigida, the blackjack oak, Quercus marylandica, the holly, Ilex opaca, Sassafras variifolium, sour-gum, Nyssa sylvatica (occasional), are found as associates of the red-cedar, together with the huckleberry, Gaylussacia frondosa, bayberry, Myrica carolinensis, Rosa carolina, Viburnum dentatum, with such herbaceous plants as Chrysopsis mariana, Gerardia purpurea, Lespedeza capitata, and Meibomia obtusa. The other constituents mentioned above remain about the same.

BLUFF FOREST ALONG GREAT EGG HARBOR BAY AT SOMERS POINT
Below Somers Point the pine forest comes down to a steep bluff.   Here, on the front slope, April 2, 1910, grew Pinus rigida, *Juniperus virginiana, Quercus stellata, and *Myrica carolinensis (Fig. 75).   The narrow, shingly beach at the foot of the bluff (Fig. 75) was tenanted by the switch-grass, Panicum virgatum, Baccharis halimifolia, and the oak, Quercus stellata, with gnarled and twisted branches, and the holly, *Ilex opaca, in the order of their occurrence.

On the level ground back from the edge of the bluff (Fig. 76) the components of the forest were, in the order of their importance: Pinus rigida, *Ilex opaca, *Juniperus virginiana, Quercus stellata, Q. marylandica, Q. alba, *Cornus florida, Hicoria glabra, with the bushy

FIGURE 76
Interior view of bluff woods, Somers Point, with pitch-pine, Pinus rigida, holly, Ilex opaca, red cedar, Juniperus virginiana. April 6, 1910. (Photograph by Henry Troth and J. W. Harshberger.)

FIGURE 77
Interior of pine woods south of salt marsh, Somers Point, April 6, 1910, with Pinus rigida, Ilex opaca, Quercus velutina, Quercus alba, etc., and oak and pine litter. (Photograph by Henry Troth and J. W. Harshberger.)

*Myrica carolinensis and the lianes, *Smilax rotundifolia and Vitis sp. This type of forest (Fig. 77) graded imperceptibly into a facies where pitch-pine, Pinus rigida (some trees 6 decimeters in diameter), was dominant (Fig. 78).   The secondary trees were Ilex opaca, *Juniperus virginiana, Quercus alba, Q. marylandica, Q. stellata, and Q. velutina.   The forest litter consisted mainly of the fallen leaves of the oaks and the pines (Fig. 79).   The shrubby undergrowth consisted of *Gaylussacia resinosa,

FIGURE 78

Interior view of pine forest with large pine, Pinus rigida, 7.62 dm. (2½ feet) in diameter, surrounded by a thicket of Ilex opaca, Quercus alba, and wintergreen, Gaultheria procumbens, with a litter of pine needles and oak leaves, Somers Point, April 6, 1910. (Photograph by Henry Troth and J. W. Harshberger.)

*Ilex glabra, *Myrica carolinensis, *Rhus copallina, and young trees of *Ilex opaca (Fig. 80), Pinus rigida, Quercus alba, Q. marylandica, and Q. stellata. All these trees were festooned with climbing vines, which add to the jungle-like growth. The lianes noted in this forest were: *Ampelopsis quinquefolia, *Rhus radicans, *Smilax rotundifolia, and *Vitis labrusca. This dense thicket has been the favorite nesting place

FIGURE 79	FIGURE 80
Open mixed forest at Somers Point, with large white-oak, Quercus alba, Pinus rigida, Ilex opaca, Juniperus virginiana, Smilax herbacea, Vaccinium pennsylvanicum, etc. (Photograph by Henry Troth and J. W. Harshberger.)	Interior of pine forest, Somers Point, April 6, 1910, with Pinus rigida, Ilex opaca, Sassafras variifolium, Quercus prinoides, and Smilax rotundifolia. (Photograph by Henry Troth and J. W. Harshberger.)

of shore and migratory birds during the summer, and in winter, owing to the shelter afforded by the evergreen trees, many birds linger on their way south, especially if the winter is an open and a mild one. The trees, shrubs, and lianes of this thicket have been bird distributed with the exception of the pines. The same type of forest thicket is characteristic of the sea islands along the coast as far south as southern Florida. Its character is largely due to the trees with avivectent fruits and seeds.*

* To emphasize this important fact, all plants with bird-carried fruits have been marked in the above account by an asterisk.

Tension Line Salt Marsh to Pine Forest at Somers Point

A sharp tension line was found at Somers Point, between the pine forest and the salt marsh (Fig. 81). The salt marsh plants meet a narrow strip, in some places 6 meters (20 feet) wide, in which Panicum virgatum, Solidago sempervirens, and Baccharis halimifolia were in association. Back of this grassy strip the outer edge of the forest consisted of Juniperus virginiana, advancing upon the grassy strip, together with

FIGURE 81

Point of pine forest projecting into salt marsh.   Pinus rigida dominant, associated with Ilex opaca, Juniperus virginiana, Myrica carolinensis, fronted by a strip of Panicum virgatum, Somers Point, April 6, 1910.   (Photograph by Henry Troth and J. W. Harshberger.)

Ilex opaca, Pinus rigida, Quercus alba (large), Q. stellata, with Myrica carolinensis, Smilax rotundifolia, and the ground herbs Chimaphila maculata and Mitchella repens (Fig. 81). The forest proper on the north side of this tongue of salt marsh consisted of the codominant Ilex opaca, Juniperus virginiana, Pinus rigida, Quercus alba (Figs. 82 and 83), while the secondary and younger trees growing in suppression to the taller trees were Ilex opaca, Quercus alba, Q. marylandica, Q. stellata, Q. velutina, and Sassafras variifolium. The forest litter consisted of

oak and pine leaves, about four centimeters deep, out of which grew
Chimaphila maculata and Gaultheria procumbens. The close association
of dark evergreen trees of medium size suggested the sclerophyllous
Mediterranean forests described by European botanists (Figs. 84 and 85).

Transition Salt Marsh to Pine Forest between Ocean Gate and Barnegat Pier

Before reaching the fresh-water marsh, which was interspersed with
wooded islands, so that the marsh ran between the islands in the form of

FIGURE 82

Large white-oak, Quercus alba, at edge of salt marsh, Somers Point, April 6, 1910,
backed by a forest of Pinus rigida, Juniperus virginiana, Ilex opaca, fronted by
Myrica carolinensis, Baccharis halimifolia, and Panicum virgatum. (Photograph by
Henry Troth and J. W. Harshberger.)

wet sloughs, a grove of Pinus rigida is entered with about 100 dominant
trees. The undergrowth consisted of Acer rubrum, Diospyros virginiana, Nyssa sylvatica, Quercus falcata, and the shrubs Ilex glabra,
Myrica carolinensis, Sassafras variifolium, and Vaccinium corymbosum,
together with the herbs Gaultheria procumbens and Trientalis americana. The smaller islands were covered with medium-sized pine trees
associated with Acer rubrum, Nyssa sylvatica, and an undergrowth of
Andromeda mariana, Ilex glabra, Myrica carolinensis, Pyrus arbutifolia,

Rhus copallina, R. radicans, Rosa carolina, together with the ferns Aspidium thelypteris, Osmunda regalis, and the herb, Baptisia tinctoria. The sloughs between the islands were fringed with Panicum virgatum, Solidago sempervirens, which grew in drier ground, together with associations of the blue-flag, Iris versicolor, in pure growth, mingling at the water's edge with Castalia odorata, Hibiscus moscheutos, Osmunda regalis, Proserpinaca palustris, and grasses and sedges, such as Eleocharis tenuis and Scirpus Olneyi. The shrub which had advanced farthest into the marsh was Myrica carolinensis.

FIGURE 83

Blending of salt marsh with Panicum virgatum at inner edge, Baccharis halimifolia, Myrica carolinensis, and forest with Pinus rigida dominant, fronted by Ilex opaca, Juniperus virginiana, Quercus alba, Q. stellata, Somers Point, April 6, 1910. (Photograph by Henry Troth and J. W. Harshberger.)

PINE-BARRENS AT NORTHERN LIMIT

Between Farmingdale and Allaire a small grove of Pinus rigida exists as a northern tongue of the main forest farther south. Here the pine trees are of all sizes, associated with Acer rubrum, Fagus grandifolia (as a low tree), Hicoria glabra, Castanea dentata, Prunus serotina, Q. alba, Q. rubra, Q. stellata, Q. velutina, and an undergrowth of such shrubs as Gaylussacia resinosa, Myrica carolinensis, Vaccinium corymbosum, V. stamineum, V. vacillans, with the herbs Chimaphila maculata and

Melampyrum americanum. On a gravelly rise of ground a slightly different facies is found, with dominant pitch-pine associated with Betula populifolia, Quercus alba, Q. ilicifolia, Robinia pseudacacia (introduced), Sassafras variifolium, together with Gaylussacia resinosa, Vaccinium pennsylvanicum, V. vacillans, and the herbs Baptisia tinctoria, Epigaea repens, and Melampyrum americanum.

FIGURE 84

Salt marsh and pine forest, Somers Point, April 6, 1910. Edge of the forest fronted with Myrica carolinensis, Juniperus virginiana (avivectent), Ilex opaca, and a group of white-oaks, Quercus alba, with Vaccinium corymbosum. (Photograph by Henry Troth and J. W. Harshberger.)

The outlying grove is in proximity to another and larger forest of pine-barren type. Here the prevailing pitch-pine, Pinus rigida, is associated with Betula populifolia (introduced), Castanea dentata, Populus grandidentata, Quercus alba, Q. marylandica, Q. phellos. The shrubs of this forest are Amelanchier intermedia, Andromeda mariana, Comptonia asplenifolia, Gaylussacia resinosa, Ilex glabra, Kalmia latifolia, Myrica carolinensis, Vaccinium pennsylvanicum, V. vacillans, together with the herbs Baptisia tinctoria, Epigaea repens, Gaultheria procumbens, Pyxidanthera barbulata, the fern, Pteris aquilina, and the reindeer-lichen, Cladonia rangiferina.

FIGURE 85

Mixed forest at Somers Point, fronted by salt marsh edged with isolated clumps of Panicum virgatum. The forest consists of Pinus rigida, Quercus alba, Ilex opaca, Juniperus virginiana. (Photograph by Henry Troth and J. W. Harshberger.)

TRANSITION FOREST AT WESTERN PINE-BARREN LIMIT

On the road between Browns Mills and New Lisbon is a swamp, west of which we enter a pitch-pine forest with pine trees 30 centimeters (1 foot) in diameter (Fig. 86), associated with Betula populifolia, Ilex opaca, Juniperus virginiana, Quercus stellata, and Q. velutina, together

FIGURE 86

Trunk and bark of pitch-pine, Pinus rigida, at Browns Mills, N. J.

with the shrubby Quercus ilicifolia. At the edge of this grove occur a few scattered trees of the transition pine, Pinus virginiana. On the opposite side of the road the pitch-pine, Pinus rigida, and the scrub-pine, Pinus virginiana, Ilex opaca, Quercus alba, are codominant (Fig. 87), and the undergrowth consists of Quercus ilicifolia, Q. stellata, Kalmia latifolia and Myrica carolinensis, with the herbs Chimaphila maculata, Panicum depauperatum and the reindeer-lichen, Cladonia rangiferina. This forest meets the deciduous forest of the Delaware Valley, composed of Castanea dentata, Fagus grandifolia, Ilex opaca, Liquidambar styraciflua, Platanus occidentalis, Quercus alba, Q. lyrata, Q. prinus.

The north bank of the Rancocas Creek is covered by a mixed deciduous forest with some Pinus rigida. The south bank has a wet pine-barren forest. The pine-barren vegetation extends west a greater distance on the south side of the North Branch of the Rancocas, at least 4.5 kilometers (3 miles) farther than on the north side, where the deciduous forest is found. Perhaps the explanation of this interesting fact is the same as the distribution of the vegetation of Mt. Fuji in Japan.* On the north slopes of that mountain the conifers predominate, while on the south side there is a most luxuriant growth of deciduous, broad-leaved trees. On the south the air is dry in winter; that does no harm to deciduous trees. In the summer, when the deciduous trees are covered with leaves, the ground is shaded and kept relatively moist, and thus, although they have a broad surface for transpiration,

* HAYATA, B.: The Vegetation of Mt. Fuji, 1911: 17.

the moisture is sufficient to supply the loss by that means. In winter they have no leaves, and, therefore, no considerable surface for evaporation. In short, deciduous trees do not require moist soil in winter, but do in spring and summer. Therefore, deciduous trees do well on slopes that are dry in winter and moist in summer.

The same type of transition forest is found near Williamstown Junction, where there is an even stand of Pinus virginiana mixed with Pinus rigida. The associated secondary trees in this forest are Ilex opaca, Quercus alba, Q. stellata, Sassafras variifolium, together with such tall

FIGURE 87
Mixed forest of codominant pines and oaks at Browns Mills Junction, October 8, 1909.
(Courtesy of N. J. Forest Commission.)

shrubs as Amelanchier intermedia and Kalmia latifolia. The smaller shrubs are Comptonia asplenifolia, Crataegus uniflora, Gaylussacia resinosa, and Myrica carolinensis. The composition of the forest at Berlin, where the pitch-pine and scrub-pine grow in association, is characterized by the associated oaks, Quercus alba, Q. marylandica, Q. prinus, Q. stellata, Q. velutina, and the shrubs Comptonia asplenifolia, Kalmia angustifolia, K. latifolia, Quercus ilicifolia, and Vaccinium pennsylvanicum. The herbs are Euphorbia ipecacuanhae, Gaultheria procumbens, Krigia virginica, and Lupinus perennis.

NOTES ON THE DISTRIBUTION OF TRANSITION PLANTS

In a swamp near Pemberton, N. J., there are standing trees of pitch-pine, Pinus rigida, associated with two non-pine-barren plants, viz., elder, Sambucus canadensis, and skunk-cabbage, Symplocarpus (Spathyema) foetidus. A cedar swamp near by, with Chamaecyparis thyoides, was characterized by an association of Amelanchier intermedia, Clethra alnifolia, Kalmia angustifolia, Liquidambar styraciflua, Magnolia glauca, Pinus strobus, and the herb Iris versicolor. Near Sumner the skunk-cabbage is associated with a few white-cedars. This association is more pronounced a few miles west of Pt. Pleasant, where Symplocarpus was found in the heart of a cedar swamp along with Decodon verti-

FIGURE 88

Typic pine forest on either side of fire strips along Central Railroad of New Jersey at Shamong, June 5, 1907. (Courtesy N. J. Forest Commission.)

cillatus and Peltandra virginica. The transition limit of the tulip-poplar, Liriodendron tulipifera, is near a small white-cedar swamp at Albion, 22 kilometers (14 miles) east of Camden. Here grew also Ilex opaca, Lindera benzoin, and the herb, Impatiens fulva, three non-pine-barren plants, associated with plants characteristic of two pine-barren swamps. At South Pemberton, Liriodendron occurs also. The extreme western limit of the white-cedar is in a grove of a few trees located in a small depression on the south shore of the lake at Clementon and in a grove 3 kilometers 2 miles east of Haddonfield, as also a small clump near the eastern edge of the town, according to Mr. James L. Pennypacker of that place. Many other interesting facts of like distribution could be determined if a botanist would follow around the border of the pine-barrens.

SUCCESSIONAL FORMATIONS

The destruction of the pine forest by fire, or the removal of the pine trees in lumbering operations, results in the replacement of the original forest by one of a different type. The composition of the second growth forest is dependent on a number of influential factors. One of the most important of them is the original character of the undergrowth of the pine forest, and another is the proximity of bodies of timber of a different character from that of the recently cleared area. The succession of species then in any particular region cannot be predicted definitely, but it varies according to the locality, the soil, and proximity to other masses of vegetation from which immigrant plants can come. The following notes, made at various times and at various places, will make clear the succession of vegetation types as it actually occurs in the pine-barren region.

NATURAL PINE-BARREN SUCCESSION.—It should be noted that when the pine-barrens (Fig. 88) have been cleared, as at Shamong (Chatsworth), without disturbing the soil conditions and the characteristic undergrowth, there has been no introduction of foreign weeds. Where the pine-barrens have been cleared, the vegetation of the unused clearings consists of trees, shrubs and herbs, which originally constituted the pine-barren undergrowth. Originally suppressed by the dominant pines, the undergrowth has a greater chance for development with the removal of the forest crown. The number of individuals of each species is increased, and there is an absolute increase of such plants as Rhus copallina and Smilax glauca, and the herbs, Aster dumosus,

FIGURE 89

Black-jack oak, Quercus marylandica, with Quercus ilicifolia, Gaylussacia resinosa, Vaccinium pennsylvanicum, Shamong, June 27, 1910. (Photograph by Henry Troth and J. W. Harshberger.)

Chrysopsis mariana, Eupatorium album, Gentiana porphyrio, Hudsonia ericoides, Liatris graminifolia, and Solidago bicolor var. concolor.

The succession, however, consists of the trees Quercus ilicifolia (=nana), Q. marylandica (Fig. 89), Q. prinoides, Q. stellata (Fig. 90), the shrubs, Comptonia asplenifolia, Sassafras variifolium, and the beardgrass, Andropogon scoparius. A few years of undisturbed conditions would restore probably the forest to its original state. A somewhat similar succession was observed at Atsion, but after fire destruction Quercus alba, the wild-indigo, Baptisia tinctoria, and the bracken-fern, Pteris aquilina, in addition to the above-mentioned plants, become of importance. If the pines do not reappear, the succession reaches its climax condition by the establishment of Quercus alba, Q. prinus, Q. velutina, and Sassafras variifolium.   A similar succession was noted on the road between Pasadena and Woodmansie, and it is mentioned here because one is enabled to add to the above list of plants that persist after the pines have been removed the following constituents of the pine-barren forest: Andromeda mariana, Comptonia asplenifolia, Crataegus uniflora, Vaccinium pennsylvanicum, V. vacillans, Tephrosia virginiana and the trailing-arbutus, Epigaea repens.   This natural succession was noted

again between Newtonville and Richland on March 23, 1910. Here Quercus coccinea occurred, together with Betula populifolia as an invader.   Kalmia angustifolia, K. latifolia, the hair-moss, Polytrichum commune, and the reindeer-lichen, Cladonia rangiferina.

OAK-COPPICE SUCCESSION IN TRANSITION REGION.—We have a succession at Williamstown Junction, at the head of Great Egg Harbor River, which is influenced largely by its proximity to the deciduous forests of West Jersey along the Delaware River.   The oak-coppice at this point consisted, on May 4, 1910, of Quercus alba, Q. ilicifolia, Q. marylandica, Q. phellos, Q. prinus, Q. stellata, and Q. velutina, as principal species, together with Acer rubrum, Hicoria glabra, Castanea dentata, Liquidambar styraciflua, Pinus echinata, P. rigida, P. vir-

FIGURE 90

Black-oak, Quercus velutina, backed by Quercus ilicifolia, Q. stellata, with an undergrowth of Gaylussacia frondosa, G. resinosa, and Hudsonia ericoides at Shamong, June 27, 1910.   (Photograph by Henry Troth and J. W. Harshberger.)

giniana, Sassafras variifolium, and the shrubs Kalmia latifolia, Vaccinium pennsylvanicum, together with the herbs Baptisia tinctoria, Comandra umbellata, Epigaea repens, Hieracium scabrum, and Lupinus perennis.   At the western border of the pine-barrens, as at Albion, whenever fields are abandoned they grow up to Pinus echinata, P. rigida, P. virginiana, Juniperus virginiana, Quercus ilicifolia, and Q. velutina.   That this forest succession has developed on an old field is indicated by the pure growth of Pinus virginiana clear of undergrowth.

OAK-COPPICE SUCCESSION IN CAPE MAY REGION.—The forest at Dennisville, between Cape May Court House and Bennett, consists of Quercus alba, Q. coccinea, Q. falcata, Q. palustris, Q. phellos, Q. prinus, Q. rubra, Q. stellata, Cornus florida, Hicoria glabra (= Carya porcina), Ilex opaca, Juniperus virginiana, Pinus rigida (occasional), P. virginiana, and between Bennett and Cape May the tulip-poplar, Liriodendron tulipifera, was found. The undergrowth in this forest consisted of Comptonia asplenifolia, Gaylussacia resinosa, Kalmia latifolia, Myrica carolinensis, Rhus copallina, Smilax rotundifolia, and the herbs Asclepias tuberosa, Baptisia tinctoria, Chimaphila maculata, Melampyrum americanum, and the lichen, Cladonia rangiferina.

MIXED PINE-OAK SUCCESSION.—Between the two arms of Shark River there is a peninsula covered with a forest of mixed pines and oaks. The edge of these woods presents a mixed character, owing to the fact that the pine has been cut and replaced by oaks mixed with pines. In some places the pine growth is dominant (Fig. 91); in other places the oak succession almost wholly so. Here is a good place at the northeastern

FIGURE 91

Pine grove on summit of Gravel Island in Shark River Bay, associated with oaks and other trees.

limit of the pine-barrens to study the succession, as influenced by the mass of deciduous forest trees lying to the north. On account of this proximity of a different type of vegetation, there are many plants, marked by an asterisk in the following account, which do not occur in the pine-barren region. At the edge of these woods, on the north bank of Shark River, with southern exposure, grow Acer rubrum, Nyssa sylvatica, Quercus alba, Q. marylandica, *Q. palustris, *Q. rubra, Sassafras variifolium, and the shrubs Alnus serrulata (= rugosa), Azalea viscosa, Gaylussacia resinosa and Myrica carolinensis. Higher upon the bluff grow Acer rubrum, Nyssa sylvatica, Pinus rigida, Quercus alba, Q. prinus, *Q. rubra, together with Chimaphila maculata, Cypripedium acaule, Gaylussacia resinosa and Ilex verticillata.

It may be said in general that the forest of Shark River peninsula consists of Acer rubrum, *Betula populifolia (rare), *Castanea dentata (rare), *Cornus florida, *Ilex opaca, *Liquidambar styraciflua, Nyssa sylvatica, Pinus rigida, Prunus serotina, Quercus alba, Q. coccinea, Q. marylandica, Q. prinus and *Q. rubra. The shrubs are Alnus serrulata (= rugosa), Andromeda mariana, Azalea nudiflora, Clethra alnifolia, Comptonia asplenifolia, Gaylussacia resinosa, Ilex glabra, Kalmia angustifolia, K. latifolia, Myrica carolinensis, Quercus ilicifolia (= nana), Q. prinoides, Rhus copallina, Sassafras variifolium, Vaccinium corymbosum, and the climbing vines Ampelopsis quinquefolium, Smilax glauca and S. rotundifolia. The herbaceous plants found here are: Chrysopsis mariana, Cypripedium acaule, Gaultheria procumbens, Osmunda cinnamomea, Melampyrum americanum, and Pteris aquilina.

Groves of pitch-pine occur in this forest where the pitch-pine is dominant. Where the forest fronts on reëntrant bays of salt marsh we find a fringe, or strip, of Alnus serrulata (= rugosa), Clethra alnifolia, Ilex glabra, Myrica carolinensis, Nyssa sylvatica, Pinus rigida, Rhus copallina, Quercus prinus, Q. stellata, Sassafras variifolium, Smilax rotundifolia and Vaccinium corymbosum.

COASTAL OAK FOREST SUCCESSION.—The oak succession forest at Belmar and Spring Lake consisted, on September 6, 1909, of such trees as Liquidambar styraciflua, Nyssa sylvatica, Pinus rigida, Quercus alba, Q. coccinea, Q. ilicifolia, Q. marylandica, Q. phellos, Q. prinus, Q. rubra, Q. stellata, Robinia pseudacacia, Sassafras variifolium, and the shrubs Clethra alnifolia, Gaylussacia resinosa, Prunus maritima, and the lianes, Smilax glauca, S. rotundifolia.

OAK-BOTTOM FORMATION.—Between the rounded pine-barren ridges, on the road from Barnegat to Warren Grove, are flat stretches with a more clayey soil, where the oak, mixing with the pine elsewhere, becomes here the dominant growth. Here the oaks, such as Quercus stellata, reach a diameter of 3 to 4 decimeters. The forest litter consists almost entirely of oak leaves. According to an old settler, these oak-bottom forests have existed always in the pine-barrens.

CHAPTER VII

PINE-BARREN PLANTS AS WEEDS

A striking fact about the pine-barren region is that the weeds are comparatively rare and easily recognized as such. They are found chiefly in the vicinity of the older and larger settlements, where the native vegetation has been damaged or destroyed by man. Spontaneous encroachment of introduced plants upon ground occupied by native plants is unknown practically, because the pine-barren formation is a closed formation and of an exclusive type. The following plants normal to the pine-barrens become, along roads and in abandoned fields, as at Whitings, more abundant, and may be reckoned as weeds, although they probably represent the natural regeneration of the vegetation, but as long as their growth opposed cultivation by man after clearing they should be classed with the weeds, as plants out of place: Andropogon scoparius, Aster spectabilis, Baptisia tinctoria, Chrysopsis mariana, Comptonia asplenifolia, Pteris aquilina, Quercus ilicifolia, Rubus cuneifolius, Sassafras variifolium, Tephrosia virginiana. Quite a few pine-barren plants do not yield to cultivation. Such are: Asclepias obtusifolia, A. tuberosa, Ionactis linariifolius, and Lupinus perennis.

INTRODUCED WEEDS

Mrs. Treat described, in 1892, a few of the weeds of southern New Jersey. Her list comprised yarrow, Achillea millefolium, shepherd's-purse, Capsella bursa-pastoris, pigweed, Chenopodium, morning-glory, Ipomoea purpurea, butter-and-eggs, Linaria vulgaris, and Japan honeysuckle, Lonicera japonica. It was several years after the land was cultivated before the dandelion, Taraxacum officinale, appeared. Now dandelions are everywhere.* Galinsoga parviflora was introduced at Vineland in 1897. Wild carrot, Daucus carota, was long established in 1897. The European hawkweed is also mentioned by Mrs. Treat. Among the troublesome grasses should be listed couch-grass, Agropyron repens, cheat, or chess, Bromus secalinus, in neglected vineyards; bur-grass,

---

* TREAT, MARY: Weeds in Southern New Jersey. Garden and Forest, v: 292; x: 313. Troublesome Grasses in Southern New Jersey. Garden and Forest, VIII: 103.

Cenchrus tribuloides, in neglected orchards; nut-grass, Cyperus rotundus, in cultivated grounds; crab-grass, Eleusine indica, in lawns.

On vacant lots at Shamong (Chatsworth) were noted Ambrosia artemisiaefolia, Diodia teres, Erigeron annuus, Euphorbia ipecacuanhae, Lepidium virginicum and Smilax glauca. The weeds found at Woodmansie on June 13, 1910, included, according to my notes, Chrysanthemum leucanthemum, Linaria canadensis, L. vulgaris, Melilotus alba, M. officinalis, Plantago lanceolata, P. Rugelii, Prunella vulgaris [Pasadena], Rudbeckia hirta [Pasadena], Taraxacum officinale, Trifolium pratense, T. repens.

The trees introduced into the pine-barrens for shade around houses and which are now fully established include Acer dasycarpum, Catalpa bignonioides [Hammonton, Newfield], Populus alba and P. deltoides.

CLASSIFICATION OF ALIEN PLANTS

The classification of alien plants, including weeds, is not an easy matter. Dr. M. Rikli, of the Polytechnicum, Zurich, has attempted such a classification.* The weeds belong to the youngest element of a flora and their existence is bound up with the activity of man. This element of a flora may be called the "Anthropophile" element, and the species belonging to it as "Anthropophytes." This anthropophile element may be divided into the

A. *Anthropochores* (Rikli), those plants which were not originally wild in the country, but which, by man's activity, were either purposely or unconsciously introduced;

B. *Apophytes* (Rikli) are species which were originally native in the country, but to some extent have abandoned their natural habitats and have gone over to the cultivated areas. Following these two heads and the several subdivisions which Rikli has made under them, the foregoing plants are arranged into—

A. ANTHROPOCHORES, brought into the country by man.

I. Intentionally introduced by man.

1. *Ergasiophytes*, exotic cultivated plants, including medicinal and ornamental plants, which have reached their habitat (field, garden, etc.) by the conscious activity of man and have been cultivated by him, *e. g.*, maize (Zea mays), sweet potato (Ipomoea batatas), white potato (Solanum tuberosum), etc.

2. *Ergasiolipophytes* (Naegeli and Thellung), relics of cultivation, were

---

* RIKLI, M.: Die Anthropochoren und der Formenkreis des Nasturtium palustre (Leyss) DC. Ber. d. Zurich botan. Gesell., 1901–03: 71–82. Review Botanisches Centralblatt, xcv. Nr. 1, 1904: 12.

originally planted in natural habitats and have maintained themselves without the intentional cultivation of man, *e. g.*, Japan honeysuckle (Lonicera japonica), white-poplar (Populus alba).

3. *Ergasiophygophytes* (Rikli), fugitives from cultivation, which have reached other habitats without the assistance of man, *i. e.*, grow wild, *e. g.*, red-clover (Trifolium pratense), white-clover (T. repens).

II. Brought into the country by the unconscious intervention of man, *e. g.*, foreign weeds.

4. *Archaeophytes* (Rikli), plants which have been in the region since prehistoric times. Probably no plant in the pine-barren region belongs to this class. The American lotus (Nelumbo lutea), introduced into South Jersey by the Indians, belongs here, but it grows at Woodstown and Sharptown, outside of the pine-barren limits.

5. *Neophytes* (Rikli), denizens, relatively frequent and constant in natural habitats, often associated with the native vegetation, *e. g.*, Achillea millefolium, Chrysanthemum leucanthemum, Daucus carota, Erigeron annuus, Linaria vulgaris, Rudbeckia hirta, etc.

6. *Epökophytes* (Rikli), colonists, of recent appearance, more or less numerous and constant in the country, but confined to artificial habitats. They are so far dependent on man for existence that their habitats require constant renewal, *e. g.*, alfalfa (Medicago sativa).

7. *Ephemerophytes* (Naegeli and Thellung), casuals, aliens, only a few and of casual occurrence, almost exclusively in artificial habitats. Owing to climatic conditions the seeds do not ripen, and the species disappear unless seeds are introduced.

B. Apophytes. Originally wild in the country in natural habitats, but later have gone over to cultivated areas.

I. Through the conscious activity of man.

8. *Oekiophytes* (Naegeli and Thellung), native cultivated plants, raised as ornamental, or economic, plants, *e. g.*, the cranberry (Vaccinium macrocarpon), lupine (Lupinus perennis), tall blackberry (Rubus argutus), dewberry (Rubus villosus), red-cedar (Juniperus virginiana).

II. Spontaneous immigrants.

9. *Spontaneous apophytes*, deserters, emigrants. Andropogon scoparius, Baptisia tinctoria, Pteris aquilina, Solidago odora, Tephrosia virginiana.

Pine-Barrens of Long Island

The only area in the Atlantic coastal plain where the pitch-pine, Pinus rigida, makes exclusive growths covering any considerable area of country is on Long Island. Nichols* states that, as at Farmington, the pitch-pine forms forests in Connecticut, but the undergrowth in that State is somewhat different, consisting of such shrubs as Ceanothus americana, Corylus americana, Comptonia asplenifolia, Quercus ilicifolia, and Rhus glabra. An intelligent comparison of the Long Island pine-barrens with those of New Jersey has been rendered possible by a visit to them on four occasions. The first visit was made with Dr. Roland M. Harper to the western end of the pine forest near Hicksville on August 25, 1909, and three later trips to the Long Island pine-barrens near Lake Ronkonkoma in the summers of 1913, 1914, and 1915, while the writer was in charge of the botanic classes at Cold Spring Harbor Marine Biological Laboratory. They have been crossed several times by trolley between Farmingdale and Amityville. The typic pine-barrens of Long Island occupy its south-central part, east of the region known as the Hempstead Plains. In fact, the eastern edge of the Hempstead Plains is invaded by a number of pine-barren plants. The pine-barrens of Long Island, according to Harper,† are confined to the southern half of Suffolk County, and extend a few miles into Nassau as isolated groves. The dry pine-barrens are fairly uniform over many square miles. The pitch-pine, Pinus rigida, is the dominant feature of the landscape as in New Jersey (Fig. 92), associated in some places with yellow pine, Pinus echinata. With the pines are associated, according to my observations, such trees as Quercus alba, Q. marylandica, Q. stellata, and in addition, according to Harper, Q. coccinea and Populus grandidentata. The shrubs of this forest include Andromeda mariana, Comptonia asplenifolia, Corylus americana, as in Connecticut, but not in New Jersey, Gaylussacia resinosa, Quercus ilicifolia, Q. prinoides, Rhus copallina, Salix humilis, Vaccinium pennsylvanicum. The herbaceous species include Aletris farinosa, Andropogon scoparius, Aster concolor, Baptisia tinctoria, Chrysopsis mariana, Tephrosia (Cracca) virginiana, Dasystoma pedicularia, Gaultheria procumbens, Ionactis linariifolius, Lespedeza capitata, Pteris aquilina, Sericocarpus asteroides, S. linifolius, Solidago bicolor, and S. odora. The bearberry, Arctostaphylos uva-ursi, and Smilax glauca are woody plants of importance in the forest vegetation. With the exception of the hazel, Corylus americana, all the other species are found in the pine-barrens of New Jersey.

* Nichols, George E.: The Vegetation of Connecticut, III, Torreya, 14: 190–1, October, 1914.

† Harper, Roland M.: The Pine Barrens of Babylon and Islip, Long Island, Torreya, 8: 1–9, January, 1908; The Hempstead Plains. Bull. Amer. Geogr. Soc., XLIII: 351–360, May, 1911; Torreya, 12: 277–287, December, 1912.

The explanation of this disjointed distribution of the pine-barren vegetation in Long Island and New Jersey has been given in some detail in a preceding section and need not be repeated here.

TENSION LINE BETWEEN PINE FOREST AND CEDAR SWAMP

The blending of the pine forest with the cedar swamp may be a gradual one if the forest is of the low, wet type, where the soil level slopes gradually to the stream along which the cedar swamp is found. The transi-

FIGURE 92

Pine-barrens of Long Island, with an even stand of pitch-pines and an undergrowth of dwarf chestnut oak, Quercus prinoides, and bear-oak, Quercus ilicifolia (nana), and associated characteristic growth near Lake Ronkonkoma, July 22, 1913.

tion may be very abrupt if the cedar swamp is situated in a depression with steep slopes on either side. The cedar swamp species will occupy the wet depression, and the pine forest vegetation will be found on the slopes and the flatter dry ground between the stream valleys. In some cases the transition is so sharp that one can stand with one foot in the cedar swamp and one foot in the pine forest. The accompanying figure (Fig. 93) illustrates the arrangement of species in abrupt tension lines at Shamong (Chatsworth). It will be seen from the illustration and from the account which follows that edaphic conditions control, *i. e.*, that the amount of water in the soil during the year determines the location of the pine forest and the location of the cedar swamps. This is illustrated characteristically in a small cedar swamp between Albion and Sumner, New Jersey, wherein the center of the swamp is a round,

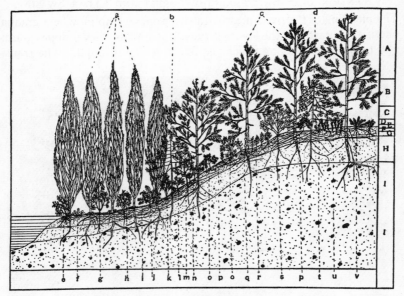

FIGURE 93

Schematic Section of Cedar Swamp and Pine-Barren Vegetation.

a, Chamaecyparis thyoides.	l, Gaylussacia frondosa.
b, Nyssa sylvatica.	m, Magnolia glauca.
c, Pinus rigida.	n, Azalea viscosa.
d, Quercus marylandica.	o, Gaultheria procumbens.
e, Alnus serrulata.	p, Pteris aquilina.
f, Sarracenia purpurea.	q, Smilax glauca.
g, Kalmia latifolia.	r, Kalmia angustifolia.
h, Clethra alnifolia.	s, Quercus ilicifolia.
i, Sphagnum cymbifolium.	t, Vaccinium pennsylvanicum.
j, Chamaedaphne calyculata.	u, Hudsonia ericoides.
k, Vaccinium corymbosum.	v, Myrica carolinensis.

A, Dominant tree layer.	F, Leaf mould.
B, Secondary tree layer.	G, Humus layer.
C, Shrub layer.	H, White sandy layer.
D, Taller herb layer.	I, Reddish, gravelly sand as subsoil.
E, Ground herb layer.	

elevated knoll of sand covered with Pinus rigida, completely surrounded by a cedar swamp filled with the white-cedar with a sphagnum basis.

At Shamong (Chatsworth) there is a cedar swamp with gradual slopes. The slope studied in particular faced the north, or was on the south side of the depression occupied with cedar swamp vegetation (Fig. 94). The upper dry slope was covered with scattered pines associated with Clethra alnifolia, Gaylussacia resinosa, Kalmia angustifolia, Quercus ilicifolia, Q. marylandica, Q. stellata, Vaccinium pennsylvanicum, V. vacillans, together with the herbs Euphorbia ipecacuanhae, Hudsonia ericoides, Pyxidanthera barbulata, and the bracken-fern, Pteris aquilina. At the foot of the slope, with a wetter soil under the pines which are still an element of the slope vegetation, appear Azalea viscosa, Clethra alnifolia (more abundant), Gaylussacia frondosa, and Ilex glabra. These plants close together to form a thicket about 1 meter high. Now appear in the association Alnus serrulata, Azalea viscosa (more common), Myrica carolinensis, Rhus vernix, Vaccinium corymbosum, and the cinnamon-fern, Osmunda cinnamomea. The massive bases of the cinnamon-fern project in some places above the wet surface of the swamp, and on the top of them young pitch-pine trees begin to grow until the swamp is invaded by scattered pines which have started on the top of the elevated fern rhizomes and stubs of old fern leaves. The wet soil of the depression at the edge of the cedar swamp proper is characterized by the presence of Acer rubrum, Clethra alnifolia, Gaylussacia frondosa, Ilex glabra, Magnolia glauca (= virginiana), and Nyssa sylvatica, which grow beneath the dense shade of the closely set white-cedar trees, Chamaecyparis thyoides. The tension line here is a broad one, because the slope is gradual. At another nearby point the tension strip is narrower, because the bank is steeper. Here, at the upper edge of the bank, were found Pinus rigida, Clethra alnifolia, Comptonia asplenifolia, Quercus marylandica, Q. stellata, and such herbs as Asclepias obtusifolia, Hudsonia tomentosa, and the bracken-fern, Pteris aquilina.

Along Deep Run, at Pancoast Mills, the tension line was studied along the north side of the cedar swamp, the slopes, therefore, facing south. The pine-barren level is low and the soil is wet. The pine trees are scattered, and the forest floor, therefore, is sunlit abundantly. Here, as a thicket undergrowth, were found Clethra alnifolia, Ilex glabra, Kalmia latifolia, Myrica carolinensis, Pteris aquilina, Smilax glauca, and the trailing wintergreen, Gaultheria procumbens. As an indication of the approach to the edge of the cedar swamp occur Acer rubrum, Azalea viscosa, Magnolia glauca, and the turkey's-beard, Xerophyllum asphodeloides. The interdigitation of the pines and cedars depends on the inequalities of the land surface, where a difference of a few centimeters in soil level makes an entire difference in the constituents of the formation. The pines are found on the drier knolls, while the white-cedars occupy the depressions between. These knolls represent probably the débris that has collected above and around the stumps of ferns, old pine and white-cedar trees, or they are hillocks of sand corresponding to unevennesses of the soil surface of the swamp.   Whatever their origin, they are characteristic of cedar swamps.   Over some of them grows the partridge-berry, Mitchella repens.

The blending of the pine forest and cedar swamp was studied at Bamber, New Jersey.   Here, on a wet slope, Pinus rigida is associated

FIGURE 94

South edge of cedar swamp blending with north-sloping pine-barren forest.   Note in the cedar swamp Chamaecyparis thyoides, Acer rubrum, Nyssa sylvatica, Azalea viscosa, and Clethra alnifolia, and in the pine forest, Pinus rigida, Quercus marylandica, Q. ilicifolia, Gaylussacia frondosa, Kalmia angustifolia, Vaccinium pennsylvanicum, Hudsonia ericoides and Pteris aquilina. Shamong (Chatsworth), June 27, 1910. (Photograph by Henry Troth and J. W. Harshberger.)

with Chamaecyparis thyoides.   The ground slopes steeply to the wetter level, where Acer rubrum, Kalmia latifolia, Pteris aquilina and Xerophyllum asphodeloides grow.   On the drier ground above the pines are small and scattered, associated with the shrubs Ilex glabra, Kalmia angustifolia, Myrica carolinensis, Quercus ilicifolia and small trees of black-jack oak, Quercus marylandica.   Two other plants are common, viz.: Gaultheria procumbens and Pteris aquilina.   A tension strip examined between Lakehurst and Whitings showed one of sharp character, the branches of the pine trees interlocking with those of the white-cedar trees. In this transition strip were found Arctostaphylos uva-ursi, Comptonia asplenifolia, Gaylussacia resinosa, Hudsonia ericoides, Quercus ilicifolia, Q. marylandica, Q. prinoides, Q. stellata, Vaccinium pennsylvanicum and V. vacillans.

CHAPTER VIII

CEDAR SWAMP FORMATION

The white-cedar trees, Chamaecyparis thyoides, form a noticeable feature of the landscape of the pine-barren region, growing as they do in close stands with their spire-shaped tops (Fig. 95). Where a pure cedar swamp can be seen in unobstructed view from a distance it presents a deeply serrated line (Fig. 95). The white-cedar trees are evergreen, and they form usually an even and dense growth. The trees grow so closely together that they form a dense shade, so dense that only the most tolerant species of shrubs and herbs will grow beneath the cedars (Figs. 1 and 96). Sometimes the trees are all of one age (Fig. 96); at other places there are trees of a second growth mingled with those that represent an older growth (Fig. 1). Formerly, the cedar swamps were more common and widely distributed than they are today (folded map, Plate I). The wood, being a most valuable one for shingles, firkins, tanks, etc., early in the history of the country, there was a great demand for white-cedar logs. Rapid inroads were made in the white-cedar swamps until only a remnant is left. Many open boggy places throughout the State owe their character to the fact that the ground they cover was originally covered with white-

FIGURE 95
Cedar swamp at Lake-hurst, August 30, 1909. Note serrate skyline of cedar trees and the gray birch, Betula populifolia.

cedars. Similarly many deciduous swamps represent the natural succession of vegetation where white-cedar has been removed in lumbering operations, and where no other attempt was made to clear the ground. Three facts are noteworthy about a cedar swamp. In the first place, a cedar growth is protective, as far as the movements of the air and the sunlight are concerned. It may be blowing a gale of wind outside, but inside a cedar grove it is calm and still, only the swaying of the tops of the trees indicating the commotion outside. This is evidenced by the persistence of the leaves of a number of deciduous trees, such as Magnolia glauca (=virginiana), on which the leaves may remain attached until the end of March, or until they are pushed off by the new leaves. Secondly, a cedar bog is cooler than the surrounding pine forest for several reasons. The ice of winter and late spring sometimes remains unmelted in the sphagnum bottom of the cedar swamp until May and occasionally later. The sphagnum, or bog moss, acts as a sponge, and therefore the water evaporates slowly from its surface and in its evaporation cools the air. Thirdly, on account of the density of the growth, a cedar swamp is warmer in winter than the pine forest. Fourthly, the dense shade and air-still condition are factors which keep the swamp cool and moist. No atmometric or photometric studies of these conditions have been made, but the environmental conditions, as above described, impress even the layman. The water held by the sphagnum is obtained from two sources of supply, viz., from rain and from the streams along which the groves of cedar usually are found. Sometimes when rains have been heavy the sphagnum is flooded with water. In exceptionally dry seasons the bog mosses are so dry that they lose their green color and become almost white and at the same time brittle, so that they easily crumble when rubbed between the fingers. The enlarged bases of the cedar trees are surrounded by raised cushions of bog mosses, consisting of Sphagnum acutifolium, S. cymbifolium, and S. squarrosum, according to the late Dr. A. F. K. Krout. Between these hummocks are depressions with standing water at certain seasons of the year and with the moss at that time supersaturated with water. Where the trees are farther apart and the depressed areas larger, small pools, or ponds, of water are formed, where a number of helophytes find congenial surroundings. At other times the pine-barren stream winds its way through these channels between the trees, uniting and separating, to unite again at places in its course through the white-cedar swamp.

The northern part of New Jersey is a glaciated region characterized by lakes and true bogs. Many of the bogs, not all, owe their origin to the invasion of the open water of a lake by bog plants. The aquatic plants begin to cover the surface of the lake; later they grow in choked masses, which are replaced by bog mosses and other bog plants. Finally shrubs and trees encroach on the surface of the bog until a climax condition is reached. The death and partial decay of these bog plants result in the formation of peat, of which there are considerable deposits in northern New Jersey. In southern New Jersey, however, we find cedar swamps with bog mosses of the genus Sphagnum, and the question naturally arises, why do these sphagnum areas, with or without white-cedar trees, occur in the coastal plain of New Jersey in an environment

FIGURE 96

Cedar swamp west of Mays Landing Station, with white-cedar, Chamaecyparis thyoides, laurel, Kalmia latifolia, as conspicuous undergrowth, sweet pepper-bush, Clethra alnifolia, sweet-bay, Magnolia glauca, etc.   (Photograph by Henry Troth and J. W. Harshberger.)

of surrounding sandy soils.   For it is an old tradition that bogs are found usually in temperate, cold, and humid climates, for as we advance toward the warmer climates, vegetable matter is more rapidly decomposed, until in tropic regions decay of animal and vegetal matter is so rapid that peat formation is unusual.   Some peat is formed in the bogs of southern New Jersey, but its depth is inconsiderable as contrasted with the thickness of peat in the glaciated parts of our country because the depressions are shallow.   To account for the presence of areas of sphagnum in the pine-barren region of New Jersey it is important to refer to the conditions under which bogs can exist with the formation of peat.

The pine-barren region of New Jersey is characterized by high summer temperatures, occasional droughts and dry summer winds, which conditions are adverse to bogs and peat formation.*   These adverse conditions are offset in part by the flat surface of the country, which is, therefore, poorly drained.   The streams have sluggish flow, and in the depressions of the land surface water is impounded.   If the rate of evaporation, which should be studied for masses of vegetation, and drainage is not too great, these depressions become filled with bodies of standing water, and if the water level does not have too great a fluctuation, such areas are invaded by aquatic plants and the margins by bog-loving species, or helophytes, and gradually the vegetation encroaches on the water surface until a bog is formed.   The formation of peat begins by a process of slow oxidation out of contact with the air, in which the amount of carbon increases as the volatile elements, like oxygen and hydrogen, decrease.   Water sufficient to cover the dead parts of the plants and prevent rapid oxidation is a prime factor in peat formation.   The presence of areas controlled by sphagnum in southern New Jersey (Map, Plate I) in a region of high summer temperature and drying winds is thus explained.

Concomitant with these conditions of bog formation and the presence of bog and cedar swamp vegetation in the coastal plain of New Jersey, we have produced characteristic features of such formations emphasized in dense shade, cool summer conditions and the protection, during summer and winter, which the white-cedar trees afford to the tolerant undergrowth of small trees, shrubs, and herbs.   The white-cedar trees in association to form a cedar swamp range in diameter from 5 centimeters (2 inches) up to 4.5 decimeters (18 inches) and 6 decimeters (2 feet) in diameter, and have a maximum height of 23 meters (75 feet).   The trunk

* Consult PARMELEE, C. W., and McCOURT, W. E.: A Report on the Peat Deposits of Northern New Jersey. Annual Report of the State Geologist, 1905: 223–307; DAVIS, CHARLES A.: The Uses of Peat.   Bull. 16, U. S. Bureau of Mines, 1911.

is tall and straight, tapering upward, so that with the slender, horizontal, short, spreading branches a narrow conic tree is formed (Fig. 93). The bark is about 2 centimeters thick, irregularly furrowed into narrow, flattish ridges, which separate into long, reddish-brown, fibrous scales (Fig. 97). The leaves are bluish-green, scale-like. Thus in serried ranks, so close together that one can hardly squeeze his way between the trunks, grow the white-cedar trees, their bases submerged at times with water, or again with the bog-moss substratum saturated with water.

FIGURE 97

Curly grass fern, Schizaea pusilla, at base of white-cedar tree, Chamaecyparis thyoides, in cedar swamp at Shamong, associated with Drosera rotundifolia and wintergreen, Gaultheria procumbens. (Photograph by Henry Troth and J. W. Harshberger.)

The oldest timber studied by Gifford Pinchot* was about eighty years old, at which age it had attained a maximum diameter of 38 centimeters (15 inches) and a height of 21 meters (70 feet). In early growth a tap-root is developed, but in later life the tree has a flat root system with strong, superficial, lateral roots. The regular growth of the tree in comparatively even stands renders the measurement of the trees simple. The following table was made by measuring 18 trees obtained during lumbering operations in Job's Swamp near Whitings and in Marigold Swamp near New Gretna.

* PINCHOT, GIFFORD: Silvicultural Notes on the White Cedar. Annual Report of New Jersey State Geologist (Report on Forests), 1899: 131–135; TAYLOR, NORMAN: A White-Cedar Swamp at Merrick, Long Island, and its Significance. Memoirs New York Botanical Garden, VI: 79–88, Plates 6–10, August, 1916.

SUMMARY OF 18 STEM ANALYSES OF WHITE-CEDAR

Tree Number	Diameter, Breast High, Inches	Height of Stump, Feet	Age, Years	Height, Feet	Total Volume, Cubic Feet
13	2.3	0.5	39	24.6	0.41
12	3.9	1.0	49	40.1	1.88
9	5.8	1.0	79	58.7	5.6
17	6.1	0.7	66	56.7	6.0
16	6.6	0.6	65	57.2	7.8
8	6.6	0.7	79	58.0	7.1
15	7.1	0.8	66	55.6	8.1
3	7.4	0.8	77	58.2	8.7
14	8.1	0.7	66	58.3	12.5
4	8.3	0.9	75	61.0	11.0
5	8.6	1.1	77	62.5	12.6
18	8.7	0.8	66	64.2	13.8
1	9.0	0.7	77	60.1	12.7
2	9.1	0.6	75	57.7	14.1
11	9.4	0.7	80	59.8	12.9
10	10.5	1.2	79	61.6	19.0
7	10.8	1.0	79	58.8	18.6
6	11.2	1.1	79	63.9	21.6

The rate of growth in diameter and height was obtained by plotting the progress of growth of each tree on cross-section paper and drawing a normal curve through the various points. The values from these average curves are as follows:

AVERAGE RATE OF GROWTH AND HEIGHT OF SEVEN WHITE-CEDARS

Age, Years	Diameter in Inches	Height in Feet
20	2.2	11
30	3.7	21
40	5.4	32
50	7.1	40.5
60	8.6	48
70	9.8	54.5
80	10.9	60

From this table it is seen that the white-cedar requires on an average sixty years to reach a height of 15 meters (50 feet), and eighty years to reach a height of 18 meters (60 feet). The trees grow more rapidly in diameter when in open places than when crowded in the forest. Valuation surveys show that at twenty years of age there were over 10,000 trees per 40 ares (1 acre), at forty years about 3,500 trees, and at eighty years in one case over 1,000 trees per 40 ares. From the accumulated data it follows: First, that it requires about sixty years to produce white-cedar lumber in paying quantities; second, that it would pay to thin the forest when it is about forty to sixty years old.

SUMMARY OF SAMPLE AREAS OF WHITE-CEDAR SWAMP

PLAT NUMBER	AREA	NUMBER OF CEDARS TWO INCHES AND OVER	NUMBER OF CEDARS UNDER TWO INCHES	AVERAGE DIAMETER, BREAST HIGH	MAXIMUM DIAMETER, BREAST HIGH	AVERAGE HEIGHT	AVERAGE AGE IN YEARS	NUMBER OF TREES TWO INCHES AND OVER PER ACRE	NUMBER OF TREES UNDER TWO INCHES
	Sq. Ft.			Inches	Inches	Feet			
57	400	107	..	0.3	..	20	21	11,663	..
58	2500	199	116	2.3	7	35	39	3,463	2,018
	Acres								
55	0.25	472	2	4.6	8	42	49	1,188	8
53	0.25	326	5	5.4	11	48.5	62	1,304	20
51	0.25	214	0	7.2	13	60	66	856	..
50	0.25	253	0	6.6	13	60	66	1,012	..
54	0.25	140	0	9.2	15	65	80	560	..
52	0.25	263	0	6.9	12	60	76	1,052	..

Two curves are presented with these figures in Pinchot's report. The first one shows the average growth in diameter of the white-cedar and the second figure depicts the curve which shows the average growth in height.

FIGURE 98

Sweet magnolia, Magnolia glauca, at edge of cedar swamp, Shamong (Chatsworth), June 27, 1910. Note fronds of bracken, Pteris aquilina. (Photograph by Henry Troth and J. W. Harshberger.)

SYNECOLOGY OF CEDAR SWAMPS

Associated with the white-cedars we find such trees as Acer carolinianum, Magnolia glauca (Fig. 98), and Nyssa sylvatica, which are more or less tolerant of the shade, together with taller shrubs, such as Azalea viscosa, Chamaedaphne (Cassandra) calyculata, Clethra alnifolia, Gaylussacia frondosa, Ilex glabra, I. verticillata (6 meters tall), Itea virginica, Kalmia latifolia (Figs. 96 and 99), Leucothoë racemosa, Rhus vernix, Sassafras variifolium, Vaccinium atrococcum, V. corymbosum, Viburnum nudum, Xolisma ligustrina, and many young cedar trees. The laurel-leaved smilax, Smilax laurifolia, is a liane clambering over the medium-sized trees, while Smilax rotundifolia is still more common, associated in places with Ampelopsis (Psedera) quinquefolia and Rhus radicans [Dennisville].   The cedar trees are draped with such lichens as Usnea barbata [One Hundred Dollar Bridge].

The undershrubs of these cedar swamps are Gaylussacia dumosa, Hypericum densiflorum, Kalmia angustifolia, while at the bog edges is found Dendrium buxifolium.   Two ferns are prominent elements of the cedar swamp vegetation, viz., the cinnamon-fern, Osmunda cinnamomea, and royal-fern, Osmunda regalis.   The grass-fern, Schizaea pusilla, is distributed locally in the New Jersey cedar swamps (Fig. 97), growing in the sphagnum at the base of the cedar trees, in which habitat I have found it at Island Heights Junction, at Forked River, at Shamong (Chatsworth) and near the Sym Place on Wading River.   The herbaceous species consist of the three sun-dews, Drosera filiformis [Warren Grove], D. intermedia and D. rotundifolia (Fig. 97), Eriocaulon decangulare, Eriophorum virginicum, Eupatorium verbenaefolium [Lakehurst], Gaultheria procumbens, Helonias bullata, Lachnanthes (Gyrotheca) tinctoria, Lophiola aurea (the latter two in open places), Lycopodium alopecuroides, Mitchella repens [Pt. Pleasant], Polygala cruciata [Lakehurst],

FIGURE 99

Cedar swamp between Whitings and Bamber, March 25, 1910, with white-cedar, Chamaecyparis thyoides, laurel, Kalmia latifolia, and sweet-bay, Magnolia glauca. (Photograph by Genji Nakahara and J. W. Harshberger.)

P. lutea, P. Nuttallii, Rhexia virginica, Rhynchospora alba, an abundance of the pitcher-plant, Sarracenia purpurea (Fig. 100), Trientalis americana and the cranberry, Vaccinium macrocarpon.

In the water of the channels between the white-cedar trees grow the white water-lily, Castalia (Nymphaea) odorata (Fig. 101), Eriocaulon decangulare, the golden-club, Orontium aquaticum, Sagittaria longirostra, Sarracenia purpurea, and the bladderwort, Utricularia clandestina. The edges of the streams that run through the cedar swamps, where the sunlight can penetrate during a part or the greater part of the day (Fig.

102), are bordered by such trees as Acer carolinianum, Magnolia glauca, Nyssa sylvatica, and such shrubs as Alnus serrulata (= rugosa), Clethra alnifolia, Ilex glabra, I. verticillata, Leucothoë racemosa and Vaccinium corymbosum.   The water's edge is marked by the presence of Castalia (Nymphaea) odorata, Dulichium spathaceum, Nuphar variegata, Rhynchospora alba, while floating submerged in the stream, with its long leaves streaming in the direction of stream flow, is found Scirpus subterminalis (Fig. 103).   An alga, Batrachospermum vagum, is attached to broken-off limbs that have fallen into the water.

FIGURE 100

Pitcher-plant, Sarracenia purpurea, in cedar swamp between Whitings and Bamber, March 25, 1910.   Note dense growth of sphagnum out of which pitcher-plants grow. (Photograph by Genji Nakahara and J. W. Harshberger.)

Basins of water occur in a number of cedar swamps where the water is impounded, especially during wet weather.   The water remains in them for some time, even during a dry spell.   Such basins of water are open to the sky above and have a rim of sphagnum, often raised into lumps. These small, almost circular pools of water are the primeval habitat of the white water-lily, Castalia (Nymphaea) odorata, Eriocaulon decangulare, E. septangulare, Rhynchospora alba, Sagittaria longirostra, the pitcher-plant, Sarracenia purpurea, while at the edges of the pool is found the golden-club, Orontium aquaticum.   Growing out of the sphagnum we find the shrubby leather-leaf, Chamaedaphne (Cassandra) caly-

FIGURE 101

Water-lily, Castalia (Nymphaea) odorata, in pool of cedar swamp at Shamong, June 27, 1910. Note Magnolia glauca, Ilex glabra, Clethra alnifolia, etc. (Photograph by Henry Troth and J. W. Harshberger.)

culata, with the herbs Drosera intermedia and Utricularia clandestina.
The outer edge, which blends with the cedar-tree-covered portion of the
swamp, is marked by Alnus serrulata, Azalea viscosa and Vaccinium
corymbosum.

In a pond at Ancora about 5 ares in extent was found Castalia (Nymphaea) odorata var. minus in the shallow water. On the wet ground around
the pond grew such shrubs as Azalea viscosa, Chamaedaphne (Cassandra)

FIGURE 102

Stream running through cedar swamp at Shamong (Chatsworth), June 27, 1910. On
right bank, looking down stream, grew Magnolia glauca, Acer rubrum, Clethra alnifolia, Alnus rugosa, and over the cedars, Smilax laurifolia. On the left side grew
Alnus rugosa, Ilex glabra, Clethra alnifolia, Magnolia glauca. (Photograph by Henry
Troth and J. W. Harshberger.)

calyculata, Gaylussacia dumosa (on rather dry knolls), Ilex glabra,
Kalmia angustifolia, with young trees of Acer carolinianum and Chamaecyparis thyoides and such herbs as Calopogon pulchellus, Drosera intermedia, Kalmia angustifolia, Lycopodium alopecuroides, Rhexia
virginica, Utricularia clandestina, and the cranberry, Vaccinium macrocarpon.

Several orchids are confined almost exclusively to cedar swamps.
They are: Arethusa bulbosa, Calopogon pulchellus, Habenaria blephari-

glottis (in flower August 17, 1912), H. clavellata, and Pogonia ophioglossoides.   The aroid, Peltandra virginica, together with Decodon verticillatus, Dulichium spathaceum, skunk-cabbage, Symplocarpus foetidus, wild-rice, Zizania aquatica, occur in a cedar swamp 3 kilometers (2 miles) west of Point Pleasant.   It is evident that the wild-

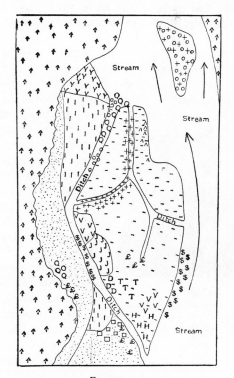

FIGURE 103

Associations in cedar-swamp stream, Davenport Branch, Toms River, August 2, 1909.

↟	Thicket.	Q	Nymphaea variegata.	D	Dulichium spathaceum.
Y	Floating sphagnum.	£	Peltandra virginica.	–	Panicularia.
I	Decodon verticillata.	V	Juncus militaris.	T	Iris prismatica.
∴	Unknown plant associa-	+	Sparganium.	$	Scirpus subterminalis.
	tions.	∧	Sagittaria longirostra.	H	Hypericum densiflorum.
o	Juncus sp.	J	Juncus canadensis.	□	Rhexia virginica.

rice has advanced up the stream from its tidal mouth and has migrated far enough inland to mingle with the typic plants of the pine-barren cedar swamps.   The sedge, Carex Collinsii, forms tussocks beneath the white-cedar trees along with Eriophorum virginicum and Xyris caroliniana.   The pitcher-plant, Sarracenia purpurea, changes its color with exposure to light.   In the dense, shaded swamp it is a bright green. In the open, sun-exposed places its leaves assume a reddish-brown color. The following plants are found at the edge of cedar swamps at Dennisville: Ilex opaca, Liquidambar styraciflua, Quercus prinus, and Solidago sempervirens. The two most western cedar swamps in New Jersey are located on the south bank of Clementon Lake, where a half-dozen trees were standing in 1914 (Fig. 104), and 1 kilometer (½ mile) east of Haddonfield, according to James L. Pennypacker of that place.

When a cedar swamp is cut over, if the conditions are favorable, young seedlings spring up in great abundance within two or three years. There is a better reproduction in dry swamps than in wet ones, chiefly because the standing water prevents the germination of the seeds. A certain amount of light is necessary for the welfare of the young plant, and when the trees grow large enough so that their crowns meet, the growth of the young trees is rendered precarious. As they grow older they become more tolerant of the shade. On an area 111 square decimeters (12 square feet), where the ground was completely seeded, there were 968 plants, which would make 3,513,480 young trees to 40 ares (1 acre). The other plat, 9 square meters (100 square feet), had 3,200 young plants, equivalent to 1,393,920 per 40 ares (1 acre). The youngest tree found by Pinchot in his studies to be bearing seed was thirteen years old.* Where the white-cedar trees have been removed without further molestation of the undergrowth, a deciduous swamp may arise in succession where Acer carolinianum, Azalea viscosa, Clethra alnifolia, Ilex glabra, Kalmia latifolia, Rhus vernix and Vaccinium corymbosum are abundant.

FIGURE 104

White-cedar, Chamaecyparis thyoides, in a small swamp of a few trees on the south shore of Clementon Lake, N. J., one of most western stations for the tree in the coastal plain of New Jersey, ten miles east of Camden.

ADDITIONAL WHITE-CEDAR SWAMP AND BOG PLANTS

The following species are found in white-cedar swamps, or in open sphagnum bogs without trees. The species of the list, arranged systematically, have not been mentioned in the foregoing account, but are given

* PINCHOT, GIFFORD: Loc. cit., 132.

with the name of one locality by way of identifying the plant with some specific habitat:

Dryopteris simulata (Forked River).
Lycopodium Chapmanii (Shamong).
   " carolinianum (Bamber).
Panicum lucidum (Atsion).
   " spretum (Lakehurst).
Sporobolus Torreyanus (Ancora).
   " serotinus (Speedwell).
Calamovilfa brevipilis (Atco).
Danthonia epilis (Forked River).
Panicularia obtusa (Egg Harbor).
Eriophorum tenellum (Speedwell).
Rhynchospora gracilenta (Bamber).
   " oligantha (Speedwell).
   " pallida (Browns Mills).
   " glomerata (Clementon).
   " axillaris (Batsto).
   " fusca (Shamong).
   " Torreyana (Atsion).
Cladium mariscoides (Dennisville).
Scleria triglomerata (Clementon).
Carex bullata (Toms River).
   " Walteriana (Cedar Brook).
   " oblita (Vineland).
   " livida (Ancora).

Carex exilis (Pt. Pleasant).
   " interior capillacea (Ancora).
   " atlantica (Sumner).
   " canescens disjuncta (Forked River).
   " trisperma (Bamber).
Juncus aristulatus (Hammonton).
   " pelocarpus (Berlin).
   " caesariensis (Shamong).
Tofieldia racemosa (Pole Branch).
Abama americana (Speedwell).
Helonias bullata (Clementon).
Blephariglottis cristata (Pt. Pleasant).
   " ciliaris (Bamber).
Sabatia lanceolata (Browns Mills).
Asclepias rubra (Bamber).
Gerardia racemulosa (Forked River).
Utricularia juncea (Ancora).
   " intermedia (Forked River).
Sclerolepis uniflora (Batsto).
Aster nemoralis (Bamber).
Eupatorium leucolepis (Speedwell).
   " resinosum (Atco).
Solidago neglecta (Clementon).

It will be noted that out of 47 plants in the above list 34 are monocotyledons, 3 are pteridophytes and the remaining 10 species are dicotyledons, showing the preponderance of monocotyledonous plants as helophytes.

Seasonal Aspects of White-Cedar Swamp Vegetation

The seasonal aspects of the vegetation of cedar swamps have been omitted purposely until this point, because a clearer expression of these aspects can be made in connection with a description of the physiognomy of such vegetation. The white-cedar swamps, having a relatively cold substratum because of the rapid evaporation there, vegetation is slow in responding to the increasing warmth of the spring days, but a notable exception to this general statement is the red-maple, which shows abundance of orange-red to crimson flowers against the unbroken green background of the white-cedar trees. The golden-club, Orontium aquaticum, with large leaves of a bluish-green, rich, velvety upper surface, that sheds the water, while the under surface is very smooth and of a light color (always wet), sends up bright yellow spadices from the dark-brown water and soil of the swamp in the early spring and lightens the small pool of cider-colored cedar water with splashes of yellow (Fig. 123). The white flowers of the leather-leaf, Chamaedaphne (Cassandra) calyculata, and swamp blueberry, Vaccinium atrococcum, are conspicuous in April. The somber green of the white-cedar swamp is almost unbroken by color until the summer aspect (Fig. 1). Then a number of shrubs and trees burst into flower. The fragrant, large white flowers of the sweet-bay, Magnolia glauca, appear against the leafy background of the cedars (Fig. 98). The laurel, Kalmia latifolia, produces its white, or pinkish white, masses of flowers, and during the year it is the most conspicuous flowering shrub of the white-cedar swamp (Fig. 96). Then, even in the deepest recesses of the swamp, masses upon masses of white to pink are seen between the closely set brown trunks of the spire-shaped cedar trees. Late summer sees the flowering of the swamp-honeysuckle, Azalea viscosa, with its glutinous white flowers, the corolla of which, as it drops off, is pendent on the long exserted style. The swamp is redolent with the perfume of the sweet pepper-bush, Clethra alnifolia, whose long spikes of white flowers almost completely cover the bush. The swamp loosestrife, Decodon verticillatus, is a handsome plant, growing sometimes in rather deep water with wand-like stems from 2 to 3 meters in length and clusters of pinkish-purple flowers in the axils of the upper leaves. The light orange heads of Polygala lutea are seen against the light green of the sphagnum background. Throughout the summer the boggy pools show the large, fragrant, starry white flowers of the water-lily, Castalia odorata (Fig. 101). The deep, reddish-brown color of the nodding flowers of the pitcher-plant, Sarracenia purpurea (Fig. 100), are also seen, while two white-flowered orchids, viz., Habenaria blephariglottis, H. clavellata, project their flower-spikes well above the mossy surface. The white flowers of Sagittaria longirostra add a dash of color along with the rusty-white heads of the cotton-grass, Eriophorum virginicum.

AUTUMN ASPECT.—The swamp-maple, Acer rubrum, holds its deep crimson leaves very late in the season, more especially the younger trees, and their reddish twigs are always conspicuous, as well as the large tresses of scarlet keys produced in late spring. This tree is always attractive. The poison-sumac, Rhus vernix, after the fall of its leaves, is noteworthy, with its stiff, hanging, axillary panicles of white berries. The tupelo, Nyssa sylvatica, in October is ablaze with bright crimson foliage, and it is easily distinguished from other swamp trees by its flat top and horizontal, spray-like branches. The cone-like fruit of Magnolia glauca is attractive in all of its stages. At first it is green, and as it matures it takes on yellow and rosy-pink hues, and when fully matured the carpels split open and reveal the beautiful, coral-red, berry-like seeds, which soon drop out, but dangle by threads. The twigs of the black-alder, Ilex verticillata, are encircled with masses of bright red berries, and the black, shining fruits of the inkberry, Ilex glabra, are also noteworthy.

WINTER ASPECT (Fig. 96).—A characteristic shrub of the white-cedar swamp is the wax-myrtle, Myrica carolinensis, with leafless branches almost covered with white, waxy fruits. The laurel, Kalmia latifolia, is evergreen, with large, glossy leaves, and its twigs of a bright yellow to red contrast strongly with the dark-green foliage. The leather-leaf, Chamaedaphne calyculata, with evergreen leaves rusty with scurfy hairs, shows racemes of flower-buds along the terminal branches and in the axils of the leathery leaves. But no plant is more attractive in winter than Azalea viscosa, the white swamp-honeysuckle. The large terminal flower-buds are variously tinted, as are also the twigs, branches, and leaf-buds. In striking contrast is the sweet pepper-bush, Clethra alnifolia, which grows beside the azalea. This shrub is a dull gray color, nearly all of its branches and twigs being surmounted with dry, gray capsules. The branches of Alnus serrulata (=rugosa) are metallic gray, terminated with clusters of male catkins, which in winter are reddish-brown and about an inch in length. The fertile spike of the same color is conspicuous. All the younger trees of Magnolia glauca hold their leaves until spring in the white-cedar swamps protected from the winds of winter. In the south it is evergreen, while in New Jersey it is only partly so. Its silky leaf-buds make the plant more interesting than in summer.

Mrs. Mary Treat, in an appreciative account written in 1894* under the caption "Christmas in the Pines," says, in her usual felicitous manner: "The Pines just now remind one of Florida rather than of New Jersey. I never remember to have seen as much freshness in a ramble here at Christmas time as now appears, and I can hardly realize that these are winter days. The foliage is so green and the berries so bright and plump that it is difficult to associate them with shrivelling frosts. Not a flake of snow has fallen, and many warm rains have served to keep up the illusion that somehow this is a belated section of summer. The woods are always fragrant, but now they do not exale the rich odor of fallen leaves, but the subtle, living aroma of growing pine trees, and the spice of shrubs and herb, which is always so refreshing."

FUNGOUS DISEASES OF THE WHITE-CEDAR.†—The white-cedar tree is not subject to any very serious disease. It is remarkably exempt from both insect and fungous enemies, and consequently it should be

* Garden and Forest, VIII: 3 (1895).
† Consult for details HARSHBERGER, J. W.: Two Fungous Diseases of the White Cedar. Proc. Acad. Nat. Sci. of Phila., 1902: 461–504, with 2 plates.

looked upon as a promising tree for systematic forestry in the eastern United States. Sydow, in his Index Universalis, gives 19 species of fungi living on Chamaecyparis thyoides. To this number one additional fungus should be added, viz., Gymnosporangium Ellisii. Of these, 10 species are found growing on the leaves, causing no material injury to them, as the fungi are usually found on dead leaves. Five fungi are confined to the branches, one is found on the trunk, two grow on the bark, two are found on the wood, and one fungus, Gymnosporangium biseptatum, occurs on both leaves and branches. The majority of these fungi are saprophytes living on the dead parts of the white-cedar. Only two fungi may be called disease-producing, viz., Gymnosporangium biseptatum and G. Ellisii.

FIGURE 105

Abnormal swellings on the white-cedar caused by a fungus, Gymnosporangium biseptatum. (Photograph by W. H. Walmsley.)

The swellings produced by Gymnosporangium biseptatum are quite characteristic (Fig. 105). The diseases may appear on trees which are 1 to 2 meters high, with stem about 2.5 centimeters in diameter. In these young trees the swelling surrounds the whole stem, being about 8 centimeters long and approximately spindle-shaped. The bark is fissured deeply by longitudinal cracks, which are somewhat wrinkled at the bottom. In a stem 1 centimeter in diameter the wood involved is quite sound, although in dried specimens it is of a more decided yellow color than the wood of the stem below, which is whitish in color. As the mycelium of the fungus is perennial, these club-shaped enlargements keep increasing in length and in diameter from year to year. The burl reached a diameter of 4 centimeters and was about 15 centimeters long in another somewhat larger specimen. The fissures become much deeper, due to abnormal formation of cork, until on one side of the stem the bark ridge is 1 centimeter high, the grooves being correspondingly deep.

The swellings produced by Gymnosporangium Ellisii are confined to the smaller branches (Fig. 106) and twigs. Near the summit of young white-cedar trees, where the branches grow upward and are thus more or less crowded together, all of such branches may be involved. The result is the formation of a fan-shaped mass of swellings. On one lateral branch of white-cedar 12 smaller branches were all massed together into a witch's-broom.

The teleutospores of Gymnosporangium biseptatum are 15–20 μ in diameter, 50–84 μ long, 3- to 4-celled or 2- to 6-celled. The intermediate hosts, species of Crataegus and Amelanchier, bear the aecial stage, known as Roestelia botryapites. The teleutospores of Gymnosporangium Ellisii are 10–16 μ in diameter, 75–190 μ long, 3- to 4- or 5-celled. The intermediate hosts, Pyrus malus and P. arbutifolia, bear the aecial stage known as Roestelia transformans.

ADDITIONAL FUNGI FOUND ON WHITE-CEDAR

In addition to the fungous diseases described above, the following fungi have been recorded as growing on the following parts of the white-cedar by various mycologists, and they are listed with synonyms in Part III, page 167, of Farlow and Seymour's Provisional Host Index of North American Fungi, 1891: Agaricus (Armillaria) melleus (trunk), Asterina cupressina

FIGURE 106

Gnarled branches (witch's-broom) of white-cedar induced by Gymnosporangium Ellisii. (Photograph by W. H. Walmsley.)

(leaf), Coniothyrium subtile (branch), Diplodia thyoides (branch), Hendersonia foliicola (leaf), H. thyoides (leaf), Hydnum Ellisianum (bark), Hypsotheca thujina (leaf), Hysteromyxa effugiens (leaf), Nectria truncata (branch), N. thujana (leaf), Nemacyclus griseus (branch), Omphalia campanella (trunk), Pestalozzia unicornis (bark), Peziza cupressi. It is improbable that all of these fungi occur on the white-cedar in New Jersey, and some of them may be accidental on that tree, but they must be considered when one wishes, by the process of exclusion, to determine the name of any particular fungus found upon this valuable tree.

CHAPTER IX

TRANSITION PINE FOREST TO DECIDUOUS SWAMP

At Mays Landing an east slope was studied where the pine forest descends into a depression filled with water (Fig. 107). Here the pitch-pine, Pinus rigida, is associated with the white-oak, Quercus alba, and such shrubs as Ilex glabra, Kalmia angustifolia, K. latifolia, Quercus ilicifolia (=nana), the wintergreen, Gaultheria procumbens, and the

FIGURE 107

Deciduous swamp between pine-barren vegetation to the right and left. Stream in foreground runs due south. Mays Landing, April 6, 1910. (Photograph by Henry Troth and J. W. Harshberger.)

bracken-fern, Pteris aquilina. The plants which are common to the two formations and which cause a blending of the formations along the tension line are Ilex glabra and Kalmia latifolia. In the swamp these two shrubs become associated with Acer carolinianum, Azalea viscosa, Clethra alnifolia, Magnolia glauca and Vaccinium corymbosum (Fig. 108).

Deciduous Swamp Formation

The successional history of the deciduous swamp may be looked for in the uninterrupted occupancy of stream depressions or valleys by an association of deciduous trees and shrubs that thrive in a wet, or swampy, soil, since the typic formations of the pine-barren region came to occupy the Atlantic coastal plain of New Jersey. Deciduous swamp vegetation has succeeded the white-cedar swamp vegetation in some places where, in the removal of the white-cedar trees, the typic deciduous-leaved undergrowth has been left undisturbed. The trees and shrubs that have been held in subjection as tolerant species by the dense shade of the dominant trees now have a chance to grow into the light and the air (Fig. 109).

FIGURE 108

Blending of deciduous swamp (left) and pine-barren (right). Here are associated Pinus rigida, white-oak, Quercus alba, laurel, Kalmia latifolia, Ilex glabra, bracken, Pteris aquilina, tall blueberry, Vaccinium corymbosum, etc. Mays Landing, April 6, 1910. (Photograph by Henry Troth and J. W. Harshberger.)

It is almost impossible at the present day to determine in all cases the origin of a particular deciduous swamp, whether it has been in undisturbed possession of a valley as a deciduous swamp for ages of forest history, or whether it represents the successional stage following the removal of white-cedar trees. Occasionally, if the succession is a recent one, there are left tall, solitary white-cedar trees along with the characteristic swamp species which indicate that the deciduous swamp represents the undergrowth of a white-cedar swamp that has grown up after the dominant white-cedar trees have been removed (Fig. 109). But the rapid destruc-

FIGURE 109

Deciduous swamp in leafless condition April 6, 1910. Large trees, the red-maple, Acer rubrum, in flower, with white-cedar. Mays Landing. (Photograph by Henry Troth and J. W. Harshberger.)

tion of the white-cedar trees, one hundred or two hundred years ago, has resulted in the spread of the deciduous swamp type of vegetation instead of the actual return to the original white-cedar swamp conditions, so that at this time we are unable to decide without historic data as to the exact successional relation of the deciduous swamps throughout the New Jersey coastal plain (Fig. 110 and folded map).

FIGURE 110
Blending of deciduous swamp and cedar swamp, Mays Landing. In deciduous swamp, a few sentinel white-cedars, also red-maple, Acer rubrum, sweet-bay, Magnolia glauca, laurel, Kalmia latifolia, blueberry, Vaccinium corymbosum, clammy azalea, Azalea viscosa. (Photograph by Henry Troth and J. W. Harshberger.)

The trees that grow to dominancy or supremacy in the deciduous swamp are Acer carolinianum, Amelanchier intermedia, Betula populifolia (introduced), Kalmia latifolia, Liquidambar styraciflua [Ocean Gate], Magnolia glauca, Nyssa sylvatica, and Sassafras variifolium (Fig. 113). Growing as secondary shrubby species we find Alnus serrulata, Azalea viscosa, Cephalanthus occidentalis, Chamaedaphne (Cassandra) calyculata, Clethra alnifolia, Ilex glabra, Leucothoë racemosa, Rhus vernix (Fig. 113), Vaccinium atrococcum, V. corymbosum and Xolisma ligustrina. The undershrubs which grow beneath and between the taller bushes are Andromeda (Pieris) mariana, Kalmia angustifolia, Myrica carolinensis, Pyrus arbutifolia. The cinnamon-fern, Osmunda cinnamomea (Fig. 114), sensitive fern, Onoclea sensibilis, royal-fern, Osmunda regalis,

FIGURE 111
Cinnamon-fern in deciduous swamp at Clementon, N. J.

marsh-fern, Dryopteris thelypteris, and chain-fern, Woodwardia virginica, are important elements of the deciduous swamp vegetation.

The lianes are Ampelopsis (Psedera) quinquefolia, Smilax glauca, S. rotundifolia, and S. tamnifolia [Woodmansie]. The blue-flag, Iris versicolor, is found in some deciduous swamps, as at Ocean Gate. In the wetter places, either intermingling with the shrubs and trees, or growing out into the open swamp or stretches of open water, we find Andropogon furcatus, Carex bullata, Dulichium spathaceum, Eriophorum virginicum, Juncus militaris, Lysimachia stricta, Orontium

FIGURE 112

Pitch-pine, Pinus rigida, cinnamon-fern, Osmunda cinnamomea, skunk cabbage, Symplocarpus (Spathyema) foetida, in association at Clementon. (Photograph by Genji Nakahara and J. W. Harshberger.)

FIGURE 113

Deciduous swamp forest, Belmar, August 24, 1910, with Acer rubrum, Quercus alba, Q. prinus, Sassafras variifolium, Magnolia glauca, Clethra alnifolia, Azalea viscosa, Gaylussacia frondosa and Smilax rotundifolia. (Photograph by E. L. Mix and J. W. Harshberger.)

aquaticum, Rhynchospora alba, Sarracenia purpurea, Sium cicutaefolium [Shark River], Symplocarpus foetidus [a transition species, Fig. 112], and Xyris fimbriata.

Floating on the water of streams that drain the deciduous swamps we find Castalia (Nymphaea) odorata, Potamogeton Oakesianus [Ancora], and submerged in the running water Scirpus subterminalis. The wet margins of the swamps are characterized by Drosera intermedia, D. linearis, D. rotundifolia, Galium pilosum puncticulosum, Hypericum densiflorum [Ancora], Lycopodium alopecuroides and Vaccinium macro-carpon.

BRANCH SWAMPS

In the southern states, where a small stream threads its way through pine woods with narrow wet banks, it is fringed on one or both banks with deciduous trees and shrubs.   Such narrow swamps are known as branch swamps.   Between Lakehurst and Whitings such a branch swamp was found.   Here grew Acer rubrum, Azalea viscosa, Clethra alnifolia, Ilex glabra, Magnolia glauca, Nyssa sylvatica, Rhus vernix, Vaccinium corymbosum, Xolisma ligustrina, and the fern, Osmunda cinnamomea.

DECIDUOUS SWAMPS OF TRANSITION REGION

These, as far as the physiognomy of the vegetation is concerned, are practically the same as the swamps in the heart of the pine-barrens, but the additional species are found that have migrated into the deciduous swamps along the edge of the pine-barren region from outside the limits of that region.   At Goshen and Dennisville, in South New Jersey, such transition swamps are found.   In a swamp south of Goshen are found in association such trees* as: Acer rubrum, *Carpinus caroliniana, *Liquid-ambar styraciflua, *Liriodendron tulipifera, Magnolia glauca, Quer-cus alba, Q. prinus, and the shrubs Alnus serrulata, Azalea viscosa, Cornus alternifolia, Ilex glabra, *I. opaca, *Lindera benzoin and Rhus vernix.   The vines are Ampelopsis (Psedera) quinquefolia and *Vitis labrusca.   The herbaceous plants of this deciduous swamp are Aralia racemosa, Carex intumescens, *Impatiens biflora (=fulva), *Lilium superbum, *Maianthemum bifolia, *Mitchella repens, Onoclea sensibilis, Osmunda regalis, *Podophyllum peltatum, Saururus cernuus, Thalic-trum polygamum and Trientalis americana.

The Dennisville Swamp is of interest because it was originally a white-cedar swamp and is replaced in part by deciduous vegetation.   Here an occasional white-cedar has been left after the lumbering operations. Here grow in association the poison-sumac, Rhus vernix, and the herbs Drosera rotundifolia, Hydrocotyle umbellata, *Lilium superbum, Os-munda cinnamomea and Pogonia ophioglossoides.   This swamp merges with a reed marsh which is fringed on the sides with Azalea viscosa, Clethra alnifolia, Ilex glabra, Osmunda regalis, Scirpus Olneyi and Vac-cinium atrococcum.

The deciduous swamps between Farmingdale and Allaire, along the north border of the pine-barrens, represent an association whose com-

* The trees and plants not found in the pine-barren region proper are marked with an asterisk.

position has depended upon the invasion of species from outside the region.   The trees noted as occurring in the deciduous swamps in this locality were Acer carolinianum, A. rubrum, *Betula populifolia, *Liquidambar styraciflua, Magnolia glauca, Nyssa sylvatica, Pinus rigida, Quercus alba, and *Q. phellos.   The shrubs which grow in these swamps are Azalea viscosa, Clethra alnifolia, Rhus vernix (Fig. 114), *Sambucus

FIGURE 114

Deciduous swamp forest with poison-oak, Rhus vernix, conspicuously as a branching shrub in the foreground, associated with Clethra alnifolia, Azalea viscosa, Vaccinium atrococcum, Acer rubrum and Pinus rigida, Belmar, August 24, 1910.   (Photograph by E. L. Mix and J. W. Harshberger.)

canadensis, Sassafras variifolium, Vaccinium corymbosum, Viburnum nudum and Xolisma ligustrina.   The woody lianes ascending the taller trees were Ampelopsis (Psedera) quinquefolia, Rhus radicans and Smilax rotundifolia.   The herbaceous swamp plants noted here include Drosera rotundifolia, Helonias bullata, *Mitchella repens, Nymphaea advena, Orontium aquaticum, *Symplocarpus foetidus, the cranberry, Vaccinium macrocarpon, and the ferns Osmunda cinnamomea and Onoclea sensibilis.

* The trees and plants not found in the pine-barren region are marked with an asterisk.

REED MARSH FORMATION

This type of formation does not exist typically in the heart of the pine-barren region, but at Dennisville, where the southern end of the pine-barrens touch the marshes in the Cape May region, typic reed swamps are found.   The ground of the reed marsh is covered with water during part of the day by a tidal flow of the streams that wind in a tortuous way and unite at places to form an intricate system of tidal channels. In these fresh-water marshes that front the pine-barren forest and blend with the white-cedar swamps (see ante) there are pure associations of the cat-tail, Typha angustifolia, covering areas of greater or less extent. There are also exclusive associations of the reed-grass, Phragmites communis, and of the sedge, Scirpus Olneyi.   The chocolate-brown water of

FIGURE 115
Marsh fronting the coastal pine forest on Shark River, July 31, 1909.   (Photograph by Roland M. Harper.)

the streams meanders between banks of Phragmites communis and Typha angustifolia.   The blue-flag, Iris versicolor, was seen in abundance along the south, or inner, edge of a reed marsh on Shark River (Fig. 115).

ADDITIONAL SWAMP PLANTS

Through the activity of the botanists of the Philadelphia Botanical Club a large collection of plants of the pine-barren region has been made, which, with those preserved at several large institutions of learning, have been consulted by Witmer Stone in the preparation of his report on the flora of southern New Jersey.   The following list of swamp plants arranged systematically has been compiled from Stone's book. The species mentioned in the descriptions above are omitted.

Sparganium americanum (Bamber).
Calamagrostis cinnoides (Speedwell).
Panicularia canadensis (Bamber).
"   obtusa (Lakehurst).
"   nervata (Speedwell).
"   pallida (Woodbine).
Cyperus strigosus (Belmar).
Eleocharis tricostata (Quaker Bridge).
Scirpus americana (Mays Landing).
"   eriophorum (Bamber).
Rhynchospora macrostachya (Manchester).
"   gracilenta (Atsion).
Scleria minor (Pasadena).
Carex folliculata (Albion).
"   intumescens (Allaire).
"   lurida (Penbryn).
"   Barrattii (Winslow).
Xyris torta (Woodbine).
"   Congdoni (Woodmansie).
"   fimbriata (Speedwell).
Eriocaulon compressum (Atsion).
"   decangulare (Atsion).
Juncus effusus (Atsion).
'   aristulatus (Bamber).
"   canadensis (Clementon).
"   acuminatus (Winslow Jc.).
Zygadenus leimanthoides (Atsion).
Melanthium virginicum (Cedar Brook).
Smilax Walteri (Atsion).
Dioscorea villosa (Belmar).
Iris prismatica (Williamstown Jc.).
Sisyrinchium atlanticum (Forked River).

Gyrostachys praecox (Atsion).
Polygonum Careyi (Manchester).
Spiraea latifolia (Hanover).
Rubus hispidus (Forked River).
Aronia arbutifolia (Atco).
Apios tuberosa (Bear Swamp).
Linum striatum (Shamong).
Polygala brevifolia (Shamong).
"   Nuttallii (Speedwell).
Ilex laevigata (Lakehurst).
Ilicioides mucronata (Atco).
Hypericum virgatum ovalifolium (Hammonton).
Viola primulifolia (Egg Harbor).
"   lanceolata (Clementon).
Rotala ramosior (Woodbine).
Rhexia mariana (Atco).
Ludvigia sphaerocarpa (Woodbine).
"   hirtella (Batsto).
"   linearis (Atsion).
"   alternifolia (Winslow).
Proserpinaca pectinata (Mays Landing).
Lysimachia terrestris (Forked River).
Sabatia lanceolata (Browns Mills).
Utricularia cornuta (Bamber).
Viburnum cassinoides (Bamber).
Eupatorium leucolepis (Atco).
Solidago fistulosa (Woodmansie).
"   uniligulata (Forked River).
Helianthus angustifolius (Atsion).
Coreopsis rosea (Browns Mills).

STREAM BANK FORMATION

Southern New Jersey is a country of low relief. There are no bold features of topography, so that the pine-barren streams wind their way to the sea, as a rule, without a strong current. The run-off is accommodated easily in ordinary rainy weather, and there are no sudden rises of water level, with consequent damaging floods, such as we find in hilly or mountainous country. The rain-water overflows the banks of these streams during times of exceptional rainfall, but the water spreads out over a flat country and over a soil peculiarly pervious and open, so that the damage from such excessive rains is always slight. The stream banks being low with but few exceptions, the vegetation that occupies these banks blends imperceptibly with that on either side, whether it be a pine forest, white-cedar forest, or deciduous swamp forest. The trees of the stream banks are those that can grow with their roots more or less in the water, and include Acer rubrum, Chamaecyparis thyoides,

Ilex opaca, Nyssa sylvatica, Pinus rigida (occasional), Quercus alba, Q. marylandica, Q. prinus, Q. stellata, and Sassafras variifolium. The most common shrubs of the stream banks are Alnus serrulata, Azalea viscosa, Cephalanthus occidentalis, Clethra alnifolia, Ilex glabra, I. verticillata, Kalmia latifolia, Leucothoë racemosa, Myrica carolinensis, Rhus copallina, Rosa lucida, Sambucus canadensis [Upper Wreck Pond], Vaccinium corymbosum, while the climbing vines, or lianes, are conspicuous at times ascending the taller trees, such as: Ampelopsis (Psedera) quinquefolia, Rhus radicans and Smilax rotundifolia. The herbaceous plants which characterize the stream banks of the coastal plain of New Jersey are Carex bullata, Cicuta maculata, Decodon (Nesaea)

FIGURE 116

Toms River, N. J., looking up stream from railroad trestle, with associations of Acer rubrum, Alnus, Dulichium, Hypericum densiflorum, Peltandra, Pontederia and Zizania. 2.27 P. M., August 2, 1909. (Photograph by Roland M. Harper.)

verticillata, Drosera intermedia, Dulichium spathaceum, Eupatorium purpureum, Galium pilosum puncticulosum, Hibiscus moscheutos, Hydrocotyle umbellata, Hypericum densiflorum, Impatiens biflora (=fulva), Juncus militaris, Lobelia cardinalis, Lysimachia stricta, Orontium aquaticum, Osmunda cinnamomea, O. regalis, Polygonum pennsylvanicum, Rhynchospora macrostachya and Xyris fimbriata.

At the edge of the stream banks, and growing in the water, occur Castalia (Nymphaea) odorata, Nymphaea (Nuphar) variegata, Peltandra virginica, Pontederia cordata, Potamogeton Oakesianus with floating leaves, and Scirpus subterminalis, almost wholly submerged and with its leaves floating in the direction of the stream flow (Figs. 116 and

117). Many of the stream bank species are typic aquatic plants and helophytes. Some of them have been described as constituents of the white-cedar swamp vegetation and some of them will be considered with the true pond and lake plants of aquatic character.

The vegetation of wet river-banks at the northern limit of the pine-barrens is studied best along the south shore of Shark River. On reaching the northern limit of the pine-barrens, where they begin to be replaced by the deciduous forest vegetation, we find an overlapping of the two types of vegetation. The distribution of the two types in such a transition region is dependent on exposure and on soil conditions. On the dry banks of Shark River, for example, where they are fully exposed to the sun, as is the north bank with southern exposure, we find the pitch-pine with associated species to be dominant. On the south bank of the river, north facing and shaded, we find that the soil is wet and springy and that the deciduous leaved vegetation is the prevalent type. In winter, with the fall of the leaves, these trees and shrubs, although exposed to trying circumstances in the north exposure, are more or less xerophytic in their leafless condition. With wet soil, the south bank vegetation partakes of the nature of a true

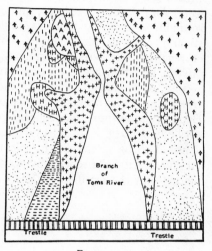

FIGURE 117

Diagram of stream shown in Figure 116.

+ Pontederia cordata at edge of water. Peltandra virginica.
H Hypericum densiflorum.
| Dulichium spathaceum.
— Zizania aquatica.
A Alnus rugosa.
⋏ Thicket with Acer rubrum.
.. Unknown plant associations.

deciduous swamp, but with this important difference that the flora is much more varied and richer and shows the presence of species not found in the pine-barrens. The composition of the vegetation of the Shark River shore reminds one of the forest flora of the Piedmont region, where the environment is more mesophytic and the soils more retentive of moisture. The following trees are characteristic of the south bank of Shark River near Belmar, New Jersey: *Acer rubrum, Betula populifolia, Hicoria glabra (=Carya porcina),

* The species marked with an asterisk are typically pine-barren ones.

Diospyros virginiana, Ilex opaca, Morus rubra, *Nyssa sylvatica, *Pinus rigida, *Prunus serotina, *Quercus alba, Q. prinus, Q. rubra, Q. stellata, Q. velutina.   The trees of secondary size are Cornus florida, Hamamelis virginiana, *Kalmia latifolia, *Magnolia glauca, *Sassafras variifolium.   The river-bank shrubs are *Alnus serrulata (= rugosa), *Azalea viscosa, Cephalanthus occidentalis, *Clethra alnifolia, *Gaylussacia frondosa, *Ilex glabra, I. verticillata, Lindera benzoin, *Myrica carolinensis, *Rhus copallina, *R. vernix, Sambucus canadensis, *Vaccinium corymbosum, Viburnum dentatum.   The lianes noted by me were Ampelopsis (Psedera) quinquefolia, *Rhus radicans, *Smilax rotundifolia and Vitis labrusca, while the commonest ferns were Polystichum acrostichoides, Onoclea sensibilis, *Osmunda cinnamomea, and O. regalis.   The herbaceous plants of these river-banks show by their mixture that they are of the pine-barrens and from the deciduous forests, swamps and marshes outside of the pineries.   They include Arisaema triphyllum, Carex stricta, *Chimaphila maculata, *Decodon verticillatus, Eupatorium purpureum, *Epigaea repens, Hibiscus moscheutos, Hydrocotyle umbellata, Impatiens biflora, Juncus canadensis, Laportea canadensis, Lobelia cardinalis, L. spicata, Maianthemum bifolia, Medeola virginica, Mitchella repens, Symplocarpus foetidus (in wettest swampy places) and the cranberry,* Vaccinium macrocarpon.

CHAPTER X

POND PLANT FORMATION

The glaciated region of the northern United States and of northern New Jersey is rich in lakes and ponds, but in the coastal plain, excepting the artificial ponds which are formed by damming a stream, or which are formed when cranberry bogs are flooded, the natural ponds or lakes are relatively few. An inspection of the maps published by the New Jersey Geological Survey shows the presence of such natural ponds as Grassy Pond, Wall Pond, Five-acre Pond, Bear's Head Pond (Tuckahoe Sheet), Savage Pond (Dennisville Sheet), and several other un-named ponds.

The heart of the East, or Lower, Plains is characterized by a depression in the low hills occupied by a tarn, or small natural lake, known as Watering Place Pond (Figs. 118 and 119), well known to old hunters who killed deer on its margin. This pond is bordered with sphagnum, out of which grow dense thickets of Azalea viscosa, Chamaedaphne (Cassandra) calyculata, Clethra alnifolia, Ilex glabra, Myrica carolinensis, and

FIGURE 118

Watering Place Pond, Lower Plains, with islands of leather-leaf, Cassandra (Chamaedaphne) calyculata. Corema Conradii is abundant on hills beyond. April 9, 1912.

Vaccinium corymbosum, with such associated herbaceous species as Lycopodium alopecuroides, Sarracenia purpurea, Vaccinium macrocarpon and Xyris Congdoni. A sedge, Dulichium spathaceum, is abundant along the pond margin. The open stretches of bog-moss border of this pond, according to Saunders,* who visited it late in the season, were characterized by Drosera filiformis, D. intermedia, Hypoxis erecta, Limo-

---

* SAUNDERS, C. F.: The Pine Barrens of New Jersey. Proc. Acad. Nat. Sci., Phila., 1900: 544–549; STONE, WITMER: Plants of Southern New Jersey, 1911: 73, 215, 485 and 802.

dorum tuberosum, Lysimachia stricta, Pogonia ophioglossoides, Polygala lutea, P. Nuttallii. The floating vegetation of this pond comprises Castalia (Nymphaea) odorata, Proserpinaca pectinata and Utricularia sp. A graceful southern sedge, Carex Walteriana, is abundant in the shallow water at the margin of the pond, along with the fern, Woodwardia virginica.

An interesting fact is mentioned by Stone with reference to the dams across pine-barren streams. The peculiar coastal flora of New Jersey will be found running up the tidewater streams and their tributaries into the pine-barrens as far as the artificial dams, which now seem to mark the limit of such coastal intrusion. On other streams the coast plants follow back to the natural limit of tide-water, and Stone thinks that some isolated colonies of such species within the pine-barrens owe their presence to the intrusion along tide-water streams that were dammed subsequently. The wild-rice, Zizania aquatica, follows the course of small streams several miles back from the coast, as in the white-cedar swamp 3 kilometers west of Point Pleasant. Dams which demarcate the pine-barren and tide-water floras sharply are found at Toms River, Batsto, Mays Landing, and Millville. Harper,* commenting on these views of Stone,

FIGURE 119

Watering Place Pond in the heart of the Lower Plain, N. J., July 29, 1913. Water covered with floating leaves of white water-lily, Castalia odorata.

adds: "This accords very well with the belief, recently expressed by the reviewer, that pioneer aquatic vegetation is commonly associated with minimum seasonal fluctuations of water, and vice versa; for seasonal fluctuations are of course least just above a dam or shoal or waterfall, and greatest just below, and these dams have probably been in existence long enough for vegetation to adjust itself pretty well to such conditions."

ADDITIONAL POND AND RIVER PLANTS

The plants noted in the foregoing description of the pond and river-bank formations were those found in a field study covering a period of several

* HARPER, ROLAND M.: Review Stone's Flora of Southern New Jersey, Torreya, 12: 220, 1912.

years. As no attempt was made to note all of the plants found in the synecologic survey, the following list of additional plants is given with one locality for each species:

Potamogeton epihydrus (Bamber).
" confervoides (Atco).
Sagittaria longirostra (Point Pleasant).
" graminea (Pleasant Mills).
Andropogon corymbosus abbreviatus (Clementon).
Panicum verrucosum (Atsion).
" longifolium (Mays Landing).
" spretum (Shamong).
Agrostis elata (Batsto).
Eleocharis Robbinsii (Shamong).

Eriocaulon septangulare (Bamber).
Elatine americana (Lakehurst).
Myriophyllum humile (Pancoast).
Limnanthemum lacunosum (Quaker Bridge).
Utricularia juncea (Ancora).
" fibrosa (Atco).
" subulata (Forked River).
" gibba (Hammonton).
" inflata (Clementon).
" purpurea (Toms River).

FIGURE 120

Savanna along North Branch of Wading River, November 13, 1912. Note the groves of white-cedar at the edge of the river and that the grasses of the savanna fringe the river-bank.

SAVANNA FORMATION

True savannas, such as exist in the southern United States, in subtropic and tropic countries, as in the orchard-scrub of East Africa, the Caatingas and Campos of Brazil, the Llanos of Venezuela, the Patanas of Ceylon, the Lalang vegetation of eastern Asia, do not exist in New Jersey. If we investigate the so-called savannas from the ecologic standpoint, considering that the vegetation of the savanna has only one resting period during the dry season, this distinction also fails, if applied to the vegetation of New Jersey. If, however, we apply the physiognomic test, then certain geographic units of the pine-barren region by extension of the term may be called savannas. A savanna, according to this physiognomic extension of the definition, is a flat, grassy area or plain, treeless or dotted over with clumps of trees or individual trees scattered over the surface. These areas may be dry throughout the year, or they may be flooded with water during a part of the year. In New Jersey it is frequently difficult to distinguish between a swamp and a savanna, which is a prairie-like area dotted over with scattered trees, but in determining whether the area is a swamp or a savanna, the decision will have to be made arbitrarily. Bearing these facts in mind, we find that there are several places in New Jersey where such savannas are found. A typic one is found in undisturbed condition in the heart of the pine-barrens along the Wading River. This one and several other savannas are indicated on the folded map of the pine-barren region (Plate I). The savanna is developed best on the south bank of Wading River, from which it rises in two terraces (Fig. 122). The first terrace, which is elevated

FIGURE 121

Savanna along North Branch of Wading River, with pine trees and white-cedar trees dotted over its surface. November 13, 1912.

scarcely above the surface of the stream, is a wet savanna, and is almost entirely destitute of trees, except a few clumps of white-cedar, Chamaecyparis thyoides, which grow along the river-bank (Fig. 120 and Fig. 121). The deeper water of the river is characterized by Castalia (Nymphaea) odorata, Eriocaulon septangulare, Juncus militaris, Orontium aquaticum (Fig. 123), Xyris Congdoni and great beds of Scirpus subterminalis.

The savanna in places comes abruptly to the water's edge (Figs. 120 and 121), while at some points low thickets of Alnus serrulata (= rugosa), Azalea viscosa, Clethra alnifolia, Ilex glabra and Osmunda regalis are at the edge of the stream between it and the savanna. The royal-fern, Osmunda regalis, grows usually in large clumps on the savanna side of such thickets. On occasional dry knolls a solitary pitch-pine is found the base of the tree being surrounded by a tall grass which covers the surface of the knoll. The dominant growth that gives character to

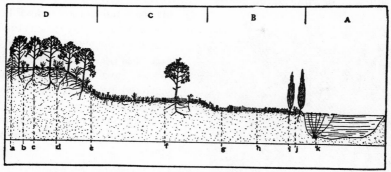

FIGURE 122
Diagram of savanna along Wading River.

a, Quercus marylandica.
b, Ilex glabra.
c, Pinus rigida.
d, Quercus ilicifolia.
e, Andropogon scoparius.
f, Sarracenia purpurea.

g, Grasses, sedges, Tofieldia racemosa, Lophiola aurea, etc.
h, Andropogon corymbosus, etc.
i, Chamaecyparis thyoides.
j, Alnus rugosa.
k, Castalia (Nymphaea) odorata

A, Wading River.
B, First and wetter terrace.

C, Second wet terrace.
D, Pine-barren level.

the wet savanna facies (Figs. 121 and 128) consists of grasses and sedges the leafy stems of which are not over three to four decimeters tall. The identifiable grasses collected by me on November 13, 1912, on the Wading River savanna, included the grasses Calamagrostis cinnoides, Eragrostis pilosa, Sporobolus serotinus, and the sedges, Dulichium spathaceum, Rhynchospora glomerata, Rh. glomerata leptocarpa, Rh. gracilenta, and the rush, Juncus acuminatus, growing on a common level, while Andropogon corymbosus abbreviatus raised its whisk-like

FIGURE 123
Golden-club, *Orontium aquaticum*, in flower in slough at edge of wet savanna along Wading River, May 27, 1916.

inflorescence to a height of 6 decimeters above the general grassy level of the wet savanna. The beard-grass, Andropogon scoparius, occurred also, but more sparingly.   The pitcher-plant, Sarracenia purpurea (Fig. 124), was found among the sedges and grasses, but the flowers of this plant and of Andromeda mariana, Eriocaulon decangulare, Iris prismatica, Lophiola aurea and Tofieldia racemosa are raised above the general surface.   Leucothoë racemosa occurred in clumps in the wet savanna (Fig. 127), and scattered here and there grew Hypericum densiflorum, Osmunda regalis and a small inconspicuous plant, Bartonia virginica.

FIGURE 124
Wet grassy savanna along Wading River, with pitcher-plant, Sarracenia purpurea, in flower on May 27, 1916.

The second terrace, elevated slightly above the first, is much drier, although it does not have a very dry soil.   On this terrace the pitch-pine trees are dotted either singly or in groups (Figs. 125 and 127).   The trees grow about 9 meters tall and bear cones of all ages.   Dotted over the grassy stretches one sees bright green clumps of the inkberry, Ilex glabra, and the bayberry, Myrica carolinensis, either in association or as single bushes.

The sheep-laurel, Kalmia angustifolia, is scattered through the coarse sedges and grasses, while the leather-leaf, Chamaedaphne (Cassandra) calyculata, grows in clumps, several 18–27 square decimeters in area (Fig. 126). The pitcher-plant, Sarracenia purpurea, is imbedded sometimes in the sphagnum and is associated with Eriophorum virginicum (Fig. 125) and clubmoss, Lycopodium alopecuroides.   The swamp fetter-bush, Leucothoë racemosa, is scat-

FIGURE 125
Small pines (invading) and Eriophorum virginicum in big savanna, south side of Wading River, November 13, 1912.

tered widely as a low shrub, becoming conspicuous in some places as an important element of the savanna vegetation beneath the scattered pine trees (Fig. 127).   Several isolated clumps of shrubs are detected.

Such are Aronia arbutifolia (with bright red fruit), A. nigra (with black fruit) and Viburnum cassinoides (with blue-black fruit). Occasional plants of Andropogon macrourus are found on the second terrace.

Found between the prevailing grasses and close to the drier ground with a sphagnum covering grow Gaultheria procumbens and Pyxidanthera barbulata. The savanna is invaded in places by dense growths of young pitch-pine trees about 6 decimeters to 1 meter tall. At another section of it clumps of Azalea viscosa, Clethra alnifolia, Ilex glabra, Magnolia glauca and Vaccinium corymbosum, with wine-red leaves and young saplings of Nyssa sylvatica, break the monotonous, prairie-like level. These savannas are covered with

FIGURE 126

Clumps of waxberry, Myrica carolinensis, Kalmia angustifolia, leather-leaf, Cassandra (Chamaedaphne) calyculata, in big savanna of Wading River, November 13, 1912.

a tall grass, Danthonia epilis, associated with a denser growth of Panicum ensifolium and several species of Rhynchospora (axillaris, oligantha, pallida). Scleria minor arises from a bed of bog-moss. Scattered about in the savanna are Lophiola aurea, pitcher-plants, Sarracenia purpurea, beds of cranberry, Vaccinium macrocarpon. One can distinguish the white tops of Eriocaulon compressum, E. decangulare. With them we find early in July the yellow spikes of Abama americana and Tofieldia racemosa, covered with rough, glutinous pubescence (Figs. 122 and 128). Two orchids are conspicuous, the pink-flowered Pogonia ophioglossoides, and the crimson-flowered Limodorum tuberosum. The shallow, iron-stained pools are bright yellow with Utricularias. Such savannas, occupying the first and second flat terraces, stretch for miles,

FIGURE 127

Big savanna, south bank of Wading River, with Leucothoë racemosa (red), Sarracenia purpurea, Lophiola aurea, November 13, 1912.

following the course of the streams and touching the third terrace, which rises abruptly several feet above the flatter levels and is covered with a typic forest of pitch-pine and associated species. During some seasons of the year the savanna terraces are overflowed with water. At other

FIGURE 128

Wet, grassy terrace of savanna along south bank of Wading River, looking toward cedar swamp on the north bank, November 13, 1912.

seasons of the year they are dry, and therefore easily crossed. The open park-like aspect of the savanna is emphasized by the groups of pine trees and cedar trees, the latter occupying vantage-points of the slow-flowing, meandering Wading River.

CHAPTER XI

PLAIN FORMATION [COREMAL]

The plains of New Jersey occupy an elevated tract of country in Burlington and Ocean counties. The two areas designated as East Plains and West Plains, or better geographically, as Lower Plains and Upper Plains, are separated from each other by the East Branch of Wading River (Oswego River), the main north tributary of which branch penetrates the Upper Plains, causing a division of the southern border of these plains into an eastern lobe and a western lobe. The eastern lobe reaches almost to Cedar Grove (now Warren Grove). The West or Upper Plains cover 3,125 hectares (7,737 acres), and the East or Lower Plains, 2,691 hectares (6,662 acres). The country is rolling. The highest elevations in the Lower Plains, which are north of Cedar (Warren) Grove are respectively 32.2, 37.6, 39.8, 41, 42.2, 43.4, 43.7, 44, 44.6, 45.9 meters (106, 124, 131, 135, 139, 143, 144, 145, 147, 151 feet). From the highest elevation on a clear day the large hotels at Atlantic City can be seen toward the south. The highest elevations of the Upper Plains that lie north of Cedar (Warren) Grove are respectively 46.8, 51.9, 53.2, 54.4, 56.2, 59.2, 60.1, 61.1, 62, 63.2 meters (154, 171, 175, 179, 185, 195, 198, 201, 204, 208 feet). The east lobe of the West Plains is separated from the main area by a bottom, or depression, with oak forest, known as "oak bottom" (see ante). Into this valley runs Sykes Branch. I have called this detached area of plains vegetation the Little Plains. If an examination of the map is made, it will be discovered that this elevated district is the water-shed of a number of important streams, such as the Wading River flowing into the Atlantic Ocean, Rancocas River, which empties into the Delaware River, Cedar Brook, Forked River, Westecunk Creek, Bass River and other smaller streams.

The hills and valleys of these plains are covered with a dwarf, elfin, or pigmy forest of pitch-pines, oaks, laurel and other shrubs with an herbaceous undergrowth. A common low shrub is the broom-crowberry, Corema Conradii, having the generic name derived from the Greek word κόρημα, a broom, from its bushy aspect. I have designated this formation of stunted, twisted and dwarfed trees and shrubs a *coremal*, which is thus aligned, as an ecologic term, with chaparral, encinal and chemisal.   Three kinds of trees may be distinguished from the standpoint of the forester, viz., first, stunted prostrate trees, from 6 to 12 decimeters (2 to 4 feet) in height, covering the greater part of the plains; second, pine trees from 18 to 42 decimeters (6 to 14 feet) in height, very stunted, branching nearly to the ground, and covering a very small fraction of the plains; third, the large pine trees, from 42 to 60 decimeters (14 to 25 feet) in height, usually very crooked and confined to the wet hollows which are found in some parts of the Lower Plains. In some places the dwarf trees of the plains are terminated abruptly by

FIGURE 129

Lower Plains, patches of Corema Conradii with low pine trees, Pinus rigida, reindeer lichen, Cladonia rangiferina, flowering moss, Pyxidanthera barbulata, dwarf oaks, April 9, 1912.

large timber, but usually there is a gradual transition from the stunted type to the tall type.   The pine trees give in near view a grayish-green ground color, broken by the lighter, glossy greens of Quercus marylandica and the lighter green of the laurel, Kalmia latifolia.   The reddish-brown of the ripening Kalmia fruits gives a dash of color wherever that bush grows.   In the distance the uneven tops of Quercus marylandica show most conspicuously with the laurel intermixed.   At places the forest line of tall trees seems quite sharp, while toward the south, from a vantage-point near the northern end of the Lower Plains, they extend as far as the eye can reach. The areas of coremal in some places touch the white-cedar swamps. Here the line of separation is a very sharp one. Dwarf vegetation (coremal) exists on one side of the line, tall, spire-shaped white-cedar trees on the other side.

The pine trees are sprouts from the stump, thus representing a coppice growth, but there are young trees (saplings) with prostrate stems and branches. Old dead stumps are found, which have sprouted again and again, until they have lost their vitality. The roots of the dwarf pitch-pine of the plains are short and poorly developed, and in some cases examined by me the main tap-root had rotted away. An actual study of the rings of a few of the prostrate, or basket-like, trees cut down on the Upper Plains shows that the number of rings rarely exceeds 18 or 20. These observations confirm the measurements of 55 stems cut on the Upper Plains, and 53 from the Lower Plains by Gifford Pinchot,* whose figures are herewith given:

AGE OF PINE SPROUTS ON THE LOWER, OR EAST, PLAINS

Height, Feet	Height, Meters	Age, Years	Height, Feet	Height, Meters	Age, Years	Height, Feet	Height, Meters	Age, Years
8	2.438	17	5.5	1.676	17	4	1.219	16
8	2.438	15	6	1.828	16	5	1.524	14
6	1.828	11	7	2.133	10	3.5	1.066	10
8	2.438	15	8	2.438	14	6	1.828	14
8	2.438	14	8.5	2.591	14	2.5	0.762	12
5	1.524	12	9	2.743	16	4	1.219	13
8.5	2.591	20	6	1.828	14	6.5	1.981	18
6	1.828	11	8	2.438	13	7	2.133	19
6	1.828	9	3.5	1.066	11	2	0.609	9

AGE OF LARGER PINES GROWING WITHIN LIMITS OF LOWER PLAINS

Height, Feet	Height, Meters	Age, Years	Height, Feet	Height, Meters	Age, Years	Height, Feet	Height, Meters	Age, Years
11	3.352	30	11	3.352	23	17	5.181	41
10	3.048	13	13	3.962	36	19	5.791	43
12	3.657	16	13	3.962	40	21.5	6.553	41
11.5	3.505	15	12	3.657	27	17	5.181	33
12	3.657	24	12	3.657	28	10	3.048	22
20	6.096	40	11.5	3.505	28	10.5	3.200	22
13	3.962	24	21	6.400	40	13	3.962	19
21	6.400	46	18	5.486	39	10	3.048	22
17.5	5.334	42	17	5.181	38	..	..	..

* Pinchot, Gifford: The Plains. Annual Report, New Jersey State Geologist, 1899, Forests: 125–130.

AGE OF SPROUTS FROM THE UPPER, OR WEST, PLAINS

Height, Feet	Height, Meters	Age, Years	Height, Feet	Height, Meters	Age, Years	Height, Feet	Height, Meters	Age, Years
4	1.219	12	3	0.914	11	7	2.133	17
3	0.914	31	2.5	0.762	7	3	0.914	7
7.5	2.286	22	4.5	1.371	8	3.5	1.066	9
6	1.828	16	5	1.524	10	5	1.524	10
5	1.524	18	5.5	1.676	16	4	1.219	11
8	2.438	13	4.5	1.371	15	3.5	1.066	9
5.5	1.676	15	6	1.828	12	4.5	1.371	15
4	1.219	13	3	0.914	8	3.5	1.066	11
2.5	0.762	9	3.5	1.066	9	4.5	1.371	10
4	1.219	8	6	1.828	13	3.5	1.066	9
8	2.438	16	7.5	2.286	16	6	1.828	15
5	1.524	13	4	1.219	7	4	1.219	10
6	1.828	15	5	1.524	12	5	1.524	13

AGE OF LARGER PINES GROWING WITHIN THE UPPER PLAINS

Height, Feet	Height, Meters	Age, Years	Height, Feet	Height, Meters	Age, Years	Height, Feet	Height, Meters	Age, Years
15	4.572	39	14	4.267	31	21	6.400	44
11	3.352	21	9	2.743	22	13	3.962	33
12	3.657	29	10	3.048	22	19	5.791	46
14	4.267	30	16	4.876	30	17	5.181	39
11	3.352	35	11	3.352	36	14	4.267	27

From the figures presented in the tables it is evident that the oldest and largest trees are under fifty years of age, and that the small growth averages about ten to fifteen years.

SYNECOLOGY OF THE COREMAL

Before we consider the cause of the present condition of the plains vegetation it is important to study that vegetation in detail from the standpoint of its synecology. We find associated with the dwarf bushy pines, loaded with cones, usually 6 to 9 decimeters (2 to 3 feet) tall (Fig. 129), Quercus marylandica, also stunted, but sometimes reaching above the pines by a growth which is 1.2 meters (4 feet) tall. The laurel, Kalmia latifolia, in full flower in early June, is sometimes 1.5 to 1.8 meters (5 to 6 feet) tall on the higher knolls of the Lower Plains. Dwarf post-oak, Quercus stellata, occurs more commonly on the Lower Plains than on the Upper, while the bear-oak, Quercus ilicifolia (= nana), is a low, prostrate, acorn-bearing bush, sometimes growing taller and more tree-like under more favorable conditions. The dwarf trees form the conspicuous elements of the vegetation (Figs. 130 and 131), and between them grow, in different associations, Andromeda mariana, Comptonia asplenifolia, Dendrium (Leiophyllum) buxifolium, Gaylussacia frondosa, G. resinosa, Hudsonia ericoides, Ilex glabra, Kalmia angustifolia, Myrica carolinensis, Pteris aquilina, Smilax glauca [Little Plain], Tephrosia (Cracca) virginiana, Vaccinium pennsylvanicum, V. vacillans, and on the ground in bare places grow Epigaea repens, Gaultheria procumbens, and Pyxidanthera barbulata (in full flower April 8, 1912). The bearberry, Arctostaphylos uva-ursi, grows in extensive radiate masses on flat, sandy stretches of the plains

FIGURE 130

View across the Lower Plain with characteristic growth of low oaks and pines. In the middle distance are the two wagons of the International Phytogeographic party which visited New Jersey on July 28–29, 1913. (Photograph by George E. Nichols.)

(Fig. 133). The herbaceous plants of the plains, as far as noted in the growing condition in the three seasons of the year, are: Andropogon scoparius, Chrysopsis mariana, Euphorbia polygonifolia and Melampyrum americanum. The reindeer-lichen, Cladonia rangiferina, grows on wide stretches of white sand.

FIGURE 131

Low growth of the pines and oaks in the Lower Plain, New Jersey, with members of the International Phytogeographic Excursion collecting the characteristic plants. July 29, 1913. (Photograph by George E. Nichols.)

The most local and peculiar plant of the Lower and Upper Plains is the broom-crowberry, Corema Conradii, which grows in two general forms (Figs. 134 and 135). The first type of plant is one which grows in dense cushions, about 3 decimeters (1 foot) tall (Fig. 135), its color varying from light green through dark green to a rich brown color, distributed in clumps between the prostrate pine trees. That this is the typic form of the plant is indicated by the fact that on the island of Nantucket, where it grows in great abundance in several widely separated localities, it is always found in the cushion form, and of the dark-green color.

On April 8, 1912, Corema, for which this formation has been designated the Coremal, was in flower, with its staminate and pistillate flowers
on different plants. The flowers, which are dioecious, or polygamous,
are in terminal clusters, each in the axil of a scaly bract, and with five or
six scarious, imbricated bractlets, but no proper calyx. The stamens
are three or four in number and exserted, with purple filaments and
brownish-purple anthers. The style is slender, usually 3-cleft. The
drupaceous fruit contains 3 (rarely 4–5) nutlets. The plants are low
shrubs, with subverticillate, narrowly linear, heath-like leaves. In
other places on the Lower Plains Corema assumes the second type of
growth, which is a diffuse one, the separate stems being scattered widely.

Such plants may be young
plants that have arisen from
scattered seeds. The local distribution of this interesting
plant early attracted the attention of systematic botanists, and
there is a copious literature
which Stone* quotes in his book
on the flora of the southern part
of the State. Lately Brown has
figured and described the plant
in Bartonia,† but neither of
these botanists describe the
broom-crowberry from the synecologic point of view. The writer
hopes a little later to emphasize,

FIGURE 132

Upper Plains (Little Plains) back of George
Cranmer's farm. The tall tree is Quercus
stellata. Note clumps of Corema Conradii
with Cladonia rangiferina, April 9, 1912.

in an especially illustrated paper,
some of the neglected points about the growth and distribution of this
striking cushion-plant.

In a basin-shaped depression in the West Plains Quercus marylandica
forms a closed association, the trees growing 2.4 meters (8 feet) tall.
The pitch-pine, Pinus rigida, is secondary to the oaks, and the undergrowth consists of Comptonia asplenifolia, Gaylussacia resinosa, Kalmia
latifolia and Quercus ilicifolia, while the low-growing species in these
depressions are Arctostaphylos uva-ursi, Gaylussacia frondosa, Pyxidanthera barbulata and Tephrosia virginiana. On an elevation nearly
covered with a coarse gravel soil the trees become dwarfed, gnarled and
twisted.

* STONE, WITMER: Loc. cit.: 530–536.
† BROWN, STEWARDSON: Bartonia, 1913 (No. 6): 1–7.

A valley-like depression in the Lower Plains was filled with water and saturated sphagnum.   The pines in these wet places were at least 4.5

FIGURE 133

Lower Plains with bearberry, Arctostaphylos uva-ursi, dwarf oaks, dwarf pitch-pines, and reindeer-lichen, Cladonia rangiferina, April 9, 1912.

to 6 meters (15 to 20 feet) tall (Figs. 136 and 137), associated with the red-maple, Acer rubrum, such shrubs as: Azalea viscosa, Clethra alnifolia, Ilex glabra, Kalmia angustifolia, Myrica carolinensis, Nyssa sylvatica, Vaccinium macrocarpon, the bracken-fern, Pteris aquilina, and the turkey's-beard, Xerophyllum asphodeloides. The gray-poplar, Betula populifolia, is an interloper. The head of this small valley was characterized by eight tall pine trees forming a small grove completely surrounded by low vegetation (Fig. 136). Here

FIGURE 134

Lower Plains showing prostrate pitch-pines and two clumps of Corema: the one to the left is green and staminate; the one to the right is chocolate-brown in color and pistillate. Flowers present on April 9, 1912.

edaphic conditions control the size and distribution of the pine trees.

The line of demarcation between the plains and the pine-barren forest seems to be a sharp one, but nevertheless there are a few scattered tall pine trees at the outer margin of the plains which become reduced in height as the Coremal is entered. Here, in the forest of tall pine trees, the oaks are subordinated to the pines, which at the edge of the plains

FIGURE 135
Corema Conradii as a cushion shrub along the road in the Lower Plain, N. J., July 29, 1913. (Photograph by George E. Nichols.)

are 6 to 9 meters (20 to 30 feet) tall, associated with stagger-bush, Andromeda (Pieris) mariana, sweet-fern, Comptonia asplenifolia, huckleberries, Gaylussacia frondosa, G. resinosa, sheep-laurel, Kalmia angustifolia, bear-oak, Quercus ilicifolia, and goat's-rue, Tephrosia (Cracca) virginiana. The vegetation of the plains joins sharply with that of a white-cedar swamp. A difference of 3 to 6 decimeters in soil level makes a marked difference in the vegetation along a stream's edge. Here the tension strip is characterized by Acer rubrum, Andromeda mariana, Chamaedaphne (Cassandra) calyculata, Clethra alnifolia and Myrica carolinensis.

MEASUREMENT OF SPECIES OF THE COREMAL

In order to determine the actual size of some of the individual trees, shrubs and herbs characteristic of the Coremal the following measurements were made by a meter stick in the field.

TREES

Pinus rigida.................................40 centimeters to 1 meter tall.
Quercus marylandica.........................+ 1 meter tall.

SHRUBS

Arctostaphylos uva-ursi......................10 centimeters tall.
Corema Conradii.............................16   "   "
Dendrium buxifolium.........................50   "   "
Gaylussacia resinosa........................40   "   "
Ilex glabra................................40   "   "
Kalmia latifolia...........................30   "   "
Quercus ilicifolia 70   "   "

HERBS

Andropogon scoparius.......................80   "   "
Chrysopsis mariana.........................40   "   "
Hudsonia ericoides.........................15   "   "

THEORIES CONCERNING THE COREMAL

Several theories have been advanced to account for the dwarf character of the pines and oaks in the Coremal. These will be given first in detail and then the experiments will be described upon which a scientific conclusion may be based. The first consistent attempt to explain the cause of the present condition of the plains is by Gifford Pinchot. We quote as follows: "Some have attributed the form of the pine to a lack of mineral constituents in the soil necessary to the growth of trees. This theory, however, is disproved by chemical analyses of the soil made by the Geological Survey, which show no greater poverty than is common to the surrounding region. The theory that fire, combined with the effect of the very poor top-soil and a hard subsoil, is the efficient cause has been advanced and certainly fire has been a very large factor in bringing about the present conditions of these areas. If the plains have been in their present condition since the country was first settled, they were probably first burned over by the Indians, who were in the habit of camping in the neighborhood, as their shell-heaps show. It is not unreasonable, therefore, to suppose that these high-rolling plains were originally stripped of their forest cover in this way. The pines, which probably returned by seed after the first fire, were burned over again and again, and their stumps sent up sprouts which became more and more feeble after successive fires.   Many old stumps, as already pointed out, became exhausted and died after repeated sprouting.   Their place was taken by seedlings and in this way the ground remained stocked with pine."

Pinchot goes on to say: "This does not explain, however, the prostrate form of the young seedlings and many older trees.   This peculiarity is not confined to the plains alone, for pine seedlings growing on bare sand in exposed situations in the neighborhood of the plains show the same tendency.   These seedlings exhibit a remarkable similarity to the forms assumed by trees near the timber line on high mountains.   It is a fair inference that the very harsh and windy situations in which they grow has an effect analogous to that of a great elevation.   Hence it is

FIGURE 136
Lower Plains showing group of tall pitch-pine trees at head of a wet hollow, April 9, 1912.

believed that exposure and poor soil are entirely sufficient to explain why the young trees are prostrate.   A large part of the pine, however, is coppice growth from old stumps which have lost their vigor to a large extent, and under the unfavorable surroundings are incapable of producing anything but straggling sprouts. Furthermore, the trees, which grow very slowly on the poor soil of the plains, are killed by fire before they have time to reach a large size."

George Cranmer, a farmer of Warren (Cedar) Grove, living on a farm at the edge of the Little Plain, showed me a portion of the plains back of his fields which he said had not been burned over for sixty years.   The pine trees and the oak trees were taller on this tract and the growth of other species in every respect thicker, and yet, contrasted with the nearby pine forests, the trees of this protected area might with propriety be termed dwarf.   This farmer advanced the idea of soil exhaustion, viz., that on account of the poverty of the soil the trees had become poorer and poorer.

The production of toxic excretions by the roots of the plants of the Coremal may be a factor of importance in inhibiting the growth of the pines and the oak trees found in this formation.   Following out a clue which the partial sterilization of the soil by chemicals or by steam gave, it was discovered that the bacteria, which are useful in ammonia-making, increased four-fold after such treatment, suggesting the presence in the soil of some agent which held them in check.   After much painstaking study it was discovered that the soil contained a living protozoan (Pleurotricha), which preyed upon the useful organisms, and that the heat and chemicals either destroyed these larger unicellular animals or inhibited their activity.   It can be said, therefore, when such cases are considered, that the fertility of the soil may be in part a biologic problem.*   That the productivity of some soils is due to biologic rather than to physic and chemic characteristics is illustrated by the attempts made to reforest Denmark.   The peninsula of Jutland was covered originally by forests, but these were destroyed, until, by the year 1500, the country had been transformed into a barren heath and sand dunes.

At various times attempts were made to reforest these heaths, but the results were disappointing until Colonel E. Delgar solved the problem.   Spruce trees (Picea alba, P. excelsa), if planted alone, did not thrive, but became sickly.   The cause of this irregularity in the growth of spruce was thought to be local conditions of the soil, but scientific investigation of such soils did not reveal any difference in the physic or chemic composition of the soil.   It was found, however, that the mountain pine

FIGURE 137

Lower Plains, showing a sedgy slough surrounded by pitch-pine trees (scattered), April 9, 1912.

(Pinus montana) acted as a nurse to spruce trees planted in its vicinity. In the same soil where spruce if planted alone would remain backward, it would, if planted close to a mountain pine, grow up vigorously.   After some years of trial it was found that the pine would hamper the growth of the spruce, and so it was cut down at an early age.   It was discovered then that even if the mountain pine was cut down at an early age, it imparted to the adjacent spruce trees the ability to grow.   The phenomenon is not understood, but it is supposed that the roots of the mountain pine are inhabited by some mycorhiza which produces the nitrogen necessary for the growth of trees, and that this organism is transferred to roots of the surrounding spruce trees.   Once this infection has taken

* HARSHBERGER, J. W.: The Soil: A Living Thing. Science, new ser., XXXIII: 741–744.

place, the presence of the mountain pine is no longer necessary and it is usually cut down.   Clearly this is a biologic relationship.

The thought arises naturally with reference to the soil of the plains, whether the tall pines can be grown there, if a nurse crop is discovered similar to the mountain pine above described.   It seems plausible that the poverty of the soil in the Upper and Lower Plains is due to the development of toxic substances which might be overcome by planting trees that are indifferent to such soil toxins, as has been accomplished so admirably in Denmark.   Future investigation of these soils will alone determine these facts.

FIGURE 138
Culture of corn in soil from Upper (Little) Plains to test character of top- and subsoils. Left, top-soil untreated; middle, subsoil untreated.   Right, garden soil as control.   (Photograph taken May 20, 1912.)

FIGURE 139
Culture of peas in soil from Upper (Little) Plains to test character of top- and subsoils. Left, top-soil untreated.   Right, subsoil untreated.   (Photograph taken May 20, 1912.)

EXPERIMENTAL TREATMENT OF COREMAL SOILS

Realizing that the problems connected with the vegetation of the plains of New Jersey and with their soils lent themselves to experimental investigation, the writer began a series of experiments to test the character of the soil: For this purpose a box of soil from the Little Plains was procured.   The samples included subsoil and top-soil.   These were placed in pans and flower-pots with and without previous sterilization by heat. In the sterilized and unsterilized top-soil and subsoil corn, peas, sunflower, wheat and okra seeds were planted on April 13, 1912, and placed in a cold-frame out-of-doors under as natural conditions as possible (Figs. 138, 139, 140, 141, 142). Nine days after the seeds were planted, on April 22, the pots and pans in the cold-frame were examined. In all of the soils, treated and untreated, the seeds had sprouted and their plumules were seen projecting above the surface in the case of the peas, sunflower and wheat. Two days afterward, on April 24, the peas, sunflower and wheat pushed their plumules well above the soil. Progress had been made in the growth of the seedlings of peas, sunflower and wheat, but at this early date there was seen an inhibition of

FIGURE 140

Culture of corn in series of soils from Upper (Little) Plains to test character of soils. Left, top-soil untreated. Second left, top-soil heated; subsoil untreated; subsoil heated. Right, garden soil untreated; garden soil heated. (Photograph May 20, 1912.)

growth in the subsoil, treated and untreated. Subsequent to April 29, when the corn-tips began to show above ground, the young plants in the pots and pans containing subsoil, treated and untreated, showed a

FIGURE 141

Culture of peas in series of soils from Upper (Little) Plains to test character of soils. Left, top-soil untreated. Second left, top-soil heated; subsoil untreated; subsoil heated; garden soil untreated. Right, garden soil heated. (Photograph taken May 20, 1912.)

constant lagging behind, due perhaps to the stiff, compact nature of the soil in which the corn was planted. The details of the experiments, as recorded in the notebooks, are too voluminous for complete description here—only the general results can be presented. The experiments ran thirty-eight days, when some of the pot-grown plants were lifted and placed between felt driers for preservation (Figs. 138, 139, 140, 141, 142). The contrast in size and growth was best seen in the shallow pan cultures. Contrasting the three pans of corn with *top-soil untreated*, the plants were 12.7 centimeters (5 inches) tall, leaves fully expanded, but slightly yellowed;

subsoil untreated, plants about 12.7 centimeters (5 inches) tall, yellowish in color, leaves suberect and inrolled; *garden soil*, plants 12.7 to 17.7 centimeters (6 to 7 inches) tall, leaves broadly expanded, dark green.

Contrasting the two pans of peas with *top-soil untreated* plants 20.3 to 25.4 centimeters (8 to 10 inches) tall, the peas were more withered and semi-prostrate, the soil less retentive of moisture; *subsoil untreated*, plants more erect, less wilted, darker green; soil stiffer and more retentive of water. It will be seen from the photographs taken of the soil cultures on April 10, 1912, that the plants grown in the stiff, heavy, impervious subsoil are of small size, more starved in appearance and show every evidence of sterile conditions of soil environment. The influence of the sterile subsoil is illustrated best in the pot cultures of corn, wheat and sunflower and least pronounced in that of the peas. The height of the plants is dwarfed, the secondary roots are shorter and the plants have an underfed appearance. The sandy top-soil of the plains, although containing some humus and relatively porous compared with the impervious, hard-baked subsoil, was less decidedly rich than the garden soil used as a control. The plants of corn, wheat and sunflower were smaller than those

FIGURE 142

Cultures of wheat and sunflowers in series of soils from Upper (Little) Plains to test character of soils. Top set, left top-soil untreated; top-soil heated; subsoil untreated; subsoil heated; garden soil untreated. Right, garden soil heated. Lower set, sunflower plants, series of soils same as in top set, and arranged in same order.

grown in ordinary garden soil, and the better growth was made in the unheated or unsterilized top-soil, as contrasted with the heated or sterilized soil (Figs. 143, 144, 145, 146, 147).

From these experiments and the series of control experiments with soils from the pine forest just over the west edge of the plains and in which normal growth was secured in top- and subsoils (see figures), we are probably safe in assuming that the inhibition of the growth of the various garden plants, including the wild plants of the plains, is due to the impervious, stiff and easily baked subsoil through which the roots can hardly penetrate, and which supplies very little plant food, but that the content of the soil in living organisms may have some influence is indicated by the difference of growth between the plants raised experimentally in heated and unheated soils (Figs. 148 and 149). That with the application of proper fertilizer, the top, or light, sandy soils of the plains could be made to produce fair crops of corn and wheat is indicated on Cranmer's farm, adjoining the plains, where corn, pumpkins, tomatoes, beets, cabbages and sweet potatoes are grown, but owing to the nature of the subsoil these attempts at cultivation will meet with very little commercial success as long as the stiff bottom soil remains as it is. The roots of the pines and the oaks, as indicated by digging them up, cannot penetrate

FIGURE 143

Culture of corn continued to July 4, 1912, in open pans with subsoil untreated.

FIGURE 144

Culture of corn continued to July 4, 1912, in open pans, when photograph was taken; top-soil untreated.

any distance into this hardpan. They decay after making futile attempts to penetrate the subsoil saturated with cold water and cemented together into a compact mass. The plants of the plains, with short, superficially formed roots, get along fairly well in the pervious sandy top-soil.

Some facts may be learned about the vegetation of the New Jersey Coremal by contrasting it with the Coremal on the Island of Nantucket. The physiognomy of the vegetation is nearly alike on the Upper and the Lower Plains of New Jersey and on the Island of Nantucket, but the Nantucket Coremal (Fig. 150) lacks the pitch-pine. In both districts there is a low vegetation of shrubby oaks, Quercus ilicifolia (= nana) and Q. prinoides; the bearberry, Arctostaphylos uva-ursi, is more abundant on the Island of Nantucket than on the plains of New Jersey, while the broom-crowberry, Corema Conradii, is found in both districts. A comparison of the Nantucket vegetation gives a clue to the origin of the plains (Coremal) and pine-barren vegetation of New Jersey. Heathland is the result of the factors which are summed up under the general term oceanic climate, and Nantucket, isolated as it is far out at sea, has an oceanic climate. The strong winds that blow and the pervious glacial soils prevent the deciduous trees of larger size from spreading out of the valleys, where they are protected, and even the pitch-pine, recently introduced into Nantucket, follows the valleys and protected slopes of the hills (Fig. 151). In all probability when the pine-barren region of New Jersey was an island an oceanic climate prevailed. The heathland, of which the Coremal is a part, was left as a relict as the plains of New Jersey. The pine-forest in those early times probably filled the valleys and later spread over the hills until all of the region was covered with pine forest except the Upper and Lower Plains, where edaphic conditions prevented the development of tall pine trees. The heathland was converted into a pine-heath. Graebner* has detailed a similar conversion of heath into pine-forest by the invasion of pines, and in such pine forests the undergrowth consists of characteristic heath plants, hence Kiefernheide.

FIGURE 145

Culture of corn continued to July 4, 1912, in open pans, with garden soil untreated.

The reason why the pine trees have been unable to grow to tall size in the New Jersey plains is because of the hard layer of soil immediately below the upper sandy layer. This layer corresponds to the caleche of Mexico, the plow-sole of agriculturists and the Ortstein of the Germans.

* GRAEBNER, P.: Die Heide Norddeutschlands. Vol. V, Die Vegetation der Erde: 239.

Klebahn* figures an Ortstein-Kiefer where the tap-root striking the hardpan is bent over, being unable to penetrate this soil layer. Graebner narrates how such pine trees grow for a time, but finally, after reaching a certain age, begin to go back until they succumb, and he describes how certain pine trees more fortunately situated by natural planting over holes through the Ortstein (Ortsteintöpfe) are able to send their tap-roots into the deeper soil layers.† Under such conditions tall thrifty pine trees will be scattered here and there, while the majority of the trees that become established in the region are dwarf and languishing. Similar conditions are found in the plains of New Jersey, where the

FIGURE 146

Culture of sunflowers continued to July 4, 1912, when photograph was taken. Left, top-soil untreated; top-soil heated; subsoil untreated; subsoil heated. Right, garden soil untreated.

low, dwarf pine trees live for a number of years and finally succumb, to be replaced by other trees that pass through a similar existence. Thus hardpan and fire are the two most important factors which have perpetuated the heath vegetation of the plains, while the surrounding region, with more pervious soil, although similarly fire swept in later years, has been preserved as a pine forest or pine heath (Fig. 151). Remove the pines and you would have the conditions as they exist on the Nantucket heathland and the vegetation would have a similar physiognomy, with such low oaks as Quercus ilicifolia and Q. prinoides forming the

* KLEBAHN, H.: Grundzüge der allgemeinem Phytopathologie: 14.
† GRAEBNER, P.: bottom of page 125.

main ground-cover (Fig. 150).   We are able, therefore, by this comparative study to connect three interesting plant formations with those of Europe.

AMERICA	EUROPE
Nantucket Heathland (with low oaks and bearberry, Arctostaphylos).	*Oak Heath* (*Eichenheide*) (with Arctostaphylos Heath, *i. e.*, Heide mit Vorherrschen von Arctostaphylos).
New Jersey Coremal (heathland with low pines, etc., in Lower and Upper Plains).	*Low Pine Heathland* (Kiefernheide).
New Jersey Pine-barrens.	*Pine Heathland* (Kiefernheide).

Summing up our study of the Coremal, or the plains vegetation, we may state that the dwarfed character of the pines, oaks, and other plants

FIGURE 147

Culture of corn and wheat continued to July 4, 1912, when photograph was taken. Left, corn in top-soil untreated; garden soil untreated. Wheat, left, top-soil untreated; subsoil untreated; subsoil heated; garden soil untreated.

of the plains is due primarily to the stiff, impervious subsoil, and the light, easily dried-out, sandy surface soil which may freeze to the depth of a meter in winter and dry out in midsummer until the subsoil is hard and stiff and the surface soil almost devoid of moisture. The elevated character of the country facilitates rapid drainage by run-off and seepage, and this influence of elevation, coupled with the strong winds that blow, has a strong dwarfing effect on the plants exposed to such conditions. These factors are sufficient to account for the origin of the Coremal as a distinct formation. Fire has had an influence on the native vegetation, but its effect has been secondary to the soil factors, for forest fires have burned over as severely and as repeatedly the nearby forest areas without so appreciably stunting the growth of the pine trees.

Undoubtedly it is true that the pines naturally of dwarf size in the Coremal, after having been burned again and again, have sent up from the charred stumps more and more feeble sprouts until many old stumps have become exhausted and have died from loss of vigor. The areas of the plains, then, are those which have edaphically an unusually impervious subsoil. In the deposit of this subsoil we can conceive of its being laid down as a sheet, or covering, of relatively impervious material of circumscribed area, surrounded by soils in which the subsoil is more favorable to the deeply penetrating roots of the pitch-pines and other pine-barren trees, as the writer has proved by ample control experiments. After deposit these impervious areas of subsoil were covered by more pervious deposits of light sands and gravels.

FIGURE 148

Culture of corn in top-soil (left) and sub-soil (right) from pine forest immediately west of the Upper (Little) Plains, started as a control of studies in the character of Upper Plains soils.

The plains are located as they are geographically because it was in parts of Ocean and Burlington counties that the materials out of which the stiff sub-soils were later formed were deposited. The factors responsible for the development of the succession which we have termed the Coremal, or plains formation, are soil, elevation, wind, and fire. Fire has played a secondary part in the development of the physiognomy of the vegetation. One of the strongest proofs that the plains have existed for many successional periods as such is the presence and persistence of Corema Conradii, which has remained as a characteristic plant of the Coremal ever since the time when, as in Nantucket, the pine-barren region of New Jersey had an

FIGURE 149

Culture of peas in top-soil (left) and sub-soil (right) from pine forest immediately west of the Upper (Little) Plains, started as a control of studies in the character of the soils of the Upper Plains.

oceanic climate when isolated as an island from the mainland by Pen-sauken Sound.

Vegetation of the Hempstead Plains, Long Island[*]

In order to have a slightly different viewpoint in our study of the plains vegetation of New Jersey a short account of the vegetation of the Hempstead Plains in western Long Island will be given.   It is the belief of the writer that the Hempstead Plains represent one element in the series of

FIGURE 150
Nantucket Heath back of Sankaty Head, covered with a dense low growth of the dwarf chestnut-oak, Quercus prinoides, August 25, 1913.

heathlands which include the plains of New Jersey and the heathland of Nantucket, for the floristic elements in all three of these formations are much the same.   In all probability they originated in the same epoch of geologic time.

The country covered by the Hempstead plains vegetation is flat and resembles the western prairie plains both in general physiognomy and in the association of plant species.   The vegetation is of the grassland

* Harper, Roland M.: The Hempstead Plains.   Bull. Amer. Geogr. Soc., XLIII: 351–360.

type rather than the heathland type. The plain is practically treeless, although dotted over with low trees. It is bisected by the valley of Meadow Brook, which flows south into the Atlantic Ocean. The scattered dwarf trees of the western half of the plains are the white birch, Betula populifolia, and the aspen poplar, Populus tremuloides, both of which are distributed by wind-carried fruits. The eastern half possesses, in addition to these two trees, occasional dwarf pitch-pine trees, Pinus rigida, and dwarf-oaks, such as Quercus ilicifolia, Q. prinoides, Q. stellata, all of which are scattered as individuals or exist in rounded clumps. The shrubs comprise, according to my notes, the stagger-bush, Andromeda (Pieris) mariana, sweet-fern, Comptonia asplenifolia, huckleberry, Gaylussacia resinosa, Rhus copallina, and a low willow, Salix humilis. The most abun-

dant and important prairie plant is the beard-grass, Andropogon scoparius, which covers many parts of the Hempstead Plain with an exclusive growth of its characteristic tufts or bunches, so that the general physiognomy is that of a western bunch-grass prairie. The most pronounced bunch-grass prairie effect is on a gravelly hillside along East Meadow Brook. Here Andropogon scoparius is undisputed possessor of the gravelly slopes, and with it a few herbs are

FIGURE 151

Inland sand-dunes, Nantucket, covered by marram-grass, Ammophila arenaria, and invaded by the pitch-pine, Pinus rigida, with Hummock Pond at the left. August 24, 1913.

found: Asclepias verticillata, Baptisia tinctoria, patches of reindeer-lichen, Cladonia rangiferina, and the rush, Juncus Greenei, and in the spring masses of Viola pedata. Several areas in western Nantucket depart from the prevailing heather vegetation and assume the physiognomy of the prairie. The beard-grass, Andropogon scoparius, is the prevailing grass in such grassland.

It will be noted, from the above description, that the physiognomy of the vegetation and the floristic character of the associated species are different from those described for the Lower and Upper Plains vegetation (Coremal) of New Jersey in the almost entire absence of low trees and in the presence of the prevailing grass, Andropogon scoparius, which gives characteristic tone and color to the Hempstead Plains of Long Island.

CHAPTER XII

CULTIVATED PLANTS OF THE PINE-BARREN REGION

A consideration of the vegetation of the New Jersey pine-barrens would not be complete if a brief consideration of the cultivated plants which have been grown successfully were omitted. The character of the successful garden and field plants is an index of the soil conditions and of the climate under which such plants are grown. A foreign phytogeographer can appreciate better the soil and the climatic conditions by referring to a list of the common cultivated plants of a region than in any other way. To indicate that certain cultivated species thrive or do not thrive in a district is a way of stating that the soil and climate in general are not suitable for such crops. Three sources of information have been used in the preparation of the following list. Through the courtesy of William F. Bassett, nurseryman and botanist of Hammonton, New Jersey, and Robert D. Maltby, Dean of the Baron de Hirsch Agricultural School, Woodbine, New Jersey, I am able to supplement my list of the cultivated plants of the pine-barrens by some additional ones.

Forest Trees

White pine (Pinus strobus).
Austrian pine (Pinus sylvestris).
Black walnut (Juglans nigra).
Butternut (Juglans cinerea).
English walnut (Juglans regia).
Pecan (Hicoria pecan).
Locust (Robinia pseudacacia).
Silver maple (Acer saccharinum).

The forest trees in the above list are cultivated either for ornament or shade, and are not spontaneous, as the trees mentioned under the head of weeds.

Fruit Trees

Apple (Pyrus malus).
Cherry (Prunus avium, P. cerasus).
Peach (Prunus persica).
Pear (Pyrus communis, P. sinensis).
Plum
Prune } (Prunus cerasifera, P. domestica, etc.).
Quince (Cydonia vulgaris).

Bush Fruits and Vines

Blackberry (Rubus nigrobaccus).
Currant (Ribes rubrum).
Fig (Ficus carica).
Gooseberry (Ribes grossularia).
Grape (Vitis—several species and hybrids).
Raspberry (Rubus idaeus, R. occidentalis (black)).

Small Fruits

Cranberry (Vaccinium macrocarpon).
Strawberry (Fragaria chiloensis).

Root Crops

Onion (Allium cepa).
Potato (Solanum tuberosum).
Radish (Raphanus sativus).
Sweet Potato (Ipomoea batatas).
Turnip (Brassica rapa).

Leaf Crops

Cabbage (Brassica oleracea).
Celery (Apium graveolens).
Lettuce (Lactuca sativa).

Fodder and Cereal Crops

Alfalfa (Medicago sativa).
Buckwheat (Fagopyrum esculentum).
Corn (Zea mays).
Cowpea (Vigna catjang).
Millet (Setaria italica).
Oat (Secale cereale).
Sorghum (Sorghum vulgare).
Wheat (Triticum vulgare).

Flower Crops

Aster (Callistephus, several species).
Cosmos (Cosmos bipinnatus, C. diversifolius, C. sulphureus).
Dahlia (Dahlia variabilis, etc.).
Gladiolus (Gladiolus, several species and hybrids).
Paeony (Paeonia, several species).
Sweet Pea (Lathyrus odoratus).

Garden Crops

Bean { Lima (Phaseolus lunatus).
{ Wax (Phaseolus vulgaris).
Canteloupe (Cucumis melo var. cantelupensis).
Cucumber (Cucumis sativa).
Okra (Hibiscus esculentus).
Pea (Pisum sativum).
Peanut (Arachis hypogaea).
Pepper (Capsicum annuum).
Pumpkin (Cucurbita pepo).
Soy (Soja hispida).
Tomato (Lycopersicum esculentum).

CHAPTER XIII

GENERAL OBSERVATIONS ON PINE-BARREN VEGETATION

Within recent years a number of European botanists have attempted to take a census of the vegetation of phytogeographic regions by counting the number of plants found in a number of limited areas, diversely situated as regards their habitats, and from the figures thus obtained to estimate the plant population of that region. Approximate values are obtained, which are useful in the comparative study of the vegetation of other regions. Clements,* in this country, advocates the use of a quadrat. A quadrat is merely a square of varying size, marked off in a formation for the purpose of obtaining accurate information as to the number and grouping of the plants present. Thus, by the use of the quadrat, it was ascertained that 259 hectares (one square mile) of alpine meadow contained approximately 1,500,000,000 plants. A more extended use of the quadrat has been to determine the relative rank of the species in a formation, the seasonal aspects of the vegetation, and by means of permanent and denuded quadrats to determine the seasonal changes in a formation and the general mode of invasion of species into a quadrat deprived of all growing plants.

Jaccard has applied somewhat the same method to a study of the flora of the Alps and the Jura. The papers received by me from this botanist in the order of their publication are:

Étude Comparative de la Distribution Florale dans une Portion des Alpes et du Jura. Bulletin de la Societé Vaudoise des Sciences Naturelle, 4e S., vol. XXXVII, No. 142 (1901).

Gesetze der Pflanzenvertheilung in der alpinen Region auf Grund statistisch-floristischer Untersuchungen, Flora, 1902, III. Heft, 90 Bd.

Nouvelles Recherches sur la Distribution Florale. Bulletin de la Societé Vaudoise des Sciences Naturelle, 5e S., vol. XLIV, No. 163 (1908).

All three of these papers are statistic discussions of the constitution of the flora of the Alps and of the Jura mountains. Special stress is laid upon the Association Coefficient (Gemeinschafts-Coefficienten = coefficient de communanté florale). For example: Of two meadows, A

* CLEMENTS, F. E.: Research Methods in Ecology: 162.

and B, Meadow A has 100 and B 120 species. Sixty species are found common to A and B. A and B have together a total number of $100 + 120 - 60 = 160$ species. The association coefficient is $\frac{60}{160} = 37\frac{1}{2}$ per cent. The generic coefficient is emphasized by Jaccard as of importance in a comparative study of vegetation. This is obtained by contrasting the number of genera and the number of species of a flora. For example: the entire alpine region of the southern Jura has 54 genera to a 100 species; the whole Swiss flora has 26 genera to a 100 species. The generic coefficient of the alpine Jura is 54 per cent., and of the Swiss flora, 26 per cent. Jaccard has studied also the individual frequence (Frequence individuelle), the floral density (Dichte eines Bestandes = densité florale), and other statistic questions concerning the alpine flora. His third paper is embellished by several important graphic tables.

Oliver and Tansley,* English botanists, have applied their methods of surveying vegetation on a large scale by a study of a salt marsh occupying the floor of an estuary called Bouche d'Erquay, on the north coast of Brittany. The method employed by them may be termed the "Method of Squares" and the "Gridiron Method" respectively. The former is suitable for the purpose of constructing a general map of an area comprising a considerable number of acres, on a scale of $\frac{1}{250}$ to $\frac{1}{500}$ or thereabouts, the latter for very detailed work restricted to small parcels of ground and giving results correct to six inches or even less. The method of squares is to peg out a square on the ground, which can be quickly and accurately mapped upon sectional paper. The gridiron method based on the first is suitable where the physic features with which the different plant formations are correlated exhibit definite variations within quite short distances.

Pine-Barren Vegetation by Quadrats

The method adopted in the statistic study of the New Jersey pine-barren vegetation is a slight modification of the method of squares, or quadrats. A selected area of pine-barren vegetation was inclosed within a square with ten-meter sides with stakes driven in at the corners, and at intervals of a meter along the sides. Six lengths of white binding tape were used, each 10 meters long. The meter intervals were indicated by a piece of red string tied around the tape. The large square was first located by stretching the four lengths of tape, and the smaller meter

---

* OLIVER, F. W., and TANSLEY, A. G.: Methods of Surveying Vegetation on a Large Scale. The New Phytologist, III: 228–237, November and December, 1904, with plate XI and Figs. 77–80.

squares were located by moving the other two pieces of tape as the survey proceeded, placing them parallel between two adjacent meter stakes on both sides of the outer and larger square.   Thus each pine-barren area surveyed included the detailed study and mapping of 100 quadrats, each a square meter.   As the method is a laborious one, requiring nearly a whole day's work on the hands and knees in a country infested by mosquitoes, only three of the large squares were finished.

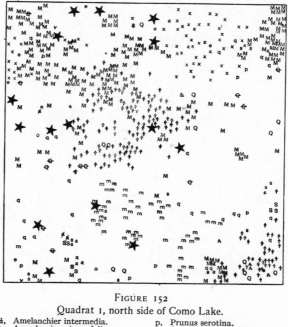

FIGURE 152

Quadrat 1, north side of Como Lake.

á,	Amelanchier intermedia.	p,	Prunus serotina.
a,	Ampelopsis quinquefolia.	&,	Quercus alba.
A,	Asclepias obtusifolia.	q,	Quercus ilicifolia.
e,	Euphorbia ipecacuanhae.	Q	Quercus marylandica.
♠,	Gaylussacia resinosa.	Š,	Quercus prinus.
M,	Melampyrum americanum.	⊖,	Quercus stellata.
x,	Moss.	S,	Smilax glauca.
m,	Myrica carolinensis.	s,	Solidago odora.
★,	Pinus rigida.	v,	Vaccinium pennsylvanicum.

The first survey was made in the pine woods on the north side of Como Lake on August 23, 1910, in the coastal pine-barrens (Fig. 152). The second was made in the pine forest 1.6 kilometers (1 mile) east of Lakehurst, in the heart of the pine-barrens (Fig. 153), and the third quadrat was located 0.8 kilometer (½ mile) east of Sumner, at the western edge of the pine-barrens (Fig. 154).   These three localities represent a wide diversity of soil, climatic, and vegetational conditions.   The data for an estimation of the plant population of the pine-barren is based on the total area of the region comprising 1,200,000 acres* out of 4,988,800 acres (7,795 square miles), the total area of the State of New Jersey. The pine-barren region is, therefore, approximately 25 per cent. the total area of the State. As one acre is equivalent to 4,046.87 square meters, the total area of the pine-barren region is 4,856,144,000 square meters.

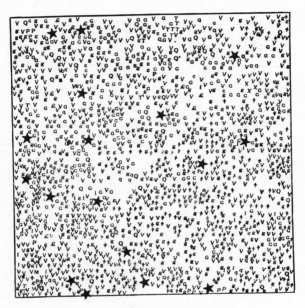

FIGURE 153
Quadrat 2, one mile east of Lakehurst.

A, Andromeda mariana.
c, Comptonia asplenifolia.
g, Gaultheria procumbens.
G, Gaylussacia frondosa.
↑, Gaylussacia resinosa.
I, Ilex glabra.
M, Melampyrum americanum.
★, Pinus rigida.
P, Pteris aquilina.

£, Quercus ilicifolia.
Q, Quercus marylandica.
q, Quercus prinoides.
X, Quercus tinctoria.
s, Sassafras variifolium.
S, Smilax glauca.
T, Tephrosia virginiana.
V, Vaccinium vacillans.

In the first (Como) ten square meter quadrat there were 15 pitch-pine trees. In the second ten square meter quadrat at Lakehurst there were 16 pitch-pine trees. In the third ten square meter quadrat at Sumner there were 15 pitch-pine trees. Taking 15 trees as the average for the three stations, and multiplying by $\frac{4,856,144,000}{10} = 485,614,400$, we get as

* VERMEULE, C. C.: The Forests of New Jersey. Annual Report of the State Geologist, 1899: 16.

the approximate total number of pine trees in the pine-barren region the astounding figure 7,284,216,000 trees of the pitch-pine, Pinus rigida. Of course these figures are only approximate, because the cleared areas have not been deducted, the areas denuded of their timber and the open country of swamps and savannas have not been subtracted, but the figures are interesting as an approximation to a total census of the pine trees in the

FIGURE 154

Quadrat 3, one-half mile east of Sumner.

C, Clethra alnifolia.
c, Comptonia asplenifolia.
E, Epigaea repens.
g, Gaultheria procumbens.
⋏, Gaylussacia resinosa.
H, Hicoria glabra.
k, Kalmia angustifolia.
K, Kalmia latifolia.
★, Pinus rigida.
P, Pteris aquilina.

p, Pyxidanthera barbulata.
£, Quercus ilicifolia.
q, Quercus prinoides.
⊕, Quercus stellata.
s, Sassafras variifolium.
S, Smilax glauca.
t, Toadstools.
v, Vaccinium pennsylvanicum.
V, Vaccinium vacillans.

region. Similarly, using the three ten square meter quadrats as a basis, we can compute the total number of plants that approximately occur in the region. Quadrat 1 (Fig. 152), located in the coastal pine forest on the north shore of Como Lake, had 680 plants, which would make on the average 6.8 plants to each secondary quadrat, one square meter in dimensions. Quadrat 2 (Fig. 153), located in the heart of the pine-barrens one mile east of Lakehurst, had 2,071 plants, or 20.071 plants to each secondary quadrat one square meter. Quadrat 3 (Fig. 154), measured in the pine forest at Sumner at the western edge of the pine-barren region, had 2,544 plants, or 25.44 plants to each secondary quadrat. The numbers 6.8, 20.071, 25.44 represent the floral density of each of the three quadrats. It is interesting to note that the smallest number of plants per secondary quadrat is found in the more sterile pervious soils of the coastal pine-barrens. As a corollary of this fact, we find that in arid districts the native shrubs and trees stand far apart, so that each individual tree has a large mass of soil from which to draw water and food. The thin character of the vegetation in the coastal pine-barrens is a direct expression of this law. The true pine-barren forest at Lakehurst has more favorable soil conditions, as reference to the discussion of soils will indicate, while the vegetation of the western edge of the pine-barrens in close proximity to the deciduous forest has still a larger number of plants per quadrat.

If we draw an average of the total number of plants in each of the three quadrats ten square meters in size, we find it to be 1,765 plants. Multiplying 1,765 by 485,614,400, the area of the pine-barrens expressed in 10 square meters, we get as the total number of pine-barren plants, including mosses, fleshy fungi, herbs, shrubs, and trees, 857,109,416,000 plants. Deducting the number of pitch-pine trees, 7,284,216,000, from the above total we get 849,825,200,000, which represents the approximate total number of all other kinds of tree, shrub, herb, and lower plant found in the region. In round numbers, the pine trees represent 0.81 per cent. of the total number of plants in the whole region, and yet, considering the size and bulk of the pine trees, they form the dominant and most conspicuous element of the vegetation.

No attempt has been made to calculate the individual frequency, but the following statement will prove of value in drawing a comparison among the three quadrats and in estimating the actual occurrence of the several plants in the pine-barren region as a whole.

PLANTS COMMON TO THE THREE QUADRATS

Pinus rigida . . . . }
Quercus stellata . . . . } 4
Smilax glauca . . . . }
Vaccinium pennsylvanicum . . . . }

PLANTS COMMON TO TWO QUADRATS (Indicated by Numbers)

Comptonia asplenifolia . . . . 2, 3 }
Gaultheria procumbens . . . . 2, 3 }
Gaylussacia resinosa . . . . 2, 3 }
Melampyrum americanum . . . . 1, 2 }
Pteris aquilina . . . . 2, 3 } 10
Quercus ilicifolia . . . . 1, 3 }
"   marylandica . . . . 1, 2 }
"   prinoides . . . . 2, 3 }
Sassafras variifolium . . . . 2, 3 }
Vaccinium vacillans . . . . 1, 2 }

PLANTS FOUND IN ONE QUADRAT ONLY (Indicated)

Amelanchier intermedia....................................1 ⎫
Ampelopsis quinquefolia....................................1 ⎪
Andromeda mariana...2 ⎪
Asclepias obtusifolia.......................................1 ⎪
Clethra alnifolia..3 ⎪
Epigaea repens...3 ⎪
Eupatorium verbenaefolium.................................1 ⎪
Euphorbia ipecacuanhae...................................1 ⎪
Gaylussacia dumosa.......................................2 ⎪
     "   frondosa....................................0 ⎪
Ilex glabra..0 ⎬ 22
Kalmia angustifolia.......................................3 ⎪
   "   latifolia.......................................3 ⎪
Myrica carolinensis.......................................1 ⎪
Prunus serotina..1 ⎪
Pyxidanthera barbulata...................................3 ⎪
Quercus alba..1 ⎪
   "   prinus.......................................3 ⎪
   "   velutina......................................2 ⎪
Solidago odora...1 ⎪
Tephrosia virginiana......................................2 ⎪
Vaccinium pennsylvanicum..................................3 ⎭

GENERIC COEFFICIENT OF THE PINE-BARREN REGION

We have called attention to the fact that Jaccard has determined the generic coefficient of the alpine Jura to be 54 per cent., and of the entire Swiss flora 26 per cent. For purposes of comparison, therefore, it is important to determine this coefficient for the pine-barren region of New Jersey. As we have ascertained, there are 555 species and 234 genera of plants native to this region. The generic coefficient is, therefore, 42.16 per cent., or in round figures 42 per cent., which is the proportion of genera to species in the pine-barren flora. A table will show how this percentage stands with respect to other regions.*

REGION	SPECIES	GENERA	GENERIC COEFFICIENT
			Per Cent.
New Jersey Pine-Barrens..................	555	234	42
Miami, Florida..........................	796	466	59
Florida Keys............................	533	346	65
Altamaha Grit Region of Georgia..........	797	404	50
Southeastern United States...............	6364	1494	23
Region of Gray's Manual.................	3413	821	24
Central Rocky Mountains.................	2733	649	24
British Flora (Druce), including aliens.......	2964	734	24
Switzerland............................	2453	659	27

* HARSHBERGER, J. W.: Vegetation of South Florida. Transactions Wagner Free Institute of Science, VII: 183, 1914.

FAMILIES OF PINE-BARREN PLANTS

It will be noted from the following two tables that the number of families of pine-barren plants is ninety-one. These families include 234 genera and 581 species of flowering plants. Of the 581 pine-barren species, 303 are known to have been collected at 6 localities and over; 411 have been gathered in from 3 to 6 localities, and 164 species in from 1 to 2 places in the pine-barrens. Ninety of the 164 have been reported from only one place, leaving 74 species collected in two localities. If we deduct 26 introduced species the total number of native pine-barren plants is reduced to 555 species. Six of the pine-barren species are of doubtful character or have been recently discovered, or rediscovered. Of the 91 families, 19 families, or 20.8 per cent., have only one species in this region. With the exception of 27 families that comprise five or over five species, 64 are small, comprising not over four species to each family.

We note the conspicuous absence* of such plant families as Equisetaceae, Selaginellaceae, Taxaceae, Chenopodiaceae, Portulacaceae, Annonaceae, Papaveraceae, Cruciferae, Saxifragaceae, Platanaceae, Malvaceae, Polemoniaceae, Solanaceae, Bignoniaceae, Caprifoliaceae, Cucurbitaceae, and Lobeliaceae. The most important families of pine-barren plants are noted in the table which follows. They are considered important, because they contain the largest number of plant species and species which, from the synecologic view, are among the most important elements which enter into and render conspicuous the several plant formations and associations described previously.

FAMILIES OF PINE-BARREN PLANTS

Family	Number of Genera	Number of Species	Family	Number of Genera	Number of Species
1. Ophioglossaceae	1	2	13. Araceae...........	5	5
2. Osmundaceae	1	2	14. Xyridaceae........	1	5
3. Schizeaceae........	2	2	15. Eriocaulonaceae	1	3
4. Polypodiaceae......	6	9	16. Pontederiaceae.....	1	1
5. Lycopodiaceae	1	4	17. Juncaceae.........	1	12
6. Pinaceae..........	3	5	18. Melanthaceae.......	8	9
7. Typhaceae.........	1	1	19. Liliaceae..........	2	3
8. Sparganiaceae......	1	2	20. Convallariaceae	2	2
9. Najadaceae........	1	3	21. Smilacaceae........	1	5
10. Alismaceae........	1	2	22. Haemodoraceae	1	1
11. Graminaceae.......	22	71	23. Amaryllidaceae.....	2	2
12. Cyperaceae.......	11	76	24. Dioscoreaceae	1	1

* HARPER, R. M.: Review of Stone's Flora of Southern New Jersey, Torreya 12: 216-225, September, 1912, especially page 220.

FAMILIES OF PINE-BARREN PLANTS.—(*Continued*)

FAMILY	NUMBER OF GENERA	NUMBER OF SPECIES	FAMILY	NUMBER OF GENERA	NUMBER OF SPECIES
25. Iridaceae	2	4	57. Hypericaceae	4	9
26. Orchidaceae	11	19	58. Elatinaceae	1	1
27. Salicaceae	2	4	59. Cistaceae	3	9
28. Myricaceae	2	2	60. Violaceae	1	6
29. Juglandaceae	2	2	61. Cactaceae	1	1
30. Betulaceae	2	2	62. Lythraceae	3	3
31. Fagaceae	1	10	63. Melastomaceae	1	3
32. Urticaceae	1	1	64. Onagraceae	6	12
33. Loranthaceae	1	1	65. Haloragidaceae	2	3
34. Polygonaceae	2	4	66. Araliaceae	1	2
35. Caryophyllaceae	4	5	67. Umbelliferae	2	3
36. Nymphaeaceae	2	3	68. Cornaceae	1	1
37. Magnoliaceae	1	1	69. Clethraceae	1	1
38. Ranunculaceae	1	1	70. Pyrolaceae	2	5
39. Lauraceae	1	1	71. Monotropaceae	2	2 (32)
40. Sarraceniaceae	1	1	72. Ericaceae	11	14
41. Droseraceae	1	3	73. Vacciniaceae	2	10
42. Iteaceae	1	1	74. Diapensiaceae	1	1
Rosaceae { 43. Rosaceae	5	10	75. Primulaceae	2	3
44. Pomaceae (10)	4	5 (17)	76. Gentianaceae	3	4
45. Drupaceae	1	2	77. Menyanthaceae	1	1
Leguminosae { 46. Caesalpiniaceae (11)	1	1 (25)	78. Apocynaceae	1	3
47. Papilionaceae	10	24	79. Asclepiadaceae	1	6
48. Geraniaceae	1	1	80. Convolvulaceae	3	3
49. Linaceae	1	4	81. Cuscutaceae	1	2
50. Polygalaceae	1	9	82. Boraginaceae	2	2
51. Euphorbiaceae	3	3	83. Verbenaceae	1	2
52. Empetraceae	1	1	84. Labiatae	8	12
53. Anacardiaceae	1	4	85. Scrophulariaceae	7	10
54. Ilicaceae	2	5	86. Lentibulariaceae	1	11
55. Aceraceae	1	1	87. Rubiaceae	5	6
56. Vitaceae	2	2	88. Caprifoliaceae	1	2
			89. Campanulaceae	2	4
			90. Cichoriaceae	4	8 (59)
			91. Compositae	18	51
			Total	234	581

RELATIVE IMPORTANCE OF FAMILIES *

CYPERACEAE .... 76 species.
GRAMINACEAE .... 71 "
OLD COMPOSITAE .... 59 "
OLD ERICACEAE .... 32 "
Old Leguminosae .... 25 "
Orchidaceae .... 19 "
Old Rosaceae .... 17 "
Juncaceae }
Onagraceae } 12 "
Labiatae }
LENTIBULARIACEAE .... 11 "

FAGACEAE } 10 species.
SCROPHULARIACEAE }
Polypodiaceae
Melanthaceae
POLYGALACEAE } 9 "
Hypericaceae }
CISTACEAE }
Violaceae
Asclepiadaceae } 6 "
Rubiaceae

* Families in small capitals are important as to number of individuals.

RELATIVE IMPORTANCE OF FAMILIES.—*(Continued)*

PINACEAE.................
Araceae..................
Xyridaceae...............
Smilacaceae.............. } 5 species.
Caryophyllaceae..........
Ilicaceae.................

Lycopodiaceae............
Iridaceae.................
Salicaceae................
Polygonaceae.............
Linaceae................. } 4 "
Anacardiaceae............
Gentianaceae.............
Campanulaceae...........

Najadaceae...............
Eriocaulonaceae ....
Liliaceae.................
NYMPHAEACEAE...........
Droseraceae..............
Euphorbiaceae............
Lythraceae............... } 3 "
Melastomaceae............
Haloragidaceae............
Umbelliferae.............
Primulaceae..............
Apocynaceae..............
Convolvulaceae...........

Ophioglossaceae...........
OSMUNDACEAE............ } 2 "
Schizaeaceae ....
Sparganiaceae............

Alismaceae ....
Convallariaceae...........
Amaryllidaceae...........
Salicaceae................
MYRICACEAE..............
Juglandaceae............. } 2 species.
Betulaceae................
Vitaceae..................
Araliaceae................
Boraginaceae.............
Verbenaceae..............
Caprifoliaceae............

Typhaceae *...............
Pontederiaceae............
Haemodoraceae...........
Dioscoreaceae............
Loranthaceae.............
Santalaceae...............
MAGNOLIACEAE...........
Ranunculaceae *..........
LAURACEAE...............
Sarraceniaceae........... } 1 "
Iteaceae..................
Geraniaceae *.............
EMPETRACEAE............
ACERACEAE...............
Elatinaceae...............
Cactaceae *...............
CORNACEAE..............
DIAPENSIACEAE...........
Menyanthaceae...........

* Introduced or relatively not abundant.

CHAPTER XIV

BIOLOGIC TYPES OF PINE-BARREN PLANTS

Raunkiaer* divides plants according to the position of the winter-buds with reference to the soil surface. The winter-buds of the *phanerophytes* are borne at least 30 centimeters above the surface of the soil. Those of the *megaphanerophytes* are borne 30 meters above the earth; those of the *mesophanerophytes*, 8 to 30 meters; *microphanerophytes*, 2 to 8 meters; *nanophanerophytes*, 0.3 to 2 meters. The *chamaephytes* are those plants whose winter buds are borne immediately above the surface of the soil. The *hemicryptophytes* have winter-buds just below the earth's surface in the upper crust of the soil. The *cryptophytes* bear buds some depth below. They are divided into the *geophytes*, which have buds under the surface of the soil. The *helophytes* and *hydrophytes* have their buds entirely covered with water. *Therophytes* are the annual plants that reproduce by seeds. Martin Vahl† proposed the terms diageic and epigeic. The diageic plants are geophytes that grow trailing under the ground, while epigaeic plants are phanerophytes and chamaephytes that develop mainly above the surface of the soil.

An attempt is made here to group the native species of the pine-barren region of New Jersey under the above growth forms, but this has not been easy, because our knowledge of many species, either described in the manuals or as herbarium specimens, is very incomplete and unsatisfactory. It is, therefore, probable that mistakes have been made. Some no doubt will neutralize others, but such errors of determination are unavoidable when the descriptions and the specimens preserved in herbaria do not enable the statistician to reach determinative conclusions.

The percentage of different growth forms in a region arranged in a series, or gamut, gives a picture or spectrum of the vegetation in the words of Raunkiaer: " Ich werde deshalb der Kürzes halben in folgenden

---

* RAUNKIAER: Types biologiques pour la geographie botanique. Bulletin de l'Academie des Sciences du Danemark, 1905.

† VAHL, MARTIN: Les Types biologiques dans quelques Formations Vegetales de la Scandinavie, Acad. Roy. des Sci. et des Lettres de Danemark, 1911: 322–323; The Growth Forms of Some Plant Formations of Swedish Lapland. Dansk. Bot. Arkiv., 1, No. 2, 1913.

eine solche statistich-biologische Uebersicht als Spektrum bezeichnen, als biologisches Spektrum oder Pflanzenklimaspektrum." He has worked out on theoretic grounds a normal spectrum, which is given in the adjoining table, where S = stem succulents; E = epiphytes; Ph = phanerophytes; Ch = chamaephytes; H = hemicryptophytes; G = geophytes; HH = helophytes and Th = therophytes.

REGION	NUMBER OF SPECIES	PERCENTAGE OF SPECIES IN EACH GROWTH FORM							
		S	E	Ph	Ch	H	G	HH	Th
Pine-Barrens of New Jersey.....	458	.21	..	14	10	38	3	24	11
Miami Florida Region..........	793	.25	3	22	16	28	6	6	18
Florida Keys..................	527	1	2	32	20	19	3	3	20
St. Thomas and St. John	904	2	1	5, 23, 30	12	9	3	1	14
Seychelles....................	258	1	3	10, 23, 24	6	12	3	2	16
Aden.........................	176	1	..	-, 7, 26	27	19	3	..	17
Transcaspian Lowlands *	768	..	..	11	7	27	9	5	41
Pamir *......................	514	..	..	1	12	63	5	5	14
Death Valley.................	294	3	..	26	7	18	2	5	42
Samos........................	400	..	..	9	13	32	11	2	33
Libyan Desert.................	194	..	..	12	21	20	4	1	42
Cyrenaica....................	375	..	..	9	14	19	8	..	50
Normal Spectrum..............	400	1	3	6, 17, 20	9	27	3	1	13

Under Ph for St. Thomas and St. John islands, for the Seychelles, Aden and the Normal Spectrum, the percentages are arranged as MM, M and N, i. e., Megaphanerophytes, Microphanerophytes and Nanophanerophytes. It will be noted that in the flora of the pine-barrens the hemicryptophytes are the most abundant, followed by the helophytes, the phanerophytes, the therophytes and the chamaephytes.

Without adopting this classification of Raunkiaer for the arrangement of plants into biologic types I will use instead the classification of plants into the following types: trees, small trees, shrubs, undershrubs, perennial herbs, annual herbs and into evergreen and deciduous-leaved species.

PINE-BARREN FORMATION

TREES

Pinus echinata.
"   rigida.
Juniperus virginiana.
Quercus alba.
"   coccinea.

Quercus marylandica.
"   prinus.
"   stellata.
"   triloba (= falcata).
Sassafras variifolium.

* PAULSEN, OVE: Studies on the Vegetation of the Transcaspian Lowlands, 1912: 135–173.

SHRUBS

Myrica carolinensis.
Quercus ilicifolia.
" prinoides.
Crataegus tomentosa (=uniflora).
Clethra alnifolia.

Dendrium buxifolium.
Gaylussacia dumosa.
" frondosa.
Kalmia latifolia.

UNDERSHRUBS

Andromeda (Pieris) mariana.
Arctostaphylos uva-ursi.
Comptonia asplenifolia.
Corema Conradii.
Gaylussacia resinosa.
Hudsonia ericoides.

Hudsonia tomentosa.
Kalmia angustifolia.
Smilax glauca.
Vaccinium pennsylvanicum.
" vacillans.

ANNUAL HERBS

Aristida dichotoma.
Agrostis hyemalis.
* Gyrostachys Beckii.
Linum medium.
Crotonopsis linearis.
Polygonella articulata.
* Bartonia paniculata.

* Bartonia virginica.
* Dasystoma pedicularia.
Gerardia Holmiana.
Melampyrum lineare.
Trichostema dichotoma.
" lineare.
Gnaphalium purpureum.

* Annual, or Biennial.

COLORLESS PARASITIC PLANTS

Cuscuta arvensis.

Cuscuta compacta.

FERNS

Pteris (Pteridium) aquilinum.

Woodwardia areolata.

PERENNIAL HERBS

Paspalum setaceum.
Aristida purpurascens.
Panicum depauperatum.
" Lindheimeri.
" meridionale.
" oricola.
" villosissimum.
" Commonsianum.
" " Addisonii.
" tsugetorum.
" columbianum.
" " thinium.
" sphaerocarpon.
" Ashei.
Amphicarpon amphicarpon.
Stipa avenacea.
Cyperus cylindricus.
" Grayi.
" filiculmis macilentus.
Scleria pauciflora.
Carex pennsylvanica.
Xerophyllum asphodeloides.
Cypripedium acaule.
Comandra umbellata.
Aletris farinosa.
Arenaria caroliniana.
Baptisia tinctoria.
Tephrosia (Cracca) virginiana.
Stylosanthes biflora.

Meibomia rigida.
" obtusa.
Lespedeza frutescens.
" hirta.
" angustifolia.
Galactia regularis.
Euphorbia ipecacuanhae.
Ascyrum hypericoides.
Helianthemum canadense.
Lechea minor.
" racemulosa.
" villosa.
Kneiffia linearis.
Pyrola americana.
Epigaea repens.
Gaultheria procumbens.
Pyxidanthera barbulata.
Gentiana porphyrio.
Asclepias tuberosa.
" amplexicaulis (=obtusifolia).
Breweria Pickeringii.
Epilobium angustifolium.
Schwalbea americana.
Monarda punctata.
Lycopus sessilifolius.
Mitchella repens.
Hieracium venosum.
" Gronovii.
Lacinaria graminifolia pilosa.

Evergreen Species

Pinus rigida.
" echinata.
Juniperus virginiana.
Xerophyllum asphodeloides.
Cypripedium acaule.
Arenaria caroliniana.
Hudsonia ericoides.
" tomentosa.
Ilex glabra.

Corema Conradii.
Arctostaphylos uva-ursi.
Dendrium buxifolium.
Epigaea repens.
Gaultheria procumbens.
Kalmia latifolia.
Pyrola americana.
Pyxidanthera barbulata.
Mitchella repens.

Woody Deciduous-leaved Species

Comptonia asplenifolia.
Myrica carolinensis.
Quercus alba.
" coccinea.
" ilicifolia.
" marylandica.
" prinoides.
" prinus.
" stellata.
" triloba (=falcata).

Sassafras variifolium.
Crataegus tomentosa (=uniflora).
Andromeda (Pieris) mariana.
Clethra alnifolia.
Gaylussacia dumosa.
" frondosa.
" resinosa.
Vaccinium pennsylvanicum.
" vacillans.

Inconspicuous.
Apetalous.
Greenish.
Yellow.
Greenish White.
White.
Pink or Red.
Purple.
Blue.
Greenish Brown.
Variegated.

FIGURE 155
Tabulation of flower colors of pine-barren plants.

Colors of Pine-Barren Flowers[*]

As the colors of flowers are important in describing the vegetation of a country, the following analysis of the plants of the pine-barren formation proper is presented. The grasses and sedges are not included.

Yellow	29	Purple	16
White	26	Blue or Violet	9
Pink or Red	10	Variegated	4

It will be noted in the above enumeration that yellow and white are the most prevalent flower colors in the pine-barren formation. The

[*] ALLEN, GRANT: The Colours of Flowers, 1882.

grasses and sedges are excluded, for although the protuberant anthers of the grass and sedge flowers give color to the vegetation, yet it can hardly be said that such plants have flower colors. For the same reason the pines and oaks have been omitted from consideration. If we take the two less specialized colors, we find that there are 55 pine-barren species with yellow and white flowers. Taking the more highly specialized flower colors, there are 39 species of pine-barren plants with pink, red, purple, blue, violet and variegated flowers, so that the unspecialized flower colors exceed the specialized by some 16 species.

WHITE-CEDAR SWAMP FORMATION

In arranging the species, according to biologic characteristics and growth forms, no reference will be made to the synecology of the white-cedar swamp formation. The association of species has been described in a former section.

TREES

Evergreen:
Chamaecyparis thyoides.
Magnolia glauca (=virginiana).
Kalmia latifolia.
Rhododendron maximum.

Alnus serrulata (=rugosa).
Myrica carolinensis.
Ilicioides mucronata.
Rhus vernix.
Azalea viscosa.

Dryopteris simulata.
Lycopodium alopecuroides.
"   carolinianum.
"   Chapmanni.

Calamovilfa brevipilis.
Danthonia exilis.
Panicularia obtusa.
Panicum ensifolium.

Carex bullata.
"   canescens disjuncta.
"   Collinsii.
"   exilis.
"   interior capillacea.
"   livida.
"   oblita.
"   trisperma.
"   Walteriana.
Cladium mariscoides.

Deciduous:
Acer carolinianum.
Nyssa sylvatica.

SHRUBS

Gaylussacia frondosa.
Vaccinium atrococcum.
"   corymbosum.
"   macrocarpon.

FERNS

Osmunda cinnamomea.
"   regalis.
Schizaea pusilla.

GRASSES

Panicum lucidum.
"   spretum.
Sporobolus serotinus.
"   Torreyanus.

SEDGES

Eriophorum tenellum.
"   virginicum.
Rhynchospora alba.
"   axillaris.
"   fusca.
"   glomerata.
"   gracilenta.
"   oligantha.
"   Torreyana.
Scleria triglomerata.

INSECT-CATCHING PLANTS

Drosera longifolia.
"   rotundifolia.
Sarracenia purpurea.
Utricularia cornuta.

Utricularia gibba.
"   intermedia.
"   juncea.

ORCHIDS

Arethusa bulbosa.
Blephariglottis blephariglottis.
" ciliaris.
" cristata.
Gymnadeniopsis integra.

Gymnadeniopsis clavellata.
Limodorum tuberosum..
Pogonia divaricata.
" ophioglossoides.

AQUATIC PLANTS

Castalia (Nymphaea) odorata.

Orontium aquaticum.

PARASITIC

Phoradendron flavescens.

PERENNIAL HERBS

Spathyema foetida.
Xyris Congdoni.
Eriocaulon decangulare.
Juncus aristulatus.
" caesariensis.
" pelocarpus
Abama americana.
Tofieldia racemosa.
Helonias bullata.
Lophiola aurea.
Hypericum boreale.
Viola lanceolata.

Decodon verticillatus.
Rhexia virginica.
Sabatia lanceolata.
Asclepias rubra.
Gerardia racemulosa.
Lycopus sessilifolius.
Aster nemoralis.
Bidens trichosperma tenuiloba.
Eupatorium leucolepis.
" resinosum.
Sclerolepis uniflora.
Solidago neglecta.

FLOWER COLORS OF WHITE-CEDAR SWAMP PLANTS

If we tabulate the colors of the flowers of the white-cedar swamp plants, we find that there are no blue flowers and no variegated flowers among the true cedar swamp plants. Excluding the white-cedar, the grasses and the sedges from consideration, the enumeration stands as follows:

Yellow....................14
White.....................21
Pink, or red...............9

Purple.........................8
Brownish-green.................3

The sum (35) of the yellow and white flowers exceeds the sum (20) of the pink, red, purple, and brown colors by 15.

BIOLOGIC TYPES FOUND IN MARSH FORMATIONS

Under this category may be enumerated the plants native to the tree swamps and open marshes of the pine-barren region. The important biologic groups of plants are given with reference to their growth forms,* as they impress the general observer of the native vegetation.

FERNS

Osmunda cinnamomea.
" regalis.

Woodwardia virginica.

GRASSES

Calamagrostis cinnoides.
Panicularia canadensis.
" nervata.

Panicularia obtusa.
" pallida.

* In connection with growth forms consult DRUDE, O.: Oekologie der Pflanzen.

SEDGES

Carex Barratti.
" bullata.
" folliculata.
" intumescens.
" lurida.
Cyperus strigosus.

Dulichium arundinaceum.
Eleocharis tricostata.
Rhynchospora gracilenta.
" macrostachys.
Scirpus americanus.
Scleria minor.

PERENNIAL HERBS WITH SEDGE- OR GRASS-LIKE LEAVES

Xyris Congdoni.
" fimbriata.
" torta.
Eriocaulon compressum.
" decangulare.
Juncus acuminatus.
" aristulatus.

Juncus canadensis.
Juncus effusus.
Zygadenus leimanthoides.
Melanthium virginicum.
Sisyrinchium atlanticum.
Gyrostachys praecox.

CLIMBING PLANTS

Apios tuberosa.
Dioscorea villosa.

Smilax Walteri.

PARASITIC PLANT

Phoradendron flavescens.

ANNUAL HERBS

Polygonum Careyi.

Rotala ramosior.

PERENNIAL HERBS

Helonias bullata.
Lilium superbum.
Iris prismatica.
Gymnadeniopsis integra.
Polygonum emersum.
Linum striatum (biennial).
Polygala brevifolia.
" Nuttallii.
Viola lanceolata.
" primulaefolia.
Hypericum virgatum ovalifolium.
Rhexia mariana.
" virginica.
Ludvigia alternifolia.
" hirtella.

Ludvigia linearis.
" sphaerocarpa.
Proserpinaca pectinata.
Oxypolis rigidior.
" rigidior longifolia.
Lysimachia terrestris.
Sabatia lanceolata.
Utricularia cornuta.
Coreopsis rosea.
Solidago fistulosa.
" neglecta.
" uniligulata.
Doellingeria humilis.
" umbellata.
Helianthus angustifolius.

WOODY PLANTS (Undershrubs, etc.)

Aronia arbutifolia.
" nigra.
Decodon verticillatus.

Rubus hispidus.
Spiraea latifolia.

SHRUBS

Capsular Fruits:
Azalea viscosa.
Chamaedaphne calyculata.
Clethra alnifolia.
Itea virginica.
Kalmia latifolia.
Leucothoë racemosa.
Xolisma ligustrina.

Baccate Fruits:
Ilex glabra.

Rhus vernix.
Vaccinium atrococcum.
" corymbosum.

Drupaceous Fruits:
Ilex laevigata.
" verticillata.
Ilicioides mucronata.

TREES

EVERGREEN:
Chamaecyparis thyoides.
Pinus rigida.

DECIDUOUS:
Drupaceous Fruits:
Nyssa sylvatica.
Sassafras variifolium.

Pomaceous Fruit:
Amelanchier intermedia.

Winged Fruit:
Acer rubrum.
Betula populifolia.

Capsular Fruit:
Liquidambar styraciflua.

Aggregate Fruit:
Magnolia glauca (= virginiana).

FLOWER COLORS OF MARSH PLANTS

The colors of the flowers of the marsh plants of New Jersey present a greater variety than do either the pineland or the white-cedar swamp plants.   Enumerated, the colors are as follows:

Inconspicuous.............. 1
No petals................. 1
Greenish................. 2
Yellow................... 22
Greenish-white............. 1

White........................ 20
Pink or red 7
Purple....................... 4
Blue......................... 2
Greenish-brown................ 3

Yellow and white, as before, are the prevailing tints of flowers among marsh plants.

Tabulating the flower colors for the pineland, white-cedar swamps, and marshes we have (Fig. 155):

	PINELAND	WHITE-CEDAR SWAMP	MARSHLAND	TOTAL
Inconspicuous...............	..	..	1	1
No petals....................	..	..	1	1
Greenish.....................	..	..	2	2
Yellow.......................	29	14	22	65
Greenish-white...............	..	..	1	1
White........................	26	21	20	67
Pink or red	10	9	7	26
Purple.......................	16	8	4	28
Blue.........................	9	..	2	11
Greenish-brown...............	..	3	3	6
Variegated...................	4	..	..	4

Consulting the table and the figure (Fig. 155) we discover that white (67), yellow 65, purple (28), pink or red (26), and blue (11) are the most frequent flower colors in the pine-barren region of New Jersey and in the order named.   The other colors are unimportant.

CHAPTER XV

PHYTOPHENOLOGY OF PINE-BARREN VEGETATION

That phase of botanic science which concerns itself with a study of the course of vegetation referred to the seasons is known as phytophenomenology, shortened to phytophenology. It is of interest to consider the period of plant activity and the period of dormancy induced by the approach of the winter cold. The bursting of the leaf and flower buds, the unfolding of the first leaf, and its final fall in the autumn are all matters of ecologic interest. The floral procession of an entire year has been studied in only a few instances, and the following account is one of the few descriptions of the procession of plant development and plant flowering, from the phytogeographic standpoint, that is, considered from the standpoint of the whole vegetation of a definite region. The recorded observations in America, heretofore, have dealt in the main with individual plants, such as the unfolding of the first leaf, the opening of the first flower, the withering of the last blossom, the fall of the leaf and the development of the fruit. These results have been tabulated for chosen species, but only a few consistent attempts have been made to study the procession of plants from early spring to late fall from the regional aspect, and few attempts have been made to make deductions based upon such a study.

The following tabulation of 548 pine-barren species, as far as the flowering and fruiting periods are concerned, is based on the field and herbarium notes of the writer and upon the work of Bayard Long, who has described succinctly his method of gathering the data in Stone's "Plants of Southern New Jersey," where the facts are recorded under each plant species. The construction of the graphic table has occupied some time. The table shows the flowering and fruiting periods of 548 pine-barren plants, seven plants less than the total number of native species, from the first of March to the end of November, a period of nine months, leaving only three months as the time when the pine-barren vegetation, taken as a whole, is entirely dormant. This period can be shortened to seven months by excluding the months of March and November, for only about four species show any activity in March and none certainly in November. It is shortened to six months if the climatic records are considered. Presumably the growing season of pine-barren plants is between the last killing frost of spring and the first killing frost of autumn. At Vineland, according to the records of the U. S. Weather Bureau, the average date of the last killing frost is April 19, and the latest recorded date May 13. The average date of the first killing frost in autumn is October 19, the earliest recorded date October 2. If we take the average dates the season of growth is exactly six months, or one hundred and eighty-three days in length. This statement was written in 1912. Since then, three years later, in January, 1915, there has appeared, while this monograph was being recopied, a work by Taylor* in which this matter is discussed with reference to the region within 100 miles of New York City. On the map, Plate 5, of Taylor's memoir, the region north of the heavy black line, which runs north of the northern part of New Jersey, has a growing season of one hundred and fifty-three days, or less, while the country south of this line has a growing season of one hundred and sixty-four days, or more, and in this part of the map the New Jersey pine-barrens fall with one hundred and fifty-nine to one hundred and eighty-two days as the length of the growing season shown on the face of the map, a close approximation to one hundred and eighty-three days, the average length of the growing season determined above from the Vineland records. This determination of the length of the growing season, as of importance in the geographic distribution of plants, is in line with the labors of Cleveland Abbe, who has published an important monograph on the subject.†

To come back to our original argument, seven months from the beginning of April to the end of October represent the grand period of activity of pine-barren vegetation, leaving five months when the vegetation, as a whole, is practically dormant. Consulting the tables, it will be seen, too, that April is a month in which very little activity is shown, so that if we exclude the early spring plants (and there are about 39 such plants, or 7 per cent. of the whole), the grand period of vegetation is shortened to six months. The principal flowering and fruiting periods are not reached until July by the majority of pine-barren plants, and the active period of flower and fruit production for this majority is crowded into the three months of July, August and September. The cul-

---

* TAYLOR, NORMAN: Flora of the Vicinity of New York. Memoirs New York Botanical Garden, v: 35–36, map, Plate V.

† ABBE, CLEVELAND: A First Report on the Relations between Climates and Crops. Bull. 36, U. S. Dept. of Agriculture, Weather Bureau, 1905, pages 1 to 386; consult also an important paper by WILLIAM G. REED and HOWARD R. TOLLEY: Weather as a Business Risk in Farming. The Geographical Review, 11: 48–53, July, 1916. The risk has been graphically and mathematically presented.

mination of this phenomenon takes place in August. So much for the general features of the phenologic table.

It should be mentioned that the several months are divided in the phenologic table into six periods of five days each, and this division of the months is sufficiently accurate for phenologic purposes, if we make each of the growing months on an average thirty days long. The flowering periods are indicated in the table by full black lines and the fruiting periods by dotted lines. It will be noted that some flowering periods are short and others are long. Some fruiting periods overlap the period of flower production, but in most cases the production of fruit follows the flower period and is separated from it by an interval of time. From a visit to the pine-barren region early in November, when every plant had entered the resting condition, it may be said that October ends the period of growth. But October is a month when the autumn plants are preparing for the period of dormancy and when very few plants flower except some fall Compositae. It may be considered to be a month of preparation rather than a month of active growth. Very few new phenologic events are begun in October. They are begun in other months and are finished in October. If this view of the month's activities is considered, it will be conceded that we could exclude October from the months of most active growth, leaving only five months out of the twelve when the pine-barren plants are in an active stage of development. These facts are of importance to a man engaged in active agricultural operations in the region.

Data from my field notes corroborate the facts displayed in the foregoing table. Excluding the smaller herbaceous plants which were found in flower and a few flowering shrubs, the pine forest on April 20, 1910, was in its winter aspect. Ten days later, on April 30, 1910, the vegetation had undergone a change. It was noted on that date that Quercus prinus was half in leaf. Quercus marylandica and Quercus ilicifolia displayed their male catkins and female flowers and were in young leafage. The leaves of Quercus stellata and Q. tinctoria were small. The leaves of the red-maple, Acer rubrum, were expanded almost entirely, but the sour-gum, Nyssa sylvatica, and Magnolia glauca were still bare of leaves with their buds unopened. The new shoots of Pinus rigida on May 4, 1910, had reached a length of four inches, and on this date the young leaves of Liquidambar styraciflua and Nyssa sylvatica were seen. By June 1 all of the above trees were in full foliage, but the leaves retained their juvenile light green color, contrasting with the dark greens of the pitch-pine, Pinus rigida. It will be seen from these data that the month of May is practically the one in which the leafage of the deciduous

FLOWERING AND FRUITING PERIODS OF PINE-BARREN PLANTS
Heavy lines, flowering periods; dotted lines, fruiting periods

Name of Plant	March	April	May	June	July	August	September	October

1 Botrychium obliquum
2 Botrychium obliquum dissectum
3 Osmunda regalis
4 Osmunda cinnamomea
5 Schiznea pusilla
6 Lygodium palmatum
7 Pteridium aquilinum
8 Woodwardia virginica
9 Woodwardia areolata
10 Asplenium platyneuron
11 Asplenium filix-foemina
12 Polystichum acrostichoides
13 Dryopteris thelypteris
14 Dryopteris simulata
15 Phegopteris dryopteris
16 Lycopodium chapmanii
17 Lycopodium alopecuroides
18 Lycopodium carolinianum
19 Lycopodium obscurum
20 Pinus virginiana
21 Pinus echinata
22 Pinus rigida
23 Chamaecyparis thyoides
24 Juniperus virginiana
25 Typha latifolia
26 Sparganium americanum
27 Sparganium americanum androcladum
28 Potamogeton oakesianus
29 " epihydrus
30 " confervoides
31 Sagittaria longirostra
32 Sagittaria graminea
33 Erianthus saccharoides
34 Andropogon scoparius
35 " corymbosus abbreviatus
36 " virginicus
37 Sorghastrum nutans
38 Paspalum psammophilum
39 " pubescens
40 " muhlenbergii
41 " setaceum
42 Amphicarpon amphicarpon

FLOWERING AND FRUITING PERIODS OF PINE-BARREN PLANTS

Name of Plant	March	April	May	June	July	August	September	October

43 Syntherisma filiformis
44 Panicum verrucosum
45 " philadelphicum
46 " virgatum
47 " " cubense
48 " agrostoides
49 " longifolium
50 " depauperatum
51 " dichotomum
52 " barbulatum
53 " lucidum
54 " clutei
55 " spretum
56 " lindheimeri
57 " leucothrix
58 " meridionale
59 " oricola
60 " villosissimum
61 " pseudopubescens
62 - commonsianum
63 " " addisonii
64 " tsugetorum
65 " columbianum
66 " " thinium
67 " ensifolium
68 " spaerocarpon
69 " ashei
70 " oligosanthes
71 " scoparium
72 " cryptanthum
73 " scabriusculum
74 Cenchrus carolinianus
75 Aristida dichotoma
76 " gracilis
77 " purpurascens
78 Stipa avenacea
79 Muhlenbergia capillaris
80 Sporobolus torreyanus
81 " serotinus
82 Agrostis maritima
83 " elata
84 " hyemalis

FLOWERING AND FRUITING PERIODS OF PINE-BARREN PLANTS

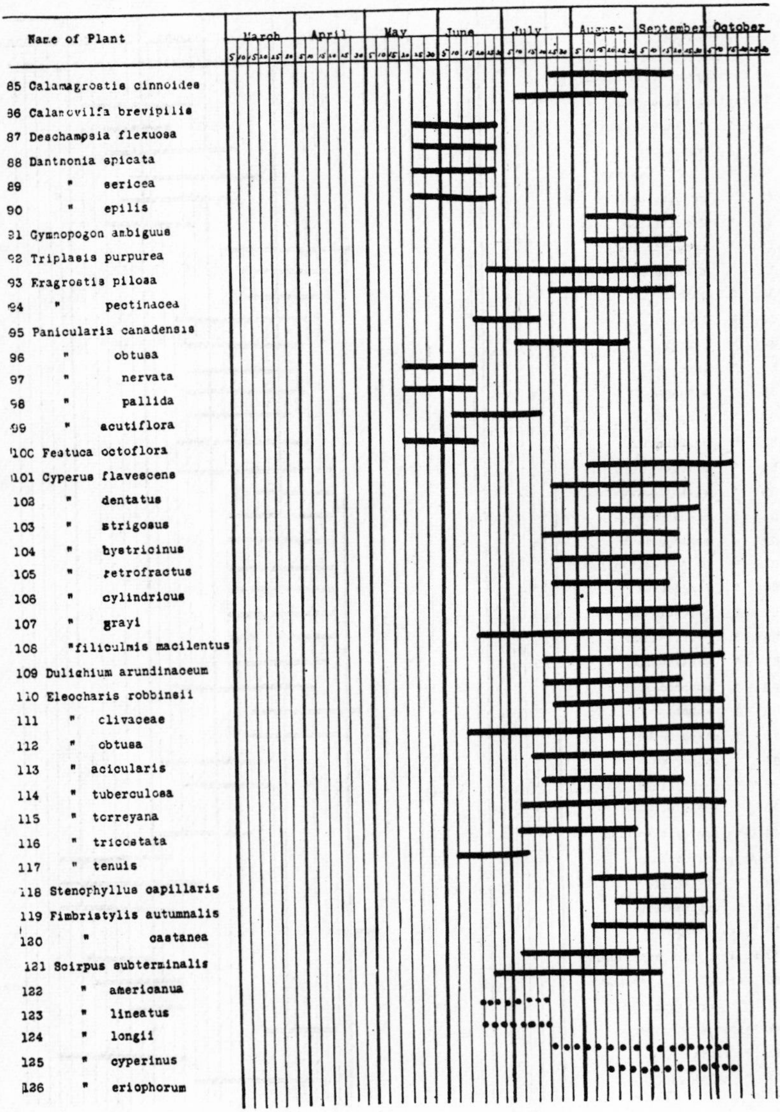

Name of Plant	March	April	May	June	July	August	September	October

85 Calamagrostis cinnoides
86 Calamovilfa brevipilis
87 Deschampsia flexuosa
88 Dantnonia spicata
89 " sericea
90 " spilis
91 Gymnopogon ambiguus
92 Triplasis purpurea
93 Eragrostis pilosa
94 " pectinacea
95 Panicularia canadensis
96 " obtusa
97 " nervata
98 " pallida
99 " acutiflora
100 Festuca octoflora
101 Cyperus flavescens
102 " dentatus
103 " strigosus
104 " hystricinus
105 " retrofractus
106 " cylindricus
107 " grayi
108 "filiculmis macilentus
109 Dulichium arundinaceum
110 Eleocharis robbinsii
111 " olivaceae
112 " obtusa
113 " acicularis
114 " tuberculosa
115 " torreyana
116 " tricostata
117 " tenuis
118 Stenophyllus capillaris
119 Fimbristylis autumnalis
120 " castanea
121 Scirpus subterminalis
122 " americanua
123 " lineatus
124 " longii
125 " cyperinus
126 " eriophorum

FLOWERING AND FRUITING PERIODS OF PINE-BARREN PLANTS

Name of Plant	March	April	May	June	July	August	September	October

127 Eriophorum tenellum
128 " virginicum
129 Rynchospora macrostachya
130 " macrostachya inundata
131 " gracilenta
132 " oligantha
133 " alba
134 " pallida
135 " knieskernii
136 " glomerata
137 " glomerata leptocarpa
138 " smallii
139 " filifolia
140 " axillaris
141 " fusca
142 " torreyana
143 Cladium mariscoides
144 Scleria triglomerata
145 " minor
146 " reticularis torreyana
147 " pauciflora
148 Carex collinsii
149 " folliculata
150 " intumescens
151 " lupulina
152 " bullata
153 " lurida
154 " lacustris
155 " vestita
156 " walteriana
157 " barrattii
158 " crinita
159 " swanii
160 " triceps
161 " oblita
162 " livida
163 " pennsylvanica
164 " umbellata tonsa
165 " exilis
166 " arnectens
167 " muhlenbergii
168 " interior capillacea

FLOWERING AND FRUITING PERIODS OF PINE-BARREN PLANTS

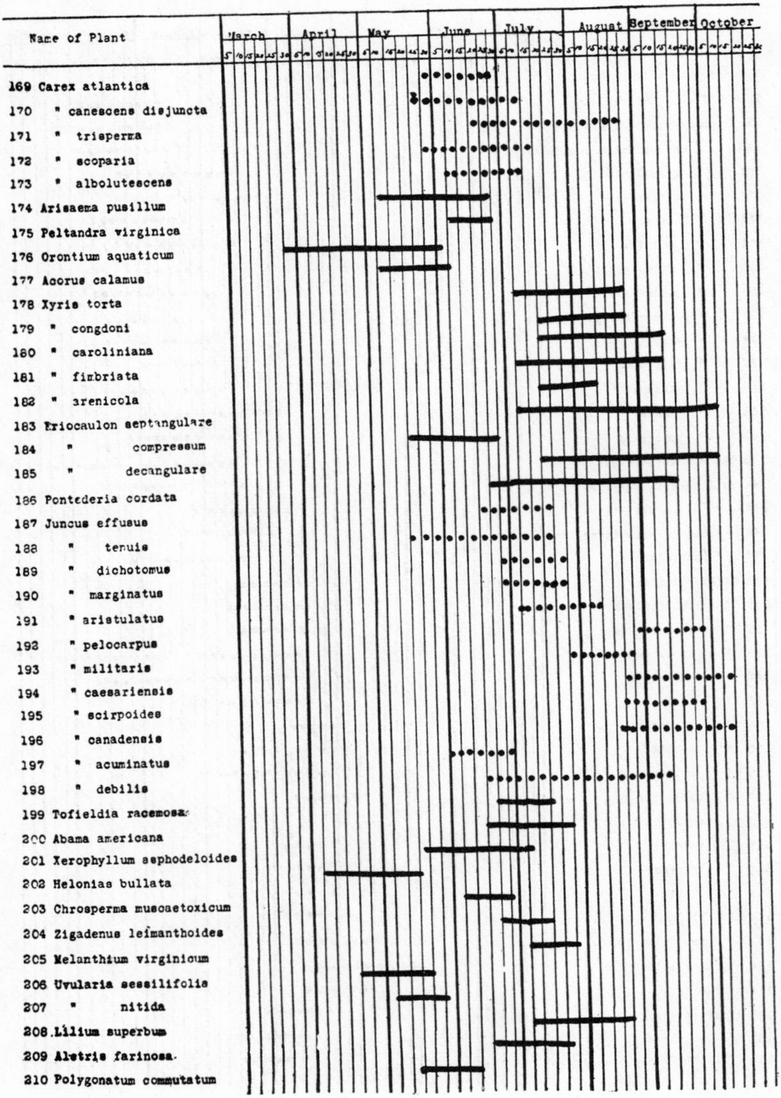

Name of Plant	March	April	May	June	July	August	September	October

169 Carex atlantica
170 " canescens disjuncta
171 " trisperma
172 " scoparia
173 " albolutescens
174 Arisaema pusillum
175 Peltandra virginica
176 Orontium aquaticum
177 Acorus calamus
178 Xyris torta
179 " congdoni
180 " caroliniana
181 " fimbriata
182 " arenicola
183 Eriocaulon septangulare
184 " compressum
185 " decangulare
186 Pontederia cordata
187 Juncus effusus
188 " tenuis
189 " dichotomus
190 " marginatus
191 " aristulatus
192 " pelocarpus
193 " militaris
194 " caesariensis
195 " scirpoides
196 " canadensis
197 " acuminatus
198 " debilis
199 Tofieldia racemosa
200 Abama americana
201 Xerophyllum asphodeloides
202 Helonias bullata
203 Chrosperma muscaetoxicum
204 Zigadenus leimanthoides
205 Melanthium virginicum
206 Uvularia sessilifolia
207 " nitida
208 Lilium superbum
209 Aletris farinosa
210 Polygonatum commutatum

FLOWERING AND FRUITING PERIODS OF PINE-BARREN PLANTS

Name of Plant	March	April	May	June	July	August	September	October

211 Medeola virginiana
212 Smilax tamnifolia
213   "   rotundifolia
214   "   glauca
215   "   laurifolia
216   "   Walteri
217 Gyrotheca tinctoria
218 Hypoxis hirsuta
219 Lophiola aurea
220 Dioscorea villosa
221 Iris prismatica
222 Sisyrinchium mucronatum
223   "   atlanticum
224 Cypripedium acaule
225 Gymnadeniopsis integra
226   "   clavellata
227 Blephariglottis cristata
228   "   ciliaris
229   "   blephariglottis
230 Pogonia ophioglossoides
231   "   divaricata
232 Isotria verticillata
233 Arethusa bulbosa
234 Limodorum tuberosum
235 Gyrostachys cernua
236   "   praecox
237   "   vernalis
238   "   beckii
239 Listera australis
240 Peramium pubescens
241 Leptorchis loeselii
242 Populus grandidentata
243 Salix nigra
244   "   humilis
245   "   tristis
246 Myrica carolinensis
247 Comptonia asplenifolia
248 Hicoria glabra
249 Betula populifolia
250 Alnus rugosa
251 Quercus coccinea
252   "   velutina

FLOWERING AND FRUITING PERIODS OF PINE-BARREN PLANTS

Name of Plant	March	April	May	June	July	August	September	October

253 Quercus triloba
254 " ilicifolia
255 " marilandica
256 " phellos
257 " alba
258 " stellata
259 " prinus
260 " prinoides
261 Phoradendron flavescens
262 Comandra umbellata
263 Polygonum eneraum
264 " careyi
265 Polygonella articulata
266 Silene antirrhina
267 Sagina decumbens
268 Arenaria caroliniana
269 Anychia canadensis
270 " polygonoides
271 Brasenia purpurea
272 Nymphaea variegata
273 Castalia odorata
274 Magnolia virginiana
275 Sassafras officinale
276 Sarracenia purpurea
277 Drosera rotundifolia
278 " longifolia
279 " filiformis
280 Itea virginica
281 Spiraea latifolia
282 " tomentosa
283 Rubus cuneifolius
284 " argutus
285 " villosus
286 " villosus enslenii
287 " hispidus
288 Aronia arbutifolia
289 " nigra
290 Amelanchier intermedia
291 Crataegus tomentosa
292 Prunus maritima
293 Baptisia tinctoria
294 Crotalaria sagittalis

FLOWERING AND FRUITING PERIODS OF PINE-BARREN PLANTS

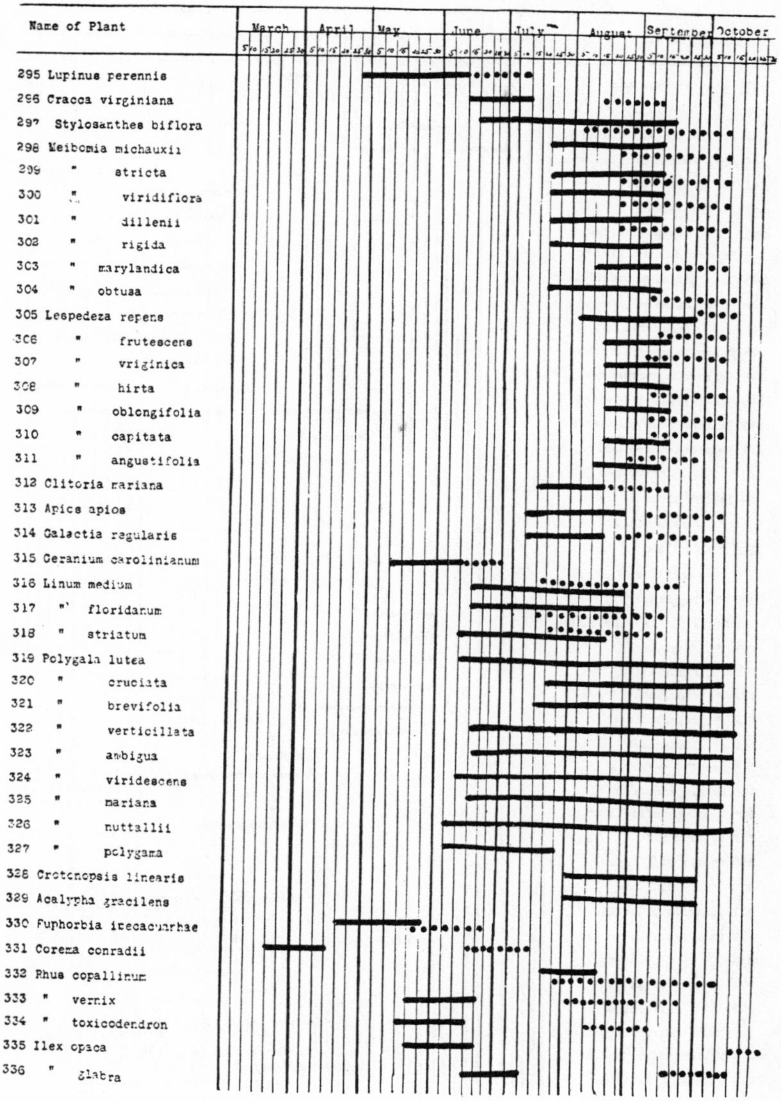

Name of Plant	March	April	May	June	July	August	September	October

295 Lupinus perennis

296 Cracca virginiana

297 Stylosanthes biflora

298 Meibomia michauxii

299   "   stricta

300   "   viridiflora

301   "   dillenii

302   "   rigida

303   "   marylandica

304   "   obtusa

305 Lespedeza repens

306   "   frutescens

307   "   vriginica

308   "   hirta

309   "   oblongifolia

310   "   capitata

311   "   angustifolia

312 Clitoria mariana

313 Apios apios

314 Galactia regularis

315 Geranium carolinianum

316 Linum medium

317   "   floridanum

318   "   striatum

319 Polygala lutea

320   "   cruciata

321   "   brevifolia

322   "   verticillata

323   "   ambigua

324   "   viridescens

325   "   mariana

326   "   nuttallii

327   "   polygama

328 Crotonopsis linearis

329 Acalypha gracilens

330 Euphorbia ipecacuanhae

331 Corema conradii

332 Rhus copallinum

333   "   vernix

334   "   toxicodendron

335 Ilex opaca

336   "   glabra

FLOWERING AND FRUITING PERIODS OF PINE-BARREN PLANTS

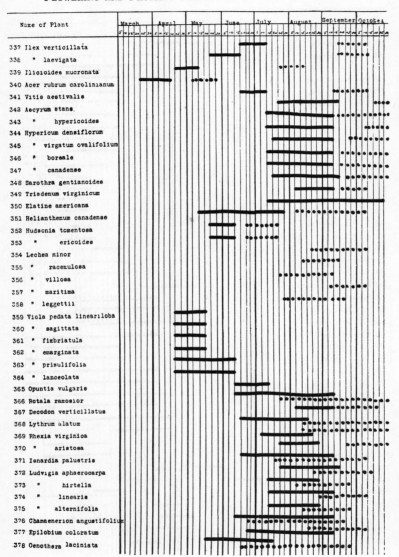

Name of Plant	March	April	May	June	July	August	September	October
337 Ilex verticillata								
338 " laevigata								
339 Ilicioides mucronata								
340 Acer rubrum carolinianum								
341 Vitis aestivalis								
342 Ascyrum stans.								
343 " hypericoides								
344 Hypericum densiflorum								
345 " virgatum ovalifolium								
346 " boreale								
347 " canadense								
348 Sarothra gentianoides								
349 Trisdenum virginicum								
350 Elatine americana								
351 Helianthemum canadense								
352 Hudsonia tomentosa								
353 " ericoides								
354 Lechea minor								
355 " racemulosa								
356 " villosa								
357 " maritima								
358 " leggettii								
359 Viola pedata lineariloba								
360 " sagittata								
361 " fimbriatula								
362 " emarginata								
363 " primulifolia								
364 " lanceolata								
365 Opuntia vulgaris								
366 Rotala ramosior								
367 Decodon verticillatus								
368 Lythrum alatum								
369 Rhexia virginica								
370 " aristosa								
371 Isnardia palustris								
372 Ludvigia aphaerocarpa								
373 " hirtella								
374 " linearis								
375 " alternifolia								
376 Chamaenerion angustifolium								
377 Epilobium coloratum								
378 Oenothera laciniata								

FLOWERING AND FRUITING PERIODS OF PINE-BARREN PLANTS

Name of Plant	March	April	May	June	July	August	September	October
379 Kneiffia linearis								
380 " longipedicellata								
381 " pumila								
382 Proserpinaca pectinata								
383 Myriophyllum humile								
384 Aralia nudicaulis								
385 Oxypolis rigidior								
386 Oxypolis rigidior longifolia								
387 Nyssa sylvatica								
388 Clethra alnifolia								
389 Pyrola americana								
390 " elliptica								
391 " chlorantha								
392 Chimaphila maculata								
393 " umbellata								
394 Monotropa uniflora								
395 Hypopitys hypopithys								
396 Azalea nudiflora								
397 " viscosa								
398 " glauca								
399 Rhododendron maximum								
400 Dendrium buxifolium								
401 Kalmia angustifolia								
402 " latifolia								
403 Leucothoe racemosa								
404 Pieris mariana								
405 Xolisma ligustrina								
406 Epigaea repens								
407 Chamaedaphne calyculata								
408 Gaultheria procumbens								
409 Arctostaphylos uva-ursi								
410 Gaylussacia frondosa								
411 " dumosa								
412 " baccata								
413 Vaccinium corymbosum								
414 " virgatum								
415 " caesariense								
416 " atrococcum								
417 " pennsylvanicum								
418 " vacillans								
419 Oxycoccus macrocarpus								
420 Pyxidanthera barbulata								
421 Lysimachia quadrifolia								

FLOWERING AND FRUITING PERIODS OF PINE-BARREN PLANTS

Name of Plant	March	April	May	June	July	August	September	October
422 Lysimachia terrestris					▬			
423 Trientalis borealis			▬▬▬					
424 Sabatia lanceolata					▬▬▬			
425 Gentiana porphyrio							▬▬▬▬	
426 Bartonia paniculata					▬▬▬			
427 " virginica					▬▬			
428 Limnanthemum lacunosum					▬▬▬			
429 Apocynum androsaemifolium				▬▬▬▬				
430 " medium				▬▬▬▬				
431 " cannabinum pubescens				▬▬▬				
432 Asclepias tuberosa				▬▬▬▬				
433 " rubra				▬▬▬				
434 " amplexicaulis				▬▬▬				
435 " variegata				▬▬▬				
436 " syrica				▬▬▬				
437 " verticillata				▬▬▬▬				
438 Breweria pickeringii				▬▬▬				
439 Ipomoea pandurata				▬▬▬▬				
440 Convolvulus sepium				▬▬▬				
441 Cuscuta arvensis						••••••••••••••••		
442 " compacta						••••••••••••		
443 Myosotis laxa			▬▬▬▬					
444 Onosmodium virginianum			▬	•••••••••••••				
445 Verbena urticifolia				▬▬▬				
446 " angustifolia				▬▬▬				
447 Trichostema dichotomum					▬▬▬			
448 Scutellaria integrifolia				▬▬▬▬▬				
449 Stachys hyssopifolia				▬▬▬▬				
450 Monarda punctata					▬▬▬			
451 Hedeoma pulegioides					▬▬▬			
452 Clinopodium vulgare					▬▬▬			
453 Koellia verticillata					▬▬▬			
454 " incana					▬▬▬			
455 " mutica					▬▬▬			
456 Lycopus virginicus						▬▬▬		
457 " sessilifolius						▬▬▬		
458 Linaria canadensis			▬▬▬▬					
459 Gratiola aurea			▬▬▬					
460 Veronica peregrina			▬▬▬					
461 Dasystoma pedicularia					▬▬▬			
462 " flava					▬▬▬			
463 " virginica					▬▬▬			

FLOWERING AND FRUITING PERIODS OF PINE-BARREN PLANTS

Name of Plant	March	April	May	June	July	August	September	October
464 Gerardia racemulosa								
465 " holmiana								
466 Schwalbea americana								
467 Melampyrum lineare								
468 Utricularia cornuta								
469 " juncea								
470 " virgatula								
471 " fibrosa								
472 " subulata								
473 " cleistogama								
474 " clandestina								
475 " gibba								
476 " intermedia								
477 " inflata								
478 " purpurea								
479 Oldenlandia uniflora								
480 Cephalanthus occidentalis								
481 Mitchella repens								
482 Diodia teres								
483 Galium pilosum								
484 " " puncticulosum								
485 Viburnum cassinoides								
486 " nudum								
487 Specularia perfoliata								
488 Lobelia inflata								
489 " nuttallii								
490 " canbyi								
491 Adopogon virginicum								
492 " carolinianum								
493 Lactuca canadensis								
494 " hirsuta								
495 Hieracium venosum								
496 " gronovii								
497 Nabalus serpentarius								
498 " virgatus								
499 Solarclepsis uniflora								
500 Eupatorium maculatum								
501 " leucolepis								
502 " album								
503 " album subvenosum								
504 " hyssopifolium								
505 " verbenaefolium								
506 " rotundifolium								

FLOWERING AND FRUITING PERIODS OF PINE-BARREN PLANTS

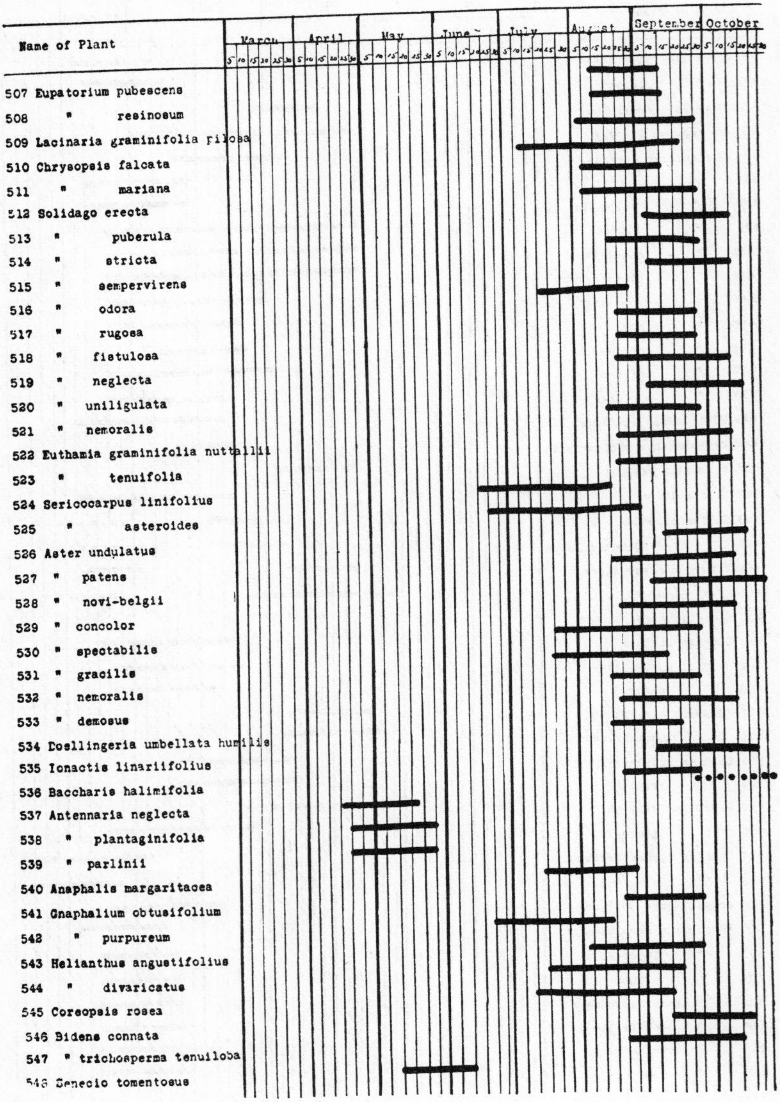

Name of Plant	March	April	May	June	July	August	September	October

507 Eupatorium pubescens

508　　　"　　　resinosum

509 Lacinaria graminifolia pilosa

510 Chrysopsis falcata

511　　　"　　　mariana

512 Solidago erecta

513　　"　　puberula

514　　"　　stricta

515　　"　　sempervirens

516　　"　　odora

517　　"　　rugosa

518　　"　　fistulosa

519　　"　　neglecta

520　　"　　uniligulata

521　　"　　nemoralis

522 Euthamia graminifolia nuttallii

523　　　"　　　tenuifolia

524 Sericocarpus linifolius

525　　　"　　　asteroides

526 Aster undulatus

527　　"　　patens

528　　"　　novi-belgii

529　　"　　concolor

530　　"　　spectabilis

531　　"　　gracilis

532　　"　　nemoralis

533　　"　　demosus

534 Doellingeria umbellata humilis

535 Ionactis linariifolius

536 Baccharis halimifolia

537 Antennaria neglecta

538　　"　　plantaginifolia

539　　"　　parlinii

540 Anaphalis margaritacea

541 Gnaphalium obtusifolium

542　　"　　purpureum

543 Helianthus angustifolius

544　　"　　divaricatus

545 Coreopsis rosea

546 Bidens connata

547　"　trichosperma tenuiloba

548 Senecio tomentosus

trees takes place, and October is the month in which the death of the foliage leaves of such trees occurs. The period of active metabolism of mature leaves is about four months and a half in length, when the active formation of new substance takes place, and when the trees get ready, by the storage of reserve materials, for the period of the winter's rest. Even on August 30, we noted the display of red autumn leaves on the red-maple, Acer rubrum, and on the sour-gum tree, Nyssa sylvatica.

A few cases of special interest should be mentioned before considering the table in detail. The pitch-pine, Pinus rigida, begins to show signs of staminate flower development on May 5. On May 20, these flowers discharged pollen. Presumably, the pollination took place on that date, or before the end of May. The pollen-grain, or microspore, is drawn into the pollen-chamber, where it remains over the summer, autumn and winter, until the next spring. The pollen-tube is probably sent out into the nucellus, as soon as the grain is deposited, but it stops active growth until April of the next year. During April the pollen-tube begins to renew its penetration of the nucellus, and the large tube-nucleus enters the pollen-tube, where it is invested completely by starch grains, and at the same time the generative cell divides into stalk-cell and body-cell. The pollen-tube consumes about two months in traversing the nucellus after its second start, entering the archegonium about July 1. Just before fertilization the body-cell divides to form the two male cells, so that at this time the pollen-tube contains the tube-nucleus, stalk-cell nucleus and the two male cells. The tip of the pollen-tube, having penetrated the overlying cells of the nucellus, reaches the wall of the embryo-sac, and either passes directly through it, or flattens out upon it in a foot-like expansion, sending out a small branch to reach the neck of the archegonium. The neck-cells are crushed and the pollen-tube comes in contact with the egg about July 1, when fertilization occurs. While the seeds are developing the cones rapidly enlarge until they reach mature size by September 6, for cones brought into a warm, dry laboratory that were gathered on that date opened and discharged their seeds after being kept indoors for a week. Outside, however, the cones do not open until the warm days of Indian summer, for a trip to the pine-barrens on November 13, 1912, showed many, but not all, of the cones of the pitch-pine fully open and discharging their winged seeds.

Some observations of Brown * are apropos. He finds that in the pinery of Cornell University the growth of the cambial layer in young twenty- to thirty-year-old specimens of Pinus rigida began in the vicinity of

* BROWN, HARRY P.: Growth Studies in Forest Trees—(1) Pinus rigida Mill. The Botanical Gazette, LIV: 386–401, November, 1912.

Ithaca as early as April 15, and it is probable also that in the vicinity of Ithaca growth begins at about the same time each spring. As the growing season is shorter at Ithaca than in the pine-barrens, it is probable that growth of the cambial layer of the pitch-pine begins early in central New Jersey. Growth at Ithaca began first in the twenty- to thirty-year-old specimens at some distance below the apical shoot, but during a period of nineteen days gradually spread upward until it reached the apex of the trees. Growth spreads down the main axis faster than it does along the lateral shoots. Except in the terminal shoot, growth in diameter was more rapid between May 25 and June 6. In the terminal shoot itself greatest rapidity of growth was manifested between June 6 and June 15. Brown finds also that there is no appreciable difference in the time of cambial awakening on the north and south side of the trees.

The evergreen flowering-moss, or pyxie, Pyxidanthera barbulata, also claims attention. The earliest date which I have of open flowers is March 23, and the last date, when only a few flowers were found, is April 30, but the herbarium records given by Bayard Long extend the flower period of this plant to May 10, so that the total period of flowering of this species is forty-eight days. On June 13 fully ripe capsules and seeds of this species were found, so that eighty-two days elapsed from the opening of the first flower until the development of mature fruit and seeds. The period of flower and fruit development in some pine-barren plants is much shorter than this, for example, Comptonia asplenifolia, eighty days, Lupinus perennis, seventy days, Acer carolinianum, sixty-five days, Hudsonia ericoides, sixty days, Salix humilis, forty-seven days. Longer periods than eighty-two days are found in quite a few plants of the pine-barrens. The white water-lily, Castalia odorata, flowers from May 23, the earliest date that I have on record, to October—about one hundred and thirty days. The production of fruit begins in this plant early in July and continues into October. The seeds may germinate immediately after escaping from the fruit, or may remain dormant in cold water until the following spring: the first is probably the case with fruits ripened before September, the second with those ripened later, according to Conard. Conard[*] found great numbers of seedlings (var. minor) in a pond south of Marlton, N. J., on August 30, 1900, with only submerged leaves. The first leaf is filiform, about 1.6 centimeters long. The second is elliptic-oblong. Intermediate leaves are ovate, cordate, rounded at apex, with rounded basal lobes and sinus. Late submerged

[*] CONARD, HENRY S.: The Waterlilies, 1905: 181.

leaves are sagittate-cordate, with deep, narrow sinus, are stiff in texture, bright emerald, and stand erect or semi-erect.

The actual period of flowering is short for the amentiferous trees and shrubs, as is shown by giving the data for the following trees, arranged systematically:

```
Populus grandidentata.................................12–15 days
Salix nigra...........................................20      "
  "   humilis........................................15      "
  "   tristis.........................................15      "
Myrica carolinensis...................................35      "
Comptonia asplenifolia................................20      "
Hicoria glabra.......................................20      "
Betula populifolia...................................25      "
Alnus rugosa.........................................25      "
Quercus coccinea.....................................15      "
  "   velutina.......................................17      "
  "   triloba........................................16      "
  "   ilicifolia......................................17      "
  "   marylandica....................................17      "
  "   phellos.........................................20      "
  "   alba............................................20      "
  "   stellata........................................15      "
  "   prinus..........................................15      "
  "   prinoides.......................................18      "
```

But in the majority of these trees and shrubs, with the exception of Populus grandidentata, which matures its fruit in May or June, and of Salix nigra, S. humilis, and S. tristis, which mature their fruits by the middle of May, all other trees and shrubs show a slow development of the fruit, which is not mature until some time has elapsed. The following facts are known about the maturation of the fruit of the remaining species in the list given above. The fruit of Myrica carolinensis matures between the end of July and the middle of August. That of Comptonia asplenifolia matures between the middle of June and early July. The fruit of the pignut, Hicoria glabra, white-birch, Betula populifolia, alder, Alnus rugosa, do not mature unti late August and September. The oaks of the pine-barrens may be divided into two groups. The annual-fruited, or white-oaks, comprising Quercus alba, Q. stellata, Q. prinus, and Q. prinoides, mature their fruits in September of the first year, i. e., the year of pollination and fertilization. The biennial-fruited, or black-oaks, comprising Quercus coccinea, Q. velutina, Q. falcata, Q. marylandica, Q. phellos, and Q. ilicifolia, do not mature their fruits until September of the next year after pollination.

For the small herbaceous plants the period of flower production of the several pine-barren species of Polygala is a long-drawn-out one. With the exception of 12 species, all of the 42 species of the family Com-

positae, represented in the pine-barren flora, flower after the middle
of July, and most of them in August and September. We will now tabulate the number of plants that come into flower at intervals of five days
throughout the year, so as to ascertain at what season the majority of
pine-barren plants come into flower. We will then be able to decide
whether the vegetation, as far as flowering is concerned, is a spring,
summer, or autumn vegetation.

TABLE SHOWING NUMBER OF PLANTS THAT COME INTO FLOWER AT
FIVE-DAY INTERVALS FROM MARCH 15 TO
SEPTEMBER 10, INCLUSIVE

March	1	0		July	1		1
	5	0			5		30
	10	0	6		10	106	21
	15	2			15		24
	20	2			20		29
	25	2			25		1
April	1	3		August	1		0
	5	5			5		24
	10	0	33		10	72	9
	15	2			15		24
	20	23			20		15
	25	0			25		0
May	1	0		September	1		1
	5	31			5		10
	10	4	91		10	15	3
	15	21			15		0
	20	32			20		1 Schizaea
	25	3			25		0
June	1	3		October	1		0
	5	30			5		0
	10	17	113		10	0	0
	15	30			15		0
	20	29			20		0
	25	4			25		0

It will be seen from the table that the blossoming of pine-barren plants
is irregular, and a graphic curve of the phenomena illustrates the same
thing. The period extends from March 15 to September 10, approximately six months. If, however, we take the number of plants that
bloom by monthly periods, we find that the majority of pine-barren
plants flower during May, June, July, and August, with the largest
number during June, July, and August. In these four months 382 species
out of 436 species come into flower. But to reach a true statement
of the number of plants in flower during any one month we must refer
again to our graphic tabulation of the pine-barren species and count the

number actually in bloom during the successive months. The following is the result of this enumeration:

Month	Number of Plants in Flower	Number that Came into Bloom
March	6	6
April	41	33
May	128	91
June	198	113
July	247	106
August	263	72
September	207	15
October	58	0

A comparison of the figures in the two columns of the preceding table indicates that although a certain number of plants may have come into flower in a certain month, yet many of these plants continue the period of flowering into the next following month, or even to two or three months, so that the number of plants actually in bloom in any one month is augmented considerably, as shown in column one. For example, many of the early spring flowers go out of flower early in June and many late summer blossoming plants begin to flower in June, so that both early spring and late summer plants help to make the total June number. This applies also to July, August and September. As no plants come into flower after September 10 and only 15 before that date, the large number of plants in flower in September is due to the continuation of the flowering periods of plants that blossomed earlier. The same applies particularly to October, a month in which no plants blossomed, but in which we find 58 species in actual flower. However, the important point which the foregoing table demonstrates is that the months of June, July, August and September are the flower months of the pine-barren vegetation season. The month of May ought to be included, perhaps, so that practically five months may be looked upon as the active flowering season of pine-barren vegetation with the culmination of the flower display in August.

From March 15 to October 30 the pine-barren region is practically never without some plant in flower. March, April, and October are the dullest months, and November, December, January, and February comprise the dormant period. Before concluding this account of the phenology of pine-barren vegetation we will construct a few figures to illustrate the procession of vegetation in each of the principal plant formations of the region, viz., pine-barrens, cedar swamps, deciduous swamps, ponds, plains, and savannas (Fig. 156). The procession of flower development as illustrated by a few plants of these formations is as follows:

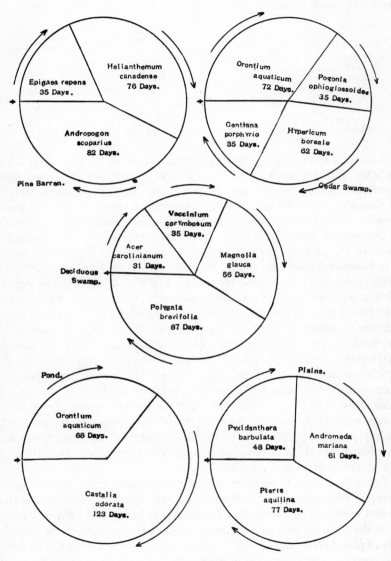

FIGURE 156

Procession of flower development in a series of plants of the pine-barren, cedar swamp, deciduous swamp, pond and plains formations.

Pine-Barren Formations:
    Epigaea repens..........................April 5–May 10.
    Helianthemum canadense..................May 10–July 25.
    Andropogon scoparius....................July 20–October 10.
Bog Formation:
    Orontium aquaticum......................March 25–June 5.
    Pogonia ophioglossoides.................June 5—July 10.
    Hypericum boreale.......................July 10–September 10.
    Gentiana porphyrio......................September 10–October 15.
Deciduous Swamp Formation:
    Acer carolinianum.......................March 20–April 20.
    Vaccinium corymbosum....................April 20–May 25.
    Magnolia glauca.........................May 20–July 15.
    Polygala brevifolia.....................July 15–October 15.
Pond Formation:
    Orontium aquaticum......................March 25–June 1.
    Castalia odorata........................June 1–October 1.
Plain Formation:
    Pyxidanthera barbulata..................March 23–May 10.
    Andromeda mariana.......................May 5–July 5.
    Pteris aquilina.........................July 5–September 20.
Savanna Formation:
    Alnus rugosa............................March 15–April 10.
    Chamaecyparis thyoides..................April 1–April 20.
    Pyrus arbutifolia.......................April 20–May 25.
    Iris prismatica.........................May 20–June 25.
    Lophiola aurea..........................June 25–July 25.
    Eriocaulon decangulare..................July 15–October 5.

It will be noted that two plants cover the flowering period of pond vegetation; three the period for the pine-barrens and plains respectively; four the white-cedar swamp and marsh periods, and five the period of flowering in the savannas. Graphically the procession in each of the formations may be indicated by a circle in which the size of each segment of the circle represents the relative length of time at which the given species is in flower (Fig. 156).

CHAPTER XVI

VEGETATIVE PROPAGATION AND STRUCTURE OF THE SHOOTS

Much attention has been given of late years to a study of those organs of the plants which are concerned with the absorption of food, which serve to anchor the plant, which store food, or which renew the plant by vegetative propagation. As the pine-barren vegetation is an ancient and exclusive one in the region where it is found and exists in climax form, it is important to study the organs by which these plants have been able, in undisturbed condition, to withstand the encroachment of other nearby formations which have reached also a climax condition. The pine-barren species occupy the soil layers so closely with their over-ground and their underground parts that few new plant introductions have succeeded in making their way into such closed formations which have occupied the region since post-Miocene times. Abroad several plant ecologists have emphasized the importance of this examination of the propagative shoots and roots. Only a few of the papers will be mentioned in passing. Tansley and Fritsch,* in a study (1905) of the flora of the Ceylon littoral, give some attention to this subject. Warming,† in 1908, published his careful investigations on the structure and biology of arctic flowering plants, and in these studies he was assisted by Petersen. Yapp,‡ in the same year, gives an excellent synopsis of the underground parts of the plants of Wicken Fen, and in a later, more-detailed paper, the results of his research on the stratification in the vegetation of a marsh and its relations to evaporation and tempera-ture. Ostenfeld, in the Botany of the Faroes (1908: 867–1018), con-siders the land vegetation of these islands with special reference to the higher plants and the vegetative propagation and structure of the shoots. Eugen Hess,§ in an inaugural dissertation, discusses the special structure

* TANSLEY, A. G., and FRITSCH, F. E.: The Flora of the Ceylon Littoral. The New Phytologist, IV: 1–17; 25–55.

† WARMING, EUGEN: The Structure and Biology of Arctic Flowering Plants. Meddelel-ser om Grönland, XXXVI.

‡ YAPP, R. H.: Wicken Fen. The New Phytologist, VII: 61–81; On Stratification in the Vegetation of a Marsh, and its Relations to Evaporation and Temperature. Annals of Botany, XXIII: 275–319, April, 1909.

§ HESS, EUGEN: Ueber die Wuchsformen der alpinen Geröllpflanzen, Universität Zurich, 1909.

of the alpine boulder plants. Sherff* has pursued much the same line of investigation in his studies of the vegetation of Skokie Marsh, with special reference to subterranean organs and interrelationship. More recently Yapp† has published a paper on a single plant, Spiraea ulmaria, and its bearing on the problem of xeromorphy in marsh plants. Cannon,‡ in two publications of considerable length from the Desert Botanical Laboratory at Tucson, has described the root habits of desert plants.

Late in my studies of pine-barren vegetation I determined to undertake a study of the underground parts of a series of pine-barren species. The result of that study is given in the following pages for 32 pine-barren plants arranged systematically. Some few of the descriptive details and some of those shown in the drawings were obtained from herbarium material which served also to check the observations made upon fresh parts while in the field.

The hapaxanthic species, or those that only flower once, propagate by seeds alone, whereas many of the perennials propagate both by seeds and vegetatively. Ostenfeld has dealt with them under three categories:

1. Spot-bound (sedentary) species, i. e., species which have no stolons, creeping rhizomes, nor bud-producing roots, or the rhizomes are so short that the plants have little, if any, power of wandering vegetatively.

2. Wandering species with epiterranean (above ground) runners.

3. Wandering species with subterranean shoots, stolons, creeping rhizomes, or bud-producing roots.

Hess,§ in the important paper mentioned above, has classified the different kinds of shoots. As his classification is applicable to the study of pine-barren plants, a translation is given by way of an introduction to what follows. To bring out the salient features of this system pine-barren species have been arranged as follows:

I. Spot-Bound Shoots

A. Single Spot-bound Shoots

1. We may designate simple, upright, terrestrial stems that bear a crown of leaves at the top, namely, caudices. The primary root persists,

---

* SHERFF, EARL E.: The Vegetation of Skokie Marsh, with Special Reference to Subterranean Organs and their Interrelationship. The Botanical Gazette, LIII: 415–435.

† YAPP, R. H.: Spiraea ulmaria and its Bearing on the Problem of Xeromorphy in Marsh Plants. Annals of Botany, XXVI: 815–870.

‡ CANNON, W. A.: The Root Habits of Desert Plants. Publ. No. 131, Carnegie Institution of Washington, 1911; Some Features of the Root-Systems of the Desert Plants. Popular Science Monthly, July, 1912.

§ HESS, EUGEN: Ueber die Wuchsformen der alpinen Geröllpflanzen, pages 38–49.

but secondary adventitious roots may be formed. Pinus rigida, Quercus marylandica, Cassia chamaecrista, Chimaphila umbellata, Melampyrum americanum.

2. Under this head may be included bulbous shoots. Cypripedium acaule, Habenaria blephariglottis, Eupatorium verbenaefolium, Liatris graminifolia, L. pilosa, Hieracium venosum.

B. Spot-bound Shoots in Clusters

3. By the branching of spot-bound shoots are formed aërial tufts or cushions. Few species form only tufted shoots without adventitious roots. Frequently, however, adventitious roots are not formed as an occasional adaptation of such species which normally grow with other plants. The rounded cushions in thick crowding display radially separated branches. Quercus prinoides, Baptisia tinctoria, Rhus copallina, Hudsonia ericoides.

4. The tufts of branches, which are provided with roots, abundant at times, are illustrated by the grasses and sedges. Such shoots with only sufficient space to grow upright develop from a short, horizontal piece. Uninterrupted growth leads to the covering of extensive surface, so that cushions are formed, but such cushions are formed of parallel shoots, not radial.

5. *Caudex multiceps.*—Uniform extended stems of mono- or sympodial branching, which grow over or under ground, were termed by Areschoug terrestrial stems (Erdstamme). If the branch grows vertically, or sharply upright, it reaches the surface in a single tuft of leaves. Hitchcock designates such a simple vertical shoot a caudex. A much-branched, upright terrestrial stem bearing tufts of leaves at each extremity is designated by Hess a branched caudex (*Caudex multiceps*). Hess distinguishes between the spot-bound form and the spread-out form, which Hitchcock calls crown (Krone), and Warming a multicipital root (radix multiceps). Whether these terrestrial stems have primary or adventitious roots or whether the flower-stem is leafy or not, is, therefore, indifferent. Examples are Euphorbia ipecacuanhae, Kalmia latifolia, Tephrosia virginiana.

II. SPREADING SHOOTS

Spreading shoots are either normal shoots or else they are elongated runners.

A. Spreading Shoots of Normal Form

6. Hess calls aërial shoots of this group which develop without roots crested shoots (Schopftriebe). Ratzeburg calls them periwigs (Perück-

en), or wig forms. In gravel, the branches are not aërial, but grow partly in the light, partly in the dark, of the soil spaces, and the small leaves on them are generally colorless. Sometimes the wig shoots wind about, between, and over the stones, a distance of 20 to 30 centimeters—Pyxidanthera barbulata.

7. Shoots which spread radially as branches of a single stem, and which thus form a crown, may be termed wheel shoots, or simply radiate shoots. These radial shoots do not develop roots except unusually long, hence they should never be classed as rhizomes. Hudsonia tomentosa, Arenaria caroliniana, Cassia chamaecrista.

8. The surface-spreading shoots which develop adventitious roots may be termed turf-forming shoots. Such are termed generally by American botanists creeping stems, and they may be orthotropic or plagiotropic.

9. Rhizome (rootstock) is any horizontal or oblique perennial stem which grows on the ground, or in it, and produces roots and shoots that are usually leafy and green. Pteris aquilina, Sparganium eurycarpum, Smilax glauca, Comptonia asplenifolia, Myrica carolinensis, Hypericum densiflorum, Nymphaea advena, Ilex glabra, Asclepias obtusifolia, Gaultheria procumbens, Andromeda mariana, Vaccinium vacillans, Arctostaphylos uva-ursi, and Sericocarpus asteroides.

B. *Extension by Elongated Slender Shoots*

10. Wandering shoots (Wandertriebe) are similar to the turf-forming shoots, but they are hidden beneath the surface and grow through the interstices of the earth, or between the stones, if the soil is a gravelly or a stony one. When they reach the surface and grow upon it, they cannot be distinguished from the turf-forming shoots. Tofieldia racemosa, Pyxidanthera barbulata.

11. The slender running shoots (Läufen) which strike root at the tip are called stolons, or runners. An offset is a short stolon.

12. A sucker is an ascending stem arising from a subterranean base usually connected with the mother plant. It is not included in the classification which Hess has given, because it does not occur among the plants which he has studied, viz., the rubble, or boulder plants (Geröllpflanzen).

Detailed Root and Stem Studies

It is only within recent years that plant ecologists have directed their attention to the study of the root and the underground stem portions of plants, as well as their physiologic condition. But the prospect is fair that in a few years we will have definite information on this important subject. The roots and underground stems of pine-barren plants are of interest because of their relation to the perennation of the species and to the physiologic activities of the shoots. There is a close relation between the character of these parts and the distribution of the plants, the association of the different species in the different plant formations, the formation of the leaves, of the flowers, and the commencement of new growth. According to Cannon, the work of Rimbach, Büsgen and Freidenfeldt, as reviewed by von Alter (Wurzel-studien, Bot. Zeit., 67: 175, 1909), is important because their researches indicate that the root system of flowering plants may be separated into two groups, according to the character of the terminal roots: they are either intensive or extensive. Intensive root systems have fine terminal roots; they are branched richly and occupy relatively small soil volume. Extensive root systems have coarse ultimate rootlets, are not richly branched and occupy a relatively large soil volume. Cannon* describes three main types of root systems to be found in the desert plants of the southwestern United States: (1) Root systems which extend horizontally from the main axis and lie, for their whole course, near the surface of the ground. (2) Root systems which are characterized by a strongly developed tap-root going down directly to a depth determined in part by the character of the soil, in part by the penetration of the rains, and in part by the character of the root itself. (3) Roots that not only reach widely, but also penetrate fairly deeply. Where the root system is an obligate type the distribution of the species is much restricted, but where it undergoes modification with changed environment the distribution of the species is much less confined. It is of interest to note especially that, as a rule, it is the latter kind of root system that is developed by such plants as occur where the soil conditions are most arid, from which it follows that the generalized type of root system (No. 3 above) is really the xerophytic type par excellence and not the type with the most deeply penetrating tap-root, as might be supposed. Reference will be made to these points as we proceed with our studies. Having classified the pine-barren plants as to their underground parts, it is important to describe next in detail the stem and root systems of each of these plants.

Pteris aquilina (Bracken-fern).—This fern develops a brown underground stem, or rhizome, which in the specimen studied reached a length of 6.7 meters (22 feet). The usual thickness is about 1.27 centimeters

---

* CANNON, W. A.: Some Features of the Root Systems of the Desert Plants. Pop. Sci. Month., July, 1912; Botanical Features of the Algerian Sahara. Publ. 178, Carnegie Institution of Washington, 1913.

(½ inch), and along each side run lighter bands called the lateral lines, which probably function in the respiration of the rootstock.   The growing-point, which is rounded, whitish, and covered with paleae, has a single apical cell, and the stem branches in two ways—first, by apparent dichotomy of the terminal growing-point, and second, by the formation of adventitious buds.   The latter are produced singly on the dorsal side of the leaf-stalks near the base.   The bases of the old leaf-stalks persist for a few years, but they ultimately disappear.   The roots, like the stem, develop by means of an apical cell which is covered with a root-cap. Root-hairs are found.   As to the structure of a single rhizome 6.7 meters (22 feet) long, the following notes are given in the metric units and the English units of measurement.

No. of Branch	Distance behind Apex		Length of Branch	
	Metric	English	Metric	English
1	5 cm.	2 ins.	2.5 cm.	1 in.
2	23 "	9 "	6.3 "	2½ ins.
3	4.6 dm.	1 ft. 6 "	. .	. .
4	6.8 "	3 " 3 "	1.0 dm.	4 ins.
5	9.1 "	3 " 0 "	. .	. .
6	1.4 m.	4 " 9 "	17.0 dm.	7 ins.
7	1.6 "	5 " 6 "	. .	. .
8	1.9 "	6 " 3 "	12.0 dm.	5 ins.
9	2.0 "	6 " 10 "	. .	. .
10	2.3 "	7 " 6 "	5.3 dm.	1 ft. 9 ins.
11	2.5 "	8 " 2 "	6.3 cm.	2½ ins.
12	2.7 "	8 " 10 "	5.0 "	2 "
13	2.9 "	9 " 6 "	7.5 "	3 "
14	3.2 "	10 " 6 "	7.5 "	3 "
15	3.3 "	11 " 0 "	7.5 "	3 "
16	3.5 "	11 " 6 "	7.5 "	3 "
17	3.7 "	12 " 4 "	3.0 dm.	1 ft.
18	3.9 "	12 " 10 "	1.7 "	7 ins.
19	4.0 "	13 " 4 "	6.3 cm.	2½ "
20	4.2 "	13 " 10 "	6.3 "	2½ "
21	4.3 "	14 " 4 "	. .	. .
22	4.5 "	14 " 10 "	. .	. .
23	4.7 "	15 " 4 "	. .	. .
24	4.8 "	15 " 10 "	. .	. .
25	5.0 "	16 " 5 "	. .	. .
26	5.2 "	17 " 1 "	. .	. .
27	5.4 "	17 " 8½ "	. .	. .
28	5.6 "	18 " 8½ "	. .	. .
29	5.7 "	19 " 1 in.	. .	. .
30	5.8 "	19 " 6½ ins.	. .	. .
31	6.0 "	19 " 9 "	. .	rhizome rotten
32	6.0 "	20 " 0 "	. .	"
33	6.2 "	20 " 6 "	. .	"
34	6.4 "	21 " 10 "	. .	"

The underground rhizome of the bracken-fern may be classified with the extensive kind of von Alten, and to the first division of Cannon's classification, and the plant is a hemicryptophyte, according to Raunkiaer.

FIGURE 157
Plotted rhizome of Pteris aquilina.
1 small square = 1 inch.
A, Level of soil.   B, Rhizome (true depth), upper dotted lines.   C, Branches.   D, Dead branches (lower dotted lines).

The depth at which the rhizome was found varied, as it accommodated itself in its growth to the larger pine and oak roots under which, or over which, it grew in its course through the soil.

At   22.8 cm. from the apex it was 11.4 cm. deep.
"   30.4   "   "   "   "   "   "   13.9   "   "
"   60.8   "   "   "   "   "   "   12.7   "   "
"   152.0   "   "   "   "   "   "   8.8   "   "
"   182.4   "   "   "   "   "   "   20.3   "   "
"   243.2   "   "   "   "   "   "   20.3   "   "
"   295.4   "   "   "   "   "   "   8.8   "   "
"   362.3   "   "   "   "   "   "   17.7   "   "
"   443.3   "   "   "   "   "   "   10.1   "   "
"   547.2   "   "   "   "   "   "   11.4   "   "
"   577.6   "   "   "   "   "   "   10.1   "   "
"   608.0   "   "   "   "   "   "   15.2   "   "

The conspicuous parts of this fern are large green fronds which arise above the ground to a height of 45 centimeters (18 inches) to 60 centimeters (2 feet), occasionally 15 to 18 decimeters (5–6 feet), and consist of a rachis with thrice pinnate divisions.   The rachis of the frond may be followed some distance into the ground.   Its buried portion is brown in color, and it is attached to the rhizome below, from which true roots are given off.   Traced in one direction from the attachment of the frond the rhizome exhibits the withered bases of the fronds, developed in former years, which have died down; while in the opposite direction it ends, sooner or later, in a rounded extremity covered with fine brown hairs, which is the growing-point of the stem.   Between the free end and the fully formed frond rounded projections, the rudiments of fronds, are seen.   These will attain their full development in the following years (Fig. 157).

Pinus rigida (Pitch-pine).—This pine is a typic phanerophyte, according to the classification of Raunkiaer (Figs. 158, 159, 160).   It develops a powerful tap-root, which persists for many years unless, as in the plains, it comes in contact with a hard, impervious subsoil, when it rots away.   Sometimes it is bifurcated, as in a tree studied along Shark River Bay.   Strong lateral roots are developed from the main primary root.   The strongest of these sec-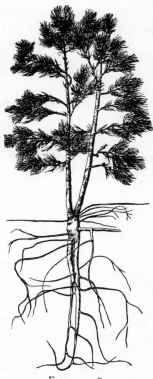ondary roots are near the soil surface, in a specimen measured 10 centimeters below the surface, and they run horizontally for some distance, finally sinking obliquely downward, or at times sharply at right angles into the deeper soil levels.   Several of these powerful lateral roots may be formed at slightly different depths.   The smaller and shorter secondary roots are developed from the lower part of the primary root, and from a physiologic standpoint may be considered to be a part of it, while the more superficial and stronger secondary roots belong in function to a slightly different category.   To give a few specific examples: A tree was studied at Como Lake.   Its first large superficial secondary root was given off 10 centimeters below the soil surface, where it soon bifurcated.   The lower division sank to a level of 23 centimeters and ran a distance of 4.5 meters; then it grew obliquely downward to a depth of 1.5 meters, reaching a distance of 6.5 meters measured at the soil surface from the base of the tree.   Another horizontal root grew down to a depth of 1.7 meters below the soil surface.

FIGURE 158

Pinus rigida, showing root system.

Some of the shorter lateral roots arose 30 centimeters below the surface of the soil.   In a low pine tree studied in the Upper Plain, which was only 30 centimeters tall, it was found that the largest roots were horizontal and superficial, gradually dipping to a gravel layer 50 centimeters below the surface.   The primary root had decayed 30 centimeters below the surface.   This decay had been hastened by the root growing to the hard pan, which has been described previously.

Sparganium americanum (Slender Bur-reed).—This marsh plant is found in the pine-barrens near Forked River, Bamber, Tuckerton, Parkdale, Bear Swamp and Clementon. It is a cryptophyte, according to Raunkiaer's classification, being included in the division of helophytes along with Typha and other marsh plants. The rhizomes of this plant are formed about two centimeters below the surface, and are strong plagiotropic shoots with large terminal buds. According to Sherff, who has studied S. eurycarpum, such plants with strong, thick rhizomes have a decided advantage over other plants of less vegetative vigor.

Dulichium spathaceum.—If we examine the underground part of Dulichium, we find a rather long horizontal rhizome with sympodial ramifications. The internodes are distinct and covered partly with rudimentary sheathing leaves. Relatively strong roots develop above the nodes, especially from the lower surface. Several stems arise from

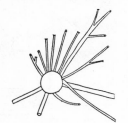

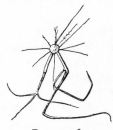

FIGURE 159
Horizontal plan of the root system of the pitch-pine, Pinus rigida. o, Sinking pine roots; –, horizontal pine roots, as shown at tips of branches.

FIGURE 160
Horizontal view of root system of pitch-pine.

the rhizome, and they show a structure very different from what we are accustomed to see in the Cyperaceae, having numerous cylindric and hollow internodes. The leaves of Dulichium are either reduced tubular sheaths upon the rhizome and at the base of the stems, or they are developed normally with a linear blade, a short, crescent-shaped ligule, and a long tubular sheath. The plant is a hemicryptophyte.

Andropogon scoparius Michx. (Beard-grass).—This grass, which has tufted culms 4 to 12 decimeters tall, is a typic hemicryptophyte. The leaves and culms arise from a rounded, woody base and the filiform secondary roots spread in every direction (Fig. 161).

Panicum depauperatum.—This erect or ascending grass, 2 to 4 decimeters tall, occurs in the driest situations in the pine-barrens. Its tufted culms and leaves arise from a plate-like stem provided with numerous filiform secondary roots. A hemicryptophyte (Fig. 162).

Abama (Narthecium) americanum.—The bog-asphodel has an aërial shoot, 2.5 to 4 decimeters tall, which arises from a horizontal rhizome covered with the fibrous remains of the old leaf-bases. The secondary roots are numerous, arising from the under surface of the rhizome. It is abundant in the wet savannas along the Wading River. A hemicryptophyte (Fig. 163).

Xerophyllum asphodeloides.—The turkey's-beard is a characteristic plant of the New Jersey pine-

FIGURE 161
Andropogon scoparius.

FIGURE 163
Abama americana.

FIGURE 162
Panicum depauperatum.

barrens. Its long, needle-like leaves form a dense hassock which represents the massing together of the leaves of several distinct plants that have a short rhizome beset with numerous secondary roots. The leaves are flattened at the base and overlapping. A hemicryptophyte (Fig. 164).

Tofieldia racemosa.—The false asphodel is a slender perennial, liliaceous plant, mostly tufted, with a slender plagiotropic rhizome growing about 1 centimeter below the surface of the boggy places in which it occurs. The rhizome shows tufts of roots at the nodes, reaching up nearly to the surface, and a few superficial, inordinately long wandering roots which develop a tuft of smaller roots at their extremity. The flowering scape of this plant reaches a height of 3 to 9 decimeters. A hemicryptophyte (Fig. 165).

Smilax glauca.—The saw-brier is an erect or climbing plant that produces a tough, wiry, plagiotropic rhizome, white in color, divided into clearly marked nodes and internodes. The nodes bear pointed buds protected partially by the sheathing leaf-base, which is the only part of the leaf found below ground. Two kinds of roots are formed, namely, long nodal roots that may reach a length of 17.7 centimeters (7 inches), and short internodal roots. A hemicryptophyte (Fig. 166).

Cypripedium acaule (Moccasin Flower).—This beautiful orchid has a short, rounded stem just below the surface of the soil. From it arises a cluster of slender roots from 3 to 25 centimeters long, produced in large numbers and of a brown color. Two large, oval leaves surround the

FIGURE 164
Xerophyllum asphodeloides.

FIGURE 165
Root system of Tofieldia racemosa.

FIGURE 166
Smilax glauca.

base of the single-flowered scape. The fibrovascular bundles of old leaves in a shredded, fibrous condition are seen frequently at the base of the stem and on the surface of the soil. A hemicryptophyte (Fig. 167).

Pogonia ophioglossoides.*—This is a terrestrial orchid with a slender rhizome, which is relatively short, vertical, or ascending, and densely covered with long, unicellular hairs, a unique feature on underground stems. The rhizome in this species, according to Holm, consists of five

---

* *Cf.* HOLM, THEO.: Pogonia ophioglossoides Nutt., a Morphological and Anatomical Study. American Journal of Science, IX: 13–19, January, 1900.

distinct internodes, the uppermost passing into a flower-bearing stem. The rhizome is vertical, bearing rudiments of two green leaves, and two scale-like and membranaceous, while a third scale-like leaf is still fresh and surrounds the base of the flower-bearing stem.

FIGURE 167
Cypripedium acaule.

The monopodial growth is characteristic of this rhizome, which ceases, however, when the flower-bearing stem dies away. After that time it becomes a sympodium, as the axillary bud will develop and continue the growth of the rhizome. This is, however, not the only bud which is visible, since another, though dormant, is developed in the axil of another leaf. This bud will grow out into a branch if the upper part of the rhizome becomes injured, but not otherwise. The main propagation of this orchid is by means of root-shoots. This striking feature makes a study of the roots of great interest. It appears that the shoot develops terminally at the end of the root, and a secondary root pushes out from the base of the young shoot and in the same direction as the mother root, as if it were a continuation of it. This secondary root repeats the development of another shoot in exactly the same manner as the first. The other secondary roots which develop higher up on the shoot grow often from the beginning in a downward direction, and do not form new shoots, but are concerned probably with absorption. Hyphae are found in the hypoderm and in the cortex of these roots. A hemicryptophyte.

FIGURE 168
Underground system of Habenaria blephariglottis.

Limodorum tuberosum (*Calopogon pulchellus*) is an orchid which has a 4- to 12-flowered scape arising from a solid bulb, which is sheathed below by the base of the solitary grass-like leaf. The cortex of the slender roots of this orchid is supplied with mycorhizal hyphae, suggesting that it and Pogonia are not entirely autophytic, but hemisaprophytic.

Habenaria blephariglottis (White Fringed Orchid).—This orchid, found in white-cedar swamps, has two kinds of roots arising from the base of the stem, namely, slender feeding roots and elongated tuberous roots, from which the buds that develop into new shoots arise. The stem which arises from these clustered roots is 4 to 6 decimeters tall, and bears several linear, lanceolate leaves on its lower part, and small linear leaves just below the crowded raceme of pure white flowers. A hemicryptophyte (Fig. 168).

Comptonia asplenifolia (Sweet-fern).—This low shrub, 3 to 6 decimeters tall, with sweet-scented, fern-like foliage, belongs to the group of nanophanerophytes of Raunkiaer. It arises from a long horizontal,

FIGURE 169
Comptonia asplenifolia.

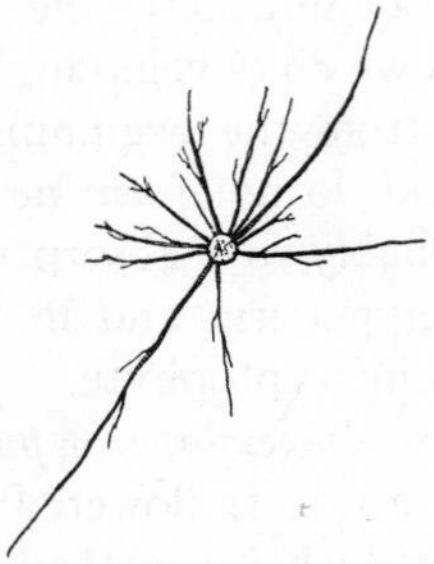

FIGURE 170
Horizontal view of the root system of
Quercus marylandica.

FIGURE 171
Quercus marylandica.

frequently forking root, sometimes three feet long, and about the thickness of a lead-pencil. It produces filiform lateral roots and is found in the surface soil just beneath the layer of pineneedles (Fig. 169).

Myrica carolinensis.—The bayberry is a nanophanerophyte. It develops ordinarily a vertically descending tap-root, but where the leaf mold, consisting of pine needles and other leaves, has been piled in masses, the bayberry develops a horizontal rhizome the thickness of a

lead-pencil. It runs a distance of about 75 centimeters away from the
main stem, and sends down short lateral roots, which become associated

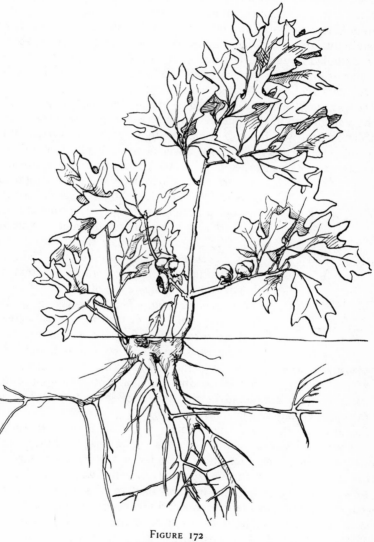

FIGURE 172
Quercus ilicifolia.

intimately with the leaf mold, so that it is almost impossible to separate
them.

Quercus prinoides (Dwarf Chestnut-oak).—This low, bushy oak develops a strong tap-root, 3 to 4 centimeters thick, that grows deep into the soil. The stronger lateral roots are found some distance below the surface. In one specimen studied the first strong lateral was 30 centimeters below the surface, the other, 42.5 centimeters. The smaller lateral roots arise from the primary root just below the soil surface. In a tree studied in undisturbed pine-barrens four stool shoots were found growing at first prostrate beneath the soil and then rising at the ends 60 centimeters above it. A nanophanerophyte.

Quercus marylandica (Figs. 170 and 171).—The black-jack oak grows to be a small tree in the pine forests of New Jersey. In a specimen 12 decimeters tall was found a conic tap-root descending vertically into the soil. From this central root developed two kinds of lateral roots. First, short roots are borne on the primary root at various depths below the surface. Second, long and thick lateral roots arise at different levels. In the tree studied there was one 9 decimeters long, 18 centimeters below the surface; a second, 5.5 decimeters long, was found 25 centimeters below the surface; a third grew 4 decimeters below the surface. These are formed usually in different ranks, so that they do not stand one above the other.

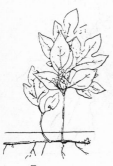

FIGURE 173
Sassafras variifolium.

Quercus ilicifolia.—The bear-oak is a dwarf tree 1 to 3, rarely 6, meters high, usually covered with acorns in the fall of the second year. The stems arise from a gnarled, knotted, and twisted crown, and the strong roots are several in number, descending obliquely into the upper subsoil. Horizontal roots are found in the upper soil layer near the surface. A mesophanerophyte (Fig. 172).

Sassafras variifolium.—The sassafras is usually a small tree in New Jersey which may reach a large size elsewhere (a phanerophyte). In the young trees in the pine forests of New Jersey the stems one meter tall arise from a horizontal rhizome sparingly provided with roots (Fig. 173).

Comandra umbellata.—The bastard-toadflax is a hemicryptophyte. It has a slender horizontal rhizome and filiform descending roots. The plant, which has upright stems 15 to 30 centimeters tall, is probably parasitic on the roots of other plants.

Hudsonia tomentosa.—This heath-like, low plant has a chocolate-brown tap-root growing downward vertically a length of 12.5 to 15 centimeters. The lateral roots which are developed from the point where the lateral branches arise are 10 to 15 centimeters long and wiry. The lateral branches which radiate outward from this central point are at first plagiotropic and at the tips orthotropic. Hence arises a cushion-like plant of a grayish-green color covered with scale-like, alternate, persistent, downy leaves. Hence we have a plant which has shoots arranged in a wheel-like or radiate manner (Fig. 174).

Hudsonia ericoides.—This perennial plant, with awl-shaped leaves, arises from a superficial rhizome about the thickness of a lead-pencil. Several strong roots and a lot of filiform ones are developed. Several stems may grow from the same rhizome. A hemicryptophyte (Fig. 175).

Hypericum densiflorum.—This is a bushy shrub, 0.5 to 2 meters tall, and branches richly above the ground, with slender branches crowded with small leaves. The root is arched or horizontal, with numerous adventitious roots growing down from

FIGURE 174
Hudsonia tomentosa.

the under surface of the horizontal portion. The main root, which is a direct continuation of the horizontal portion, grows downward obliquely, where it divides into a number of strong branches. A short stub on the upper surface of the horizontal portion is that of last year's leafy stem. A hemicryptophyte (Fig. 176).

Nymphaea (Nuphar) advena.—The cow-lily, or spatterdock, is a hydrophytic cryptophyte with strong horizontal rhizomes, 5 to 10 centimeters thick, and found mostly 8 to 25 centimeters below the soil surface. Where this plant mixes with Scirpus and Typha, it may be classed as a true helophyte, but above it has been classed as a true hydrophyte, the distinction between hydrophytes and helophytes not being defined always sharply. From the rhizome, the surface of which has a checkered appearance owing to the prominence of the old leaf scars, arise the floating (*variegata*), or emersed (*advena*), erect leaves, which are round to ovate, or almost oblong, with an open sinus and subtriangular lobes. The bases of these leaves overlap where they join the rhizome. A large number of strong roots grow down from the under side of this rhizome.

FIGURE 175
Hudsonia ericoides.

Castalia (Nymphaea) odorata.—The rhizome of this hydrophyte and cryptophyte is horizontal and much smaller than that of the preceding plant, but found at a similar depth, where a few stout persistent branches may be found.   They are 2.2 to 3.2 centimeters thick and 30 centimeters to 1 meter long, pale in color, covered with a dense, short, black pubescence.   The apex is rounded, clothed with stipules and fine hairs.   The roots, according to Conard, remain uninjured after the evaporation of water from ponds in the fall.   In this species 3 to 5 or 6 roots occupy a triangular area on the cushion behind each petiole; the apex of the triangle is farthest from the petiole and is occupied by the first formed root of medium size.   Next to it is a second and much larger root, and

FIGURE 176
Hypericum densiflorum.

FIGURE 177
Tephrosia virginiana.

the base of the triangle has a few smaller ones.   A root-cap is present on the tip of every root and large intercellular air-passages in the cortex of the roots.

Tephrosia (Cracca) virginiana.—The goat's-rue is a perennial herb with a rootstock that may branch several times.   These branches are as thick as a lead-pencil and sink downward obliquely to a depth of 10 centimeters before uniting.   The stubs of old aërial roots indicate that some of the rhizome branches have produced at least 5 to 6 annual shoots which are erect and 3 to 6 decimeters tall.   The lateral roots, produced 10 centimeters below the surface, are yellowish-white, cord-like, and at least 25 centimeters long.   A hemicryptophyte (Fig. 177).

Baptisia tinctoria.—The wild-indigo has a strong primary root which descends vertically into the soil and reaches at the soil surface a thickness of 2.5 centimeters. In a measured specimen the first lateral root was found 14 centimeters below the soil surface, and it reached a length of 1 meter, growing downward obliquely into the soil. The shorter lateral roots were found in general at twice that depth, namely, 28 centimeters. From the crown of this root arise the aërial shoots, usually only one being produced annually, but a number of stubs, at least two, sometimes 4, indicate where former shoots have arisen. A hemicryptophyte (Fig. 178).

FIGURE 178
Baptisia tinctoria.

FIGURE 179
Lupinus perennis.

Lupinus perennis.—The blue lupine is a perennial plant with palmately compound leaves of 7–11 leaflets. Its stems arise from an underground rhizome that may be buried 10 centimeters under the soil surface. A hemicryptophyte (Fig. 179).

Galactia regularis.—The milk-pea is a prostrate perennial with short, four- to eight-flowered racemes. From six to eight stems may arise from one deeply descending taproot. Hence its root is of the multicipital kind. A cryptophyte (Fig. 180).

Cassia chamaecrista.—This annual plant, or therophyte of Raunkiaer's classification, has a slender primary root, about 8 to 10 centimeters long, developed below a subterranean, hypocotyledonary portion, 2 centimeters long. The secondary roots, which are produced at intervals along the primary root, bear scattered tubercles. The longest measured lateral root was 22 centimeters long. The upright portion of the plant,

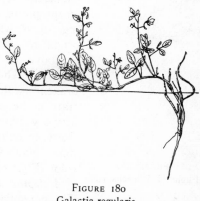

FIGURE 180
Galactia regularis.

which is at least 30 centimeters tall, is divided into a number of branches which show at first a plagiotropic tendency, later assuming an upright or orthotropic growth. Hence the branching system is candelabra-like (Fig. 181).

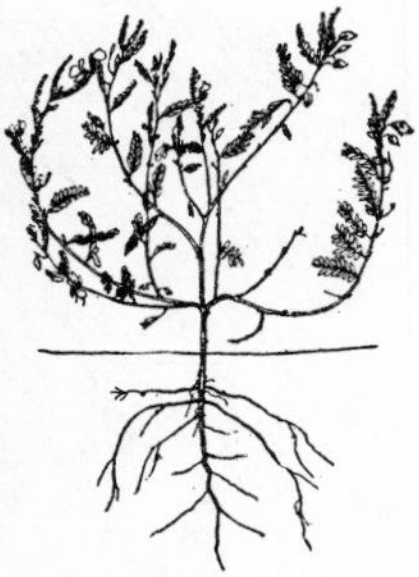

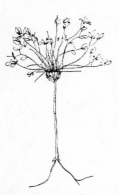

FIGURE 181
Cassia chamaecrista

Ilex glabra.—The ink-berry is a shrub 6 to 9 decimeters high, arising from a horizontal root-stock 2 centimeters thick and at least 6 to 7 deci-meters long. Filiform secondary roots arise from this rather thick rootstock. A nanophan-erophyte.

Euphorbia ipecacuanhae.—This spurge has a long, vertically directed primary root, which may be over 80 centimeters in length. From the crown of this root, which may be 1 to 1.5 centimeters thick, arise numerous stems which grow in a diffusely spreading manner. The lateral roots are given off some distance below the surface. One was found at 20 centimeters below and another at 40 centimeters below. A chamaephyte, or a hem-icryptophyte, depend-ing upon whether the origin of the branches is above or below the surface. The radiating branches, which are pros-trate on the ground, bear leaves varying from ob-ovate, or oblong, to narrowly linear, and from a light, glaucous-green to a deep reddish-brown color (Fig. 182).

FIGURE 183
Asclepias amplexicaulis.

FIGURE 182
Euphorbia ipecacuanhae.

Asclepias amplexicaulis (= *A. obtusifolia*).— This perennial herb has an upright stem from 3 to 8 decimeters tall, arising from a perpendicular rhizome at least 20 to 30 centimeters in length. Along the rhizome at intervals are borne opposite lateral buds separated by distinct internodal regions. From the crown of this vertical rootstock may arise two leafy shoots, and a stub is seen frequently

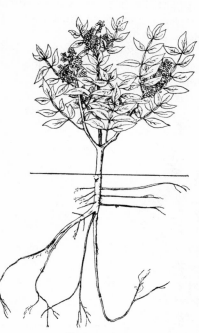

FIGURE 184
Rhus copallina.

FIGURE 185
Clethra alnifolia.

where the stem of last year was formed. No lateral roots were observed in this plant. A cryptophyte (Fig. 183).

Rhus copallina.—The dwarf sumac is rarely over two meters tall. It arises from a primary root which probably reaches a total The secondary lateral roots begin

depth of 7 decimeters to 1 meter. to grow horizontally at a depth of 6 to 7 centimeters below the surface of the soil. A nanophanerophyte (Fig. 184).

Clethra alnifolia (Sweet Pepperbush).—This is a shrub which varies greatly in size and grows under a great variety of soil conditions. In the pine woods several stems arise from a thick horizontal rhizome, which is provided sparingly with secondary roots. A nano- to a meso-phanerophyte (Fig. 185).

FIGURE 186
Chimaphila maculata.

Chimaphila maculata (Spotted Wintergreen).—The spotted, leathery leaves of this plant arise from an upright, low stem developed with a rhizome about the thickness of a lead-pencil, buried 5 to 10 centimeters below the surface.   A hemicryptophyte (Fig. 186).

Dendrium (Leiophyllum) buxifolium.—The sand-myrtle grows in a tufted manner, many upright, leafy stems arising from a thick, obliquely growing rhizome.   A nanophanerophyte (Fig. 187).

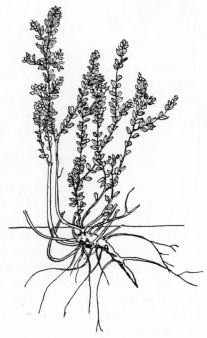

FIGURE 187
Dendrium buxifolium.

Gaultheria procumbens.—The wintergreen is a chamaephyte, or hemicryptophyte, with evergreen leaves that arise on short, upright branches that are also flower-bearing.   These leafy shoots, 5 to 15 centimeters tall, are distributed at intervals along the slender, horizontal, creeping stems found on, or as much as 5 centimeters below, the surface. It is a dark-brown color.   The filiform roots are not over 8 centimeters long and are found on the under surface of the rootstock.

A specimen of wintergreen rhizome was measured 48 centimeters long, and another shorter one was 36 centimeters (Fig. 188).

Andromeda (Neopieris) mariana.—The stagger-bush is a nanophanerophyte, 5 to 10 decimeters tall, with oblong to oval leaves and fascicles of nodding flowers in racemes on leafless shoots in the early spring. The upright stems arise from a horizontal rhizome 1.5 centimeters thick, at least 60 to 65 centimeters long. It is found 7 to 8 centimeters below the surface of the soil. The filiform lateral roots are branched and 8 centimeters long (Figs. 189 and 190).

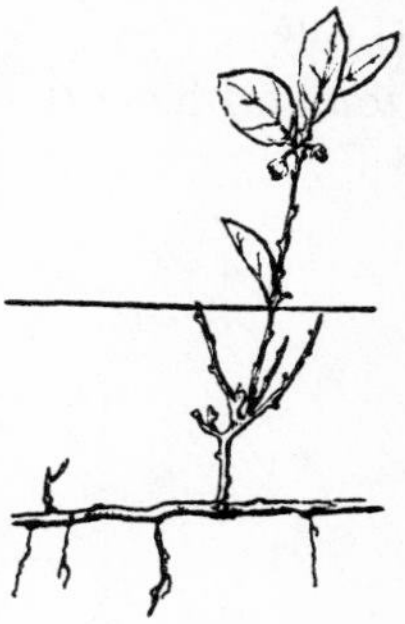

Kalmia angustifolia (Sheep Laurel).—This low evergreen shrub is extremely common in the pine forests. Its leafy stems arise from a much-branched rhizome provided with strong secondary roots. A nanophanerophyte (Fig. 191).

Kalmia latifolia.—The laurel is a mesophanerophyte. Its irregularly shaped, knobby rootstock is chocolate-brown in color and 5 centimeters thick. It grows downward obliquely and joins the rootstock branches of equal size. The lateral roots are much branched, 7 to 10 centimeters long, and produced most numerously at the crown of the plant (Fig. 192).

Epigaea repens.—The trailing-arbutus is a chamaephyte. Its stems, which trail over the ground, arise from a rhizome which merges gradually with the roots, one of which is a tap-root. This is an evergreen plant which it is difficult to grow out of a peaty soil, where its roots, covered with an ectotrophic mycorhiza, find the best conditions

FIGURE 188
Gaultheria procumbens.

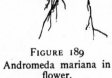

FIGURE 189
Andromeda mariana in flower.

FIGURE 190
Andromeda mariana in fruit.

for their development.   Coville* has succeeded in growing it from seed (Fig. 193).

Gaylussacia resinosa.—The black huckleberry is a low, much-branched shrub, 0.3 to 1 decimeter high.   It arises from a woody branched rhizome with short secondary roots.   A nanophanerophyte (Fig. 194).

FIGURE 191
Kalmia angustifolia.

Vaccinium pennsylvanicum.—The low, sweet blueberry is characterized easily by its low habit, bright, warty green twigs, its narrow, lanceolate leaves, and its blue-black berries.   Its tufted stems arise from a woody rhizome that grows downward obliquely in the humous layer of

* COVILLE, F. V.: Experiments in Blueberry Culture. Bull. 193, Bureau of Plant Industry.

the soil and produces several strong horizontal secondary roots. A nanophanerophyte (Fig. 195).

Vaccinium vacillans.—This abundant blueberry has a long, creeping rhizome 8 millimeters thick, which is produced 2.5 to 4 centimeters below the soil surface, and grows at least 90 centimeters long. The chocolate-brown lateral roots are tufted and 6 to 8 centimeters long. The upright, aërial stems that arise from this rhizome are 3 to 9 decimeters tall. A nanophanerophyte (Fig. 196).

Chamaedaphne (Cassandra) calyculata.—The leather-leaf is a low, erect shrub with coriaceous leaves that remain for more than a year on the branches and turn reddish-brown in autumn. During the early part of the growing season the shoots of the year develop large leaves, and toward the end of the growing period

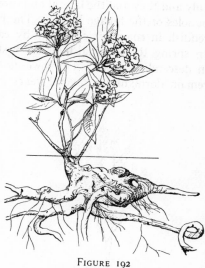

FIGURE 192
Kalmia latifolia.

small leaves are formed in whose axils are flower-buds that remain closed over winter. Where the tops of the plants project beyond the cover of

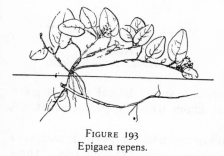

FIGURE 193
Epigaea repens.

snow, they may be winter-killed seriously, so that there may be dead leaves and buds at the extremities of the shoots.*

Arctostaphylos uva-ursi.—The bearberry is a prostrate, creeping shrub with its branches outspread on the surface of the ground, but in the course of time the older portions are buried by leaf mold. The last of the year's shoots curve upward slightly, but afterward become more horizontal (plagiotropic). The plant grows centrifugally, and its younger

* WARMING, EUG.: The Structure and Biology of Arctic Flowering Plants, 1908: 32; GATES, FRANK C.: The Relation of Snow Cover to Winter Killing in Chamaedaphne calyculata. Torreya, 12: 257–262, November, 1912.

branches have relatively few adventitious roots, but these secondary roots become more abundant upon the older parts of the branches, especially if covered with soil.   The leaves remain green upon the stems from two to three years.   The terminal buds on the branches are small in July and they and the lateral buds are protected by the basal part of the petioles of the foliage leaves.   The buds bear scale-leaves, greenish or reddish in color, which merge by easy gradations into foliage leaves. In spring the terminal buds are first to develop and the others follow in descending order.   The lowermost buds on the year's shoots may remain dormant for several years and may then develop.   After the

FIGURE 194
Gaylussacia resinosa.

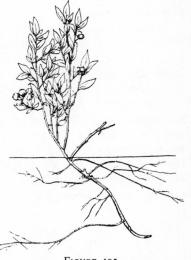

FIGURE 195
Vaccinium pennsylvanicum.

terminal flower-cluster has blossomed, a strong lateral shoot is given off below it, often forming a sympodium with the parent shoot.   A chamaephyte (Fig. 197).

Pyxidanthera barbulata.—The flowering-moss, or pyxie, is usually a prostrate, or creeping plant, and it shows sun and shade forms.   Three different kinds of shoots giving rise to three distinct growth forms may be distinguished in this chamaephyte.   First, we have a creeping stem, or rhizome, produced in part under the surface of the soil and in part above the surface.   The older parts may be covered with humus to a depth of 1.5 centimeters, and are provided richly with short, brown, filiform roots, while the newer portion is running along on the surface of the soil and is without the filiform rootlets.   This form of shoot produces an extended mat- or cushion-plant.   Some of the prostrate shoots are at least 20 centimeters long, bearing numerous upright leafy branches which are from 1.5 to 2 centimeters tall and beset with small, linear leaves.

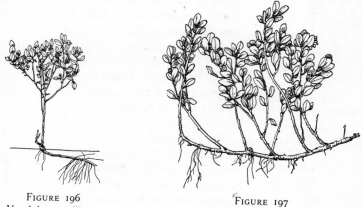

FIGURE 196
Vaccinium vacillans.

FIGURE 197
Arctostaphylos uva-ursi.

The second kind of shoot is the periwig type.   The shoots are short, and are branched in an upward direction from a common base.   They may be covered by humus to a depth of 3 to 4 centimeters, and such buried shoots are covered with brown rootlets.   Such a method of branching gives rise to rounded, circumscribed cushions which are usually found in gravel soils or in the bright sunlight.

The third kind of shoot occurs on plants that are found in places where there is a considerable depth of unconsolidated leaf mold, where the shoots, to reach the surface, must find their way through the interstices of such

FIGURE 198
Pyxidanthera barbulata.

superficial deposits.   The plant produces in such situations the wandering shoots, which bear tufts of loosely aggregated green branches whenever they reach the surface.   The filiform rootlets are found on the older parts of such wandering shoots (Figs. 61, 198, 279, F and G).

Melampyrum lineare (= *M. americanum*).—The cow-wheat is an erect, branching annual, or therophyte, growing from a primary root which descends obliquely into the soil to a depth of 5 to 6 centimeters, and produces a number of lateral roots which vary in length from 3 to 5 centimeters. It is a common plant in the pine-barrens of New Jersey, growing about 30 centimeters tall, and with opposite, linear, lanceolate leaves (Fig. 199).

FIGURE 199
Melampyrum lineare.

Chrysopsis mariana.—The golden-aster is a perennial herb with silky stems which arise from a weak horizontal rhizome. It grows about 1 to 3 decimeters tall. A hemicryptophyte (Fig. 202).

Hieracium venosum.—The rattlesnake-weed is a perennial herb with an almost leafless scape, bearing a loose corymb of capitula. The rosette of lanceolate-spatulate, radical leaves arises from the top of a vertical rhizome that is scarcely 2 centimeters long and bears numerous secondary roots all over its surface. These roots spread in all directions and are at least 10 to 12 centimeters long. The crown of leaves almost meets the upper roots, so that not over 5 millimeters of actual rhizome surface is bare either of roots or attached leaf bases. A hemicryptophyte (Fig. 200).

FIGURE 200
Hieracium venosum.

FIGURE 201
Eupatorium verbenæ-folium.

Sericocarpus asteroides (= *S. conyzoides*).—The white-topped aster is a perennial herb, 35 centimeters tall, and leafy to the top. The underground parts of this plant are not over 2 centimeters below the surface of the soil. There is a short horizontal rhizome that is branched usually, from which arises an intricate mass of strong secondary roots, some of

FIGURE 202
Chrysopsis mariana.

FIGURE 203
Sericocarpus asteroides.

them 15 to 20 centimeters long, growing 2 centimeters below the surface. Usually only one upright stem is developed from a single branch of

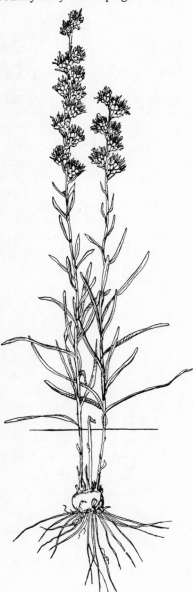

the rhizome. In addition the plant develops a series of suckers. One plant in the field had six of these suckers, bearing pinkish colored scale-leaves with adventitious secondary roots along their course of 8 centimeters, and a tuft of petiolate, spatulate leaves at their extremity. Such a plant may be described as surculose, and according to Raunkiaer's classification, it is a hemicryptophyte (Fig. 203).

Eupatorium   verbenaefolium.— This plant is an erect cryptophyte with aërial stems 50 centimeters tall, bearing opposite, ovate-lanceolate, c o a r s e l y toothed, sessile leaves. This upright stem arises from a corm-like base, buried 5 centimeters below the soil surface. Numerous roots arise from this rounded base, and the longest one measured was 30 centimeters long, and it grew down obliquely into the soil. The majority of the upper roots grew almost horizontally (Fig. 201).

Liatris graminifolia.—The button-snakeroot is a hemicryptophyte with an upright stem 3 to 9 decimeters tall, arising from a corm-like base from 1 to 2 centimeters in diameter. Numerous white, filiform roots 3 to 4 centimeters long arise from its under surface, and two to several upright leaf- and flower-bearing stems from its upper side about 4 centimeters below the soil surface. Several buds arise at the top of the corm (Fig. 204).

FIGURE 204
Liatris graminifolia.

CHAPTER XVII

GENERAL REMARKS ON ROOT DISTRIBUTION

The distribution of the roots and underground parts of pine-barren species is determined by several important conditions which must be considered in any detailed account of the ecology of pine-barren vegetation. The first important factor is the living plant considered specifically, as to the structural characteristics of its underground system. Each plant has specific root and underground stem characters, which are ancestral, and which are subject only to differences of length, thickness, color, and other minor modifications due to the environment (Fig. 205). The second important factor in considering the subject of the distribution of the underground parts of pine-barren plants is the actual association of the species together where the growth of one kind of root system excludes the growth of another kind. Thirdly, the soil structure, or texture, has great influence in determining the distribution of the underground systems of these plants, as also the ground water, the temperature of the soil and its aëration. The bacterial and fungous floras of such soils also have an important bearing upon the distribution through the soil of pine-barren roots and root-stocks.

In the preceding pages we have described the subterranean organs of different species of pine-barren plants, and in this study we have emphasized the specific differences of structure and growth in the soil. The soil structure, or texture, its permeability or impermeability, its color and other peculiarities, have been discussed in detail, but the following notes, taken in the field, may be inserted at this point as directly applicable to the discussion. The following is a series of measurements of the strata of pine-barren soils taken at a number of widely separated localities in the region:

Pine Forest, Como Lake:
Leaf Mould Layer...................... 6 centimeters
White Sand Layer...................... 27   "
Red Gravel Layer...................... 33   "

Pine Forest, Avon:
Upper Humous Layer.................... 16   "
White Sand Layer...................... 30   "
Red Gravel Layer...................... 29   "

Pine Forest, Shamong:
> Upper Humous Layer...................... 8 centimeters
> White Sand { Root-penetrated 16 cm. }..... 32   "
>   { Rootless .... 16 cm. }
> Red Sand Layer........................Unknown depth*

Pine Forest, Sumner:
> Upper Humous Layer..................... 20 centimeters
> White Sand Layer....................... 40   "
> Reddish Sand Layer.....................Unknown depth

Upper Plain:
> Upper Humous Layer..................... 15 centimeters
> White Sand Layer....................... 30   "
> Yellow Sand with Limonite................ 1.2 meters
> Light Yellow Sand with Ground-water.

Deciduous Forest, Clementon:
> Upper Humous Layer..................... 30 centimeters
> Middle Soil Layer........................ 25   "
> Water Table.............................

We have discussed the successive layers found in a pine-barren forest, beginning with the tall pine trees, then the undergrowth of oaks, then the shrubs, then the undershrubs, the herbaceous plants, and those prostrate on the surface of the soil. Similarly, we find strata of the underground parts of the plants, which ramify through the layers of the soil above described. We wish to ascertain now the layer of the soil where most of the roots and rhizomes occur and after we have definitely located them to draw conclusions therefrom. Of these 55 species, 4 are aquatic, 5 are bog plants, leaving 46 as true pine forest species. Of this 46, we find that 5, or one-ninth, are surface growers, or chamaephytes, such as Gaultheria procumbens, Epigaea repens, Arctostaphylos uva-ursi, and the flowering-moss, Pyxidanthera barbulata (Fig. 198). The demands then that these plants make on the soil is only on the most superficial part of the upper humous layer. Twenty-six species occupy the hemicryptophyte layer and their roots and root-stocks live in part in the upper humous layer and in the topmost part of the white sandy layer. Such are arranged systematically as follows: Andropogon scoparius, Xerophyllum asphodeloides, Smilax glauca, Cypripedium acaule, Comptonia asplenifolia, Helianthemum canadense, Hudsonia ericoides, Arenaria caroliniana, Hypericum densiflorum, Comandra umbellata, Tephrosia (Cracca) virginiana, Cassia chamaecrista, Andromeda mariana, Vaccinium vacillans, V. pennsylvanicum, Gaylussacia resinosa, Kalmia

---

* At Lakewood the red-gravel layer of virgin soil in a gravel pit was found to be 9.1 meters deep, and in a railroad cut, 4.5 to 6 meters thick.

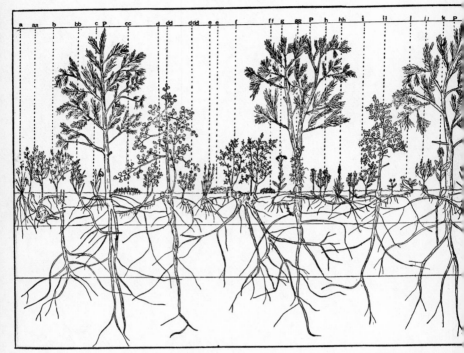

FIGURE 205

Diagrammatic representation of the aërial and root systems of the plants of the pine-barren formation

a,   Panicum depauperatum.
aa, Andromeda (Pieris) mariana.
b,   Baptisia tinctoria.
bb, Vaccinium pennsylvanicum.
c,   Cypripedium acaule.
cc, Pyxidanthera barbulata.
d,   Gaylussacia resinosa.
dd, Quercus marylandica.
ddd, Hypericum densiflorum.
e,   Andropogon scoparius.

ee, Epigaea repens.
f,   Quercus ilicifolia.
ff, Cladonia rangiferina.
g,   Sericocarpus asteroides.
gg, Smilax glauca.
h,   Kalmia angustifolia.
hh, Xerophyllum asphodeloides.
i,   Comptonia asplenifolia.
ii, Quercus stellata.
j,   Chimaphila maculata.

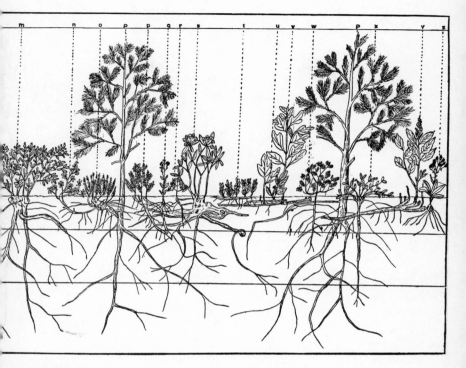

(The horizontal lines below the soil surface represent successive soil strata. See pages 245 and 246.)

jj, Vaccinium vacillans.
k, Asclepias obtusifolia.
kk, Liatris graminifolia pilosa.
l, Hudsonia ericoides.
ll, Andropogon scoparius.
m, Quercus prinoides.
n, Lupinus perennis.
o, Dendrium buxifolium.
p, Tephrosia (Cracca) virginiana.
q, Eupatorium verbenaefolium.

r, Melampyrum americanum.
s, Kalmia latifolia.
t, Arctostaphylos uva-ursi.
u, Gaultheria procumbens.
v, Sassafras variifolium.
w, Cassia chamaecrista.
x, Euphorbia ipecacuanhae.
y, Clethra alnifolia.
z, Hieracium venosum.
P, Pinus rigida.

angustifolia, Dendrium buxifolium, Melampyrum lineare, Eupatorium verbenaefolium, Liatris graminifolia, Chrysopsis falcata, Solidago odora, Aster spectabilis and Hieracium venosum.

The next deeper layer of soil, which comprises practically the white sandy layer and the topmost part of the reddish sand, or gravel beneath, is occupied by the roots and rhizomes of Pteris aquilina, horizontal roots of Pinus rigida, Myrica carolinensis, Quercus prinoides, Q. marylandica, Q. ilicifolia, Q. stellata, Sassafras variifolium, Baptisia tinctoria, Ilex glabra, Euphorbia ipecacuanhae, Asclepias amplexicaulis, Rhus copallina and Kalmia latifolia.

In some cases the upper lateral roots of these 14 pine-barren species are found in the upper humous layer, entering into competition with the complex of roots of the plants listed above. The reddish sand and gravel layer beneath the sandy layer is penetrated by few roots, and the roots that penetrate are deep-growing, primary roots of the principal trees, such as: Pinus rigida, Quercus marylandica, Q. stellata, and such shrubs as Q. ilicifolia, Q. prinoides, Kalmia latifolia, and Rhus copallina. From this enumeration it is clear that the roots of the majority of pine-barren plants are superficial ones found in the first 30 centimeters of the surface soil, which would include the compact humous layer below the loose leaf mould layer, which need not be included, and the white sandy layer, which includes a considerable amount of humus, and therefore ought to be included strictly as part of the humous layer (Fig. 205). The roots and root-stocks of the perennial plants of the pine-barrens make a continuous demand upon the soil for water and food substances. The annual plants, or therophytes, such as Melampyrum lineare, Cassia chamaecrista, and others absorb materials from the soil only during the summer months, when they make a rapid growth. The same classification of root systems can be made for pine-barren species that Cannon has made for the desert plants of the southwest: (1) Root-systems, which extend horizontally from the main plant and lie for their whole course near the surface of the ground; (2) root systems which are characterized by a strongly developed tap-root going down directly to a depth determined in part by the character of the soil, in part by the penetration of rain-water, and in part by the character of the root itself; (3) and roots that reach not only widely, but penetrate also fairly deeply. The majority of the pine-barren plants rooted in the humus and with horizontal subterranean organs belong to the first category. The pines belong to the second category and the oaks to the third (Fig. 205).

With these facts in mind the distribution of the rainfall throughout the year must be considered in relation to the growth and successful competition of the various pine-barren species of plants. It is important to emphasize the statement of Kraus* that, "under natural conditions, one may find in the smallest areas an endless diversity of chemically and physically distinct habitats," for there can be no criticism, if one cites the general conditions that prevail in the region. An estimate of the amount of rain and melted snow throughout the year expressed in inches† gives the following for two localities in the pine-barrens.

	ATCO	VINELAND
January	3.73	4.60
February	3.36	4.06
March	4.25	4.43
April	2.85	3.12
May	2.93	3.76
June	4.15	3.52
July	4.07	4.25
August	5.87	5.09
September	4.76	4.38
October	2.96	3.33
November	3.93	3.72
December	3.90	4.01
Total	46.76	48.27

But that this precipitation is not always uniformly distributed in the months of the year is indicated from the table of extreme monthly precipitation in inches and hundredths for Vineland:

	GREATEST	LOWEST
January	6.30	1.30
February	6.25	1.73
March	6.84	1.22
April	8.32	1.73
May	8.45	0.77
June	5.59	0.60
July	9.82	1.85
August	10.64	1.28
September	9.91	0.69
October	6.75	1.08
November	7.24	1.49
December	7.52	1.88

At Vineland, in 1881, the rainfall for July, August, and September amounted to 6 inches, as compared with the average of 14.2 inches. And there were two periods of twenty-one and twenty-two days respectively when no rain fell. As a general thing it may be stated that, with ordinary rains, the rain-water does not penetrate the soil to any

* KRAUS, G.: Boden und Klima auf kleinstem Raum. Versuch einer exakter Behandlung des Standorts auf den Wellenkalk, pp. 184, pls. 7, Jena, 1911; Reviewed in The Plant World, 15: 300, December, 1912.

† All the United States Weather Bureau records are in inches.

considerable depth, first, because there is not enough of it, and second, the thirsty roots of the first 30 centimeters (1 foot) of sandy humous soil absorb it and very little percolates into the lower soil layers. The water-table lies usually so deep that it is not available as a source of water supply to the superficially rooted plants, for the capillary rise of water is extremely slow.* With heavy rains, or a steady one of several days' duration, the rain-water drains into the subsoil, which is replenished thereby. I have no meteorologic records which give the number of such rains throughout the year, but they are few and far between. The length of time that the water in available amounts remains in the soils is a variable one. But from the first table above it is seen that the rainfall is sufficient during the year for the superficially rooted annuals and perennials which can utilize all of the water, which falls as rain, whether the showers be heavy or light. But there are critical periods when no rain falls and then the xerophytic structures of such plants as Kalmia latifolia, Gaultheria procumbens, Epigaea repens, Xerophyllum asphodeloides, Ilex glabra, Arctostaphylos uva-ursi, tide them over the period of drought. With the deep-rooted trees, it is otherwise, for during the critical period of dry weather and the cold dry period of winter, their deep vertical and superficial horizontal root systems enable them to get a supply of water under all climatic conditions quite sufficient to supply the loss by transpiration. During dry weather the deeper roots are depended upon to secure the water lost in transpiration; during wet weather, or when the surface soil is wet, the superficial roots are active. The amount of transpiration is under control in our northern pines† in which the surface exposed by the evergreen, acicular leaves, as a whole, renders it necessary for the individual leaves to be xeromorphic and xerophytic. This structure enables these coniferous trees to live in regions where there is a season of physiologic drought. The tracheidal structure of their wood is well suited to their xerophytic evergreen leaves. With the horizontal root system the species of pine growing in the New Jersey pine-barrens can secure additional supplies of water from the surface layers after a light shower. The deciduous trees, as trophophytes, are potential xerophytes, for in winter with the fall of their leaves the rate of transpiration is materially reduced. As the pine-barren soils have a fair capacity for retaining water and occupy an intermediate position with respect to the rate of percolation of water

* Ante, under description of soils.
† GROOM, PERCY: Remarks on the Oecology of Coniferae. Annals of Botany, xxiv, 241–269, April, 1910.

through them, there is probably no period of the ordinary year when some water is not available to the native plants of the region.*

The pine-barren soils, being open and pervious to water, the sunlight being intense and the sand at times heated highly by the sun's rays, the plants that live under such conditions must have been modified structurally by the influence of their environment, so as to become drought resistant. This implies adaptations for enduring drought, adaptations for evading drought and adaptations for escaping drought. In the presence of water-storage tissues, as in succulent roots, stems and leaves and in the ability to become dormant during the critical season of the year pine-barren plants show adaptations for enduring drought. The various structural peculiarities by which transpiration is controlled, or by which the root system has an excessive development are adaptations for evading drought. Finally, when a plant is annual and grows only during the favorable season and then dies to the ground, having liberated its seeds, it escapes the period of cold, or physiologic dryness. In the pine-barrens the first and third methods of resisting drought are rather infrequent, but structural peculiarities of leaf and exceptional root development are not uncommon.

Another matter of interest should be mentioned at this juncture. The conductive power of the pine-barren soils is such that they become heated intensely by the sun's rays falling directly upon them.   In short, they are warm soils. They often become so hot in the heat of the day as to be unbearable to the touch of the bare foot or hand. The influence of such sandy soil is both heating and drying, and somewhat like that of a desert, making the days hot and the nights, owing to the rapid radiation of heat, cool. These factors have a decided influence on the character of the vegetation of the region. Having determined the depth at which the subterranean organs of pine-barren plants are found, we can appreciate why this type of vegetation of superficially rooted perennials has been able to occupy southern New Jersey, as a climax formation, since very early times (see ante). The matted and interlaced condition of the roots and root-stocks of the herbs, shrubs and trees is so intricate that no alien plant has a chance of establishing itself in an area where the original vegetal covering has remained unbroken. There is such an interrelation and reciprocity of underground parts that each is found at the soil level where its functions can be performed best and where there is soil space enough to grow. This adjustment of the root systems of the pine-barren plants which grow in association has taken ages to bring to its present state, and such plants as enter into the different formations

* See ante under Soils.

grow together because their root systems are in a complicated reciprocity of growth, so that the soil space is filled after a long period of competition. Such competition by which many species were replaced by others has led to an adjustment in which, instead of competition, the root organs each fill a particular soil space.   Through the survival of the fittest the struggle for existence among the different native species has become an almost negligible factor.   Such plants naturally exclude other species that have subterranean parts that are unable to be adjusted to the complex of the roots of the long-established pine-barren plants.

The moisture conditions under which a species grows may be those which prevent the encroachment of competing species.   The non-entry of some species is determined probably during the early stages of germination and growth of the seedling.   Obviously, says Yapp,* if the conditions are such as are very unfavorable to the young growth, there is a poor chance of its surviving to the adult stage.   (Compare Fig. 205.)

With the few water plants to which we have drawn attention it is otherwise.   During the entire year there is water sufficient for their normal growth, and hence, although we have classed them as cryptophytes, yet it is mainly with reference to the depth below the surface of the muddy bottom of the ponds and the streams where their roots and root-stocks are found.   Such plants as Tofieldia racemosa, Pogonia ophioglossoides, Calopogon pulchellus, Habenaria blephariglottis, and the shrub, Chamaedaphne calyculata, have usually sufficient water throughout the year, except in cases of unusual drought, when the sphagnum dries up.

* YAPP, R. H.: Sketches of Vegetation at Home and Abroad, IV. Wicken Fen, The New Phytologist, VII: 76.

CHAPTER XVIII

LEAF FORMS OF PINE-BARREN PLANTS

The older text-books of botany, for the sake of training the botanic student in the identification of species by the use of manuals, or floras, emphasized the descriptive terms used in describing the shapes, margins, apices, and other parts of leaves. These terms are still in common use, but latterly considerable emphasis has been given to the study of leaf forms from the ecologic and morphogenetic points of view. Goebel was one of the first of modern botanists to break the leashes of the older morphology. In his classic Pflanzenbiologische Schilderungen (1889–1893) he considers the form of the leaf from its morphogenesis. Experimental methods are used, which are nowhere better illustrated than in his discussion of water plants, where the influence of land conditions and water conditions· are determined experimentally for a number of water plants. Sir John Lubbock, in his little book, "Flowers, Fruits and Leaves" (1894), approaches the subject from much the same standpoint, and Henslow, in his illuminating book, the "Origin of Plant Structures" (1895), gives in Chapter XII a modern statement of the form and structure of leaves. L. H. Bailey devotes 52 pages of "Lessons with Plants" (1898) to a study of leaves and foliage. Warming (1901), in a brochure entitled "Om Lovbladformer,"* gives an interesting résumé of the subject. Sir John Lubbock gives a chapter to leaves in his "Notes on the Life History of British Flowering Plants" (1905), and Goebel, in Part II of his monumental work, the "Organography of Plants," presents in the English translation of his work (1905) a full synopsis of the views of modern morphologists on the origin, shape, and structure of various leaf forms. If we analyze the leaf forms of the 555 native pine-barren species, our analysis shows that the following leaf forms exist in the region. As a number of duplications occur in the following list, and as it was found difficult to classify exactly some of the leaf forms, the total number of leaves listed in the following arrangement does not total the entire number of 555 pine-barren species of plants. This discrep-

---

* WARMING, EUG.: Oversigt over det Kgl. Danske. Videnskabernes Selskabs Forhandlinger, 1901, No. 1.

ancy does not detract, however, from the value of the enumeration of all the different kinds of leaf forms.   In the groups the species are arranged systematically.

1. LEAVES WITH CIRCINATE PTYXIS, SOME AS SPOROPHYLLS.—The leaves of this group show a blade that is characterized by apical growth and a manner of unfolding that is called circinate.   The group includes the ferns: Osmunda regalis, O. cinnamomea, Schizaea pusilla, Lygodium palmatum, Pteris (Pteridium) aquilinum, Woodwardia virginica, W. areolata, Asplenium platyneuron, A. filix-foemina, Polystichum acrostichoides, Dryopteris thelypteris, D. simulata, Phegopteris dryopteris. = 13.

2. LYCOPODIUM, OR CEDAR FORM.—This type is exhibited in the clubmoss, genus Lycopodium, with an axis beset with numerous small lanceolate or subulate leaves, arranged in two to many ranks.   In the pine-barrens this type is represented by four species of Lycopodium, and by the trees Chamaecyparis thyoides, Juniperus virginiana, and the herbs, Arenaria caroliniana, Pyxidanthera barbulata.   = 8.

3. NEEDLE LEAVES IN FASCICLES.—The acicular leaves of the pines are found in fascicles of two in Pinus echinata and in threes in Pinus rigida.   They are evergreen.   = 2.

4. ALGOID LEAVES.—These are narrow submerged leaves which resemble the thallus of certain algae in their long and narrow blades.   Such leaves are found in Potamogeton confervoides and Zannichellia palustris. = 2.

5. LINEAR, OR GRASS FORM.—This group includes 142 species of grasses and sedges.   Although the sedges do not always have the generalized type of linear leaf, yet it is unnecessary for our purpose to make a separation of this form into a number of different categories.   Here we may include the leaves of Schizaea pusilla (a fern), Sagittaria graminea, 5 species of Xyris, 4 species of Eriocaulon, 12 species of Juncus, Tofieldia racemosa, Abama americana, Xerophyllum asphodeloides, Zygadenus leimanthoides, Gyrotheca tinctoria, Hypoxis hirsuta, Iris prismatica, Sisyrinchium angustifolium, S. atlanticum, Euthamia graminifolia (radical leaves), E. tenuifolia (radical leaves), Sericocarpus linifolius (radical leaves).   = 177.

6. HEATHER FORM.—The leaves are small and closely crowded on the shoots of plants of this type.   Here may be mentioned Corema Conradii, Hudsonia ericoides, H. tomentosa.   = 3.

7. MINUTE SUBULATE LEAVES.—These leaves are small, but they are scattered on the stem, otherwise types 6 and 7 might have been combined.   The following plants have small leaves: Sarothra gentianoides,

Lechea Leggettii, Bartonia virginica, B. paniculata, Utricularia cornuta, U. juncea, U. subulata, U. cleistogama. =8.

8. LINEAR LEAVES, NOT GRASS-LIKE.—The linear leaves of the following plants are usually cauline and without overlapping bases and sheaths, as in grasses and sedges: Potamogeton epihydrus (submerged leaves), Polygonella articulata, Linum medium, L. floridanum, Polygala viridescens, P. mariana, P. Nuttallii, Crotonopsis linearis, Linaria canadensis, Ludwigia linearis, Kneiffia linearis, K. longipedicellata, Trichostema lineare, Stachys hyssopifolia, Diodia teres, Lobelia Nuttallii, L. Canbyi, Sclerolepis uniflora, Lacinaria graminifolia, Euthamia graminifolia Nuttallii, E. tenuifolia, Sericocarpus linifolius, Ionactis linariifolius, Helianthus angustifolius, Coreopsis rosea, Chrysopsis falcata, Anaphalis margaritacea, Gnaphalium obtusifolius. The specific names of a considerable number of these plants refer to the linear leaf form. =28.

9. IRIS, OR TYPHA LEAF FORM.—The erect, sword-shaped blade of leaves of this type is familiar in the Iris, or blue-flag. Relatively few pine-barren plants show this type and all that have this form of leaf are aquatic: Acorus calamus, Typha latifolia, Sparganium americanum, S. americanum androcladum, and Iris versicolor.

10. SIMPLE LANCEOLATE LEAVES (Willow and Knot-weed Type).— This group includes a diverse number of plants of no close systematic affinity, which display a lance-shaped stem leaf. For convenience we will divide them into trees, shrubs, and herbaceous plants.

(a) *Trees and Shrubs.*—Salix nigra, S. humilis, S. tristis, Myrica carolinensis, Comptonia asplenifolia, Quercus phellos, Itea virginica, Spiraea latifolia, S. tomentosa, Malus angustifolia, Aronia arbutifolia, Amelanchier intermedia, Prunus serotina, Ilex verticillata, I. laevigata, Ilicioides mucronata, Nyssa sylvatica, Viburnum nudum. = 18.

(b) *Herbs.*—Polygonum emersum, P. Careyi, Crotalaria sagittalis, Acalypha gracilens, Ludvigia alternifolia, Chamaenerion angustifolium, Epilobium coloratum, Asclepias tuberosa, Myosotis laxa, Onosmodium virginicum, Verbena urticifolia, V. angustifolia, Lycopus virginica, L. sessilifolius, Lobelia inflata, Solidago erecta, S. puberula, S. sempervirens, S. odora, S. rugosa, S. fistulosa, S. neglecta, S. uniligulata, S. nemoralis, Aster novi-Belgii, A. nemoralis, A. dumosa, Helianthus divaricatus and Bidens connata. =29.

11. YUCCA FORM.—This type is a sword-like blade, which is somewhat leathery in texture. It is represented by one pine-barren plant, viz., Eryngium yuccifolium.

12. RADICAL LANCEOLATE OR SPATULATE LEAVES.—There are quite

a few plants with clustered leaves arising as a rosette from the crown of the root or root-stock. Besides the radical leaves, there are cauline ones. Here are included: Adopogon virginicum, Hieracium venosum, H. Gronovii, Solidago erecta, S. puberula, S. stricta, S. sempervirens, S. rugosa, S. neglecta, S. uniligulata, Sericocarpus asteroides, Aster spectabilis, A. gracilis, Antennaria neglecta, Gnaphalium purpureum, Senecio tomentosus. = 16.

13. SIMPLE BROAD DECIDUOUS LEAVES.—(a) *Unlobed*.—Populus grandidentata, Betula populifolia, Alnus rugosa.

(b) *Lobed*.—Quercus coccinea, Q. velutina, Q. digitata, Q. ilicifolia, Q. marylandica, Q. alba, Q. stellata, Q. prinus, Sassafras variifolium, Acer rubrum carolinianum, Vitis aestivalis. = 14.

14. BAY-LIKE LEAVES.—These leaves are broad with almost entire margins, evergreen and leathery, resembling those of the true laurel, Laurus nobilis. The group includes: Smilax laurifolia, Magnolia glauca, Kalmia latifolia, Rhododendron maximum, Ilex opaca, I. glabra. = 6.

15. SMILAX-IPOMOEA LEAF FORM.—The leaves of these two genera of climbing plants are broad approaching heart-shaped. The group includes Smilax tamnifolia, S. rotundifolia, S. glauca, S. Walteri, Dioscorea villosa, Ipomoea pandurata, Convolvulus sepium. = 7.

16. VIOLET LEAF FORM.—This group was made to include the pine-barren species of the genus Viola, such as: Viola pedata, V. sagittata, V. fimbriatula, V. emarginata, V. primulifolia, V. lanceolata. = 6.

17. BROAD SAGITTATE LEAVES.—Only two plants have typic arrow-shaped leaves, viz., Sagittaria longirostra and Nymphaea variegata.

18. DANDELION FORM.—The leaves of the dandelion, Taraxacum officinale, are incised-pinnatifid. A few pine-barren plants have leaves of this type: Oenothera laciniata, Dasystoma pedicularia, D. flava, D. virginica, Adopogon carolinianum, Lactuca canadensis, L. hirsuta. = 7.

19. NABALUS LEAF FORM.—The genus Nabalus, or Prenanthes, includes herbs with alternate, mostly petioled, dentate, lobed, or pinnatifid leaves. The upper are sometimes auriculate and clasping. In the pine-barrens of New Jersey there are two species, Nabalus serpentarius and N. virgatus. = 2.

20. LEATHERY LEAF.—This type of leaf may be described as coriaceous or ericaceous, because common in that family.

(a) *Deciduous*.—Clethra alnifolia, Azalea nudiflora, A. viscosa, Leucothoë racemosa, Andromeda (Neopieris) mariana, Xolisma ligustrina, Gaylussacia frondosa, G. dumosa, G. baccata, Vaccinium corymbosum, V. virgatum, V. caesariense, V. atrococcum, V. pennsylvanicum, V. vacillans. = 15.

(b) *Evergreen.*—Pyrola americana, P. elliptica, P. chlorantha, Chimaphila maculata, C. umbellata, Dendrium buxifolium, Kalmia angustifolia, K. latifolia, Epigaea repens, Chamaedaphne calyculata, Gaultheria procumbens, Arctostaphylos uva-ursi, Vaccinium macrocarpon. = 13.

21. COMPOUND PINNATE LEAVES.—This heading is a term of the older morphology. Warming has emphasized the importance of enumerating the plants that have simple and compound leaves in any particular geographic region, as that of the pine-barrens of New Jersey. The following plants have compound pinnate leaves: Hicoria alba, H. glabra, Drymocallis arguta, Rosa humilis, Cassia nictitans, Tephrosia (Cracca) virginiana, Apios tuberosa, Rhus copallina, R. vernix, Aralia nudicaulis, A. spinosa, Oxypolis rigidior. = 12.

22. COMPOUND TRIFOLIATE LEAVES.—The compound leaves with three leaflets are more common in the pine-barrens than the preceding type. The plants with trifoliate leaves are: Rubus cuneifolius, R. argutus, R. villosus, R. hispidus, Baptisia tinctoria, Stylosanthes biflora, Meibomia Michauxii, M. sessilifolia, M. stricta, M. viridiflora, M. Dillenii, M. rigida, M. marylandica, M. obtusa, Lespedeza repens, L. frutescens, L. virginica, L. hirta, L. oblongifolia, L. capitata, L. angustifolia, Clitoria mariana, Galactia regularis and Rhus radicans. = 24.

23. COMPOUND PALMATE LEAVES.—A single plant has leaves of this type, viz., Lupinus perennis. = 1.

24. VARIABLE LEAF FORMS.—Euphorbia ipecacuanhae belongs to this group. The wild-ipecac or ipecac spurge, has leaves that vary from green to red and in outline from linear to orbicular. It is a remarkable plant, because these leaf forms are found on plants that grow in close proximity to each other. = 1.

25. FLOATING LEAF FORMS.—The water plants with floating leaves are Brasenia purpurea, Nymphaea variegata, Castalia odorata, Limnanthemum lacunosum, Utricularia inflata, Potamogeton Oakesianus, P. epihydrus, Orontium aquaticum. = 8.

26. DISSECTED SUBMERGED LEAVES.—In aquatic plants the submerged leaves have frequently a different conformation from those which stand above the surface of the water, or which float upon it. The influence of water, illumination, and temperature, is to produce a divided or dissected leaf and in some types a ribbon-shaped leaf is found. The pine-barren plants with dissected leaves are Proserpinaca pectinata, Myriophyllum humile, M. heterophyllum, Utricularia intermedia, U. inflata, U. clandestina, U. purpurea, U. gibba, U. fibrosa, Potamogeton Oakesianus. = 10.

27. MONOTROPA LEAF FORM.—As the Indian-pipe, Monotropa uniflora, is a saprophyte, the reduced leaves of this and other plants are related to a saprophytic or a parasitic life. The leaves of the following plants are colorless, scale-like and closely appressed: Monotropa uniflora, Monotropa hypopitys. = 2.

28. DODDER FORM.—The dodders (Cuscuta) are parasitic plants attached by haustorial roots to various host plants. The leaves, as a consequence of this parasitic life, are reduced to a few scale leaves around the clusters of flowers. Here belong Cuscuta arvensis, C. compacta.

29. FLY-TRAP LEAF FORM.—The leaves of the Droseras vary from linear to orbicular. The surface of the blade is covered with gland-bearing filaments or tentacles. Each gland is surrounded by a drop of viscid secretion, which glistens in the sunlight, hence the name, sundew. If a small insect alights on the leaf, it is caught by the viscid secretion. The tentacles bend over the insect until it is firmly held and digestion proceeds until the soft parts of the insect are dissolved and absorbed. In the pine-barrens there are three species: Drosera rotundifolia, D. longifolia, D. filiformis. = 3.

30. PITCHERED LEAF FORM.—The ascidial leaves of the pitcher-plant, Sarracenia purpurea, are well known to catch insects, but it is doubtful whether any digestion takes place. It is more probable that the pitcher filled with insects decays and the putrid organic material, in part the dead bodies of insects, in part plant tissue, fertilizes the soil in which the plant grows. = 1.

31. BLADDER LEAF FORM.—The insect-trapping bladders of Utricularia are modified leaves. Not all of the pine-barren species have these bladders. The following five species are provided with them: Utricularia fibrosa, U. clandestina, U. gibba, U. inflata, U. purpurea. = 5.

32. CLASPING LEAVES.—Two plants have leaves attached to the stem in this manner: Specularia perfoliata and Aster patens. = 2.

33. LEAVES OPPOSITE (HERBACEOUS).—The following plants have opposite leaves: Silene antirrhina, Sagina decumbens, Arenaria caroliniana, Acer rubrum caroliniana, Ascyrum stans, A. hypericoides, Hypericum densiflorum, H. maculatum, H. boreale, Triadenum virginianum, Elatine americana, Rhexia virginica, R. aristosa, R. mariana, Isnardia palustris, Lysimachia terrestris, Sabatia lanceolata, Gentiana porphyrio, Apocynum androsaemifolium, A. cannabinum pubescens, Asclepias rubra, A. variegata, Verbena urticifolia, V. angustifolia, 11 species of the family Labiatae, Veronica peregrina, Dasystoma pedicularia, D. flava, D. virginica, Oldenlandia uniflora, Cephalanthus occi-

dentalis, Mitchella repens, Diodia teres, Viburnum cassinoides, V. nudum, 7 species of Eupatorium, Coreopsis rosea and Bidens connata. = 54.

34. VERTICILLATE LEAVES.—The following are some of the pine-barren plants with verticillate leaves: Lilium superbum, Medeola virginica, Polygala cruciata, P. brevifolia, P. verticillata, P. ambigua, Decodon verticillatus, Lysimachia quadrifolia, Trientalis borealis, Asclepias verticillata, Monarda punctata (bracts included), Galium pilosum, Eupatorium maculatum, E. hyssopifolium. = 14.

GENERAL REMARKS ON LEAF FORMS

We have enumerated in the preceding 34 groups 518 species of plants, some of which have been included twice and occasionally 3 times. However, this can be said, that the whole pine-barren flora of 555 native species has been pretty well covered in the above classification. Only those forms of leaves have been omitted which have been considered relatively unimportant or which it has been impossible to classify satisfactorily. In order to compare the leaf forms of the pine-barren region with two other phytogeographic regions, we will divide all pine-barren plants into two groups: those with simple leaves and those with compound leaves. By making this distinction, we can correlate the figures thus obtained with those given by Warming for Denmark and Lagoa Santa in South America. Of the 555 native pine-barren species, 50 species have compound leaves and 505 have simple leaves. The list of 50 species with compound leaves includes 13 ferns with pinnately compound fronds. Excluding the ferns, there are 37 species with compound leaves and 542 species, less 37 species, or 505 species with simple leaves, or as 1 : 13.6. Warming estimated that there were 410 species of Danish vascular plants comprising the forest flora of the country and 1000 species in the open country. In the woods of Denmark there exist approximately 180 species with leaves divided or compound, and 230 species with simple leaves. The ratio is then as 3 : 4. In the open places, he estimated that there were 750 species with simple leaves and 250 species with divided or compound leaves. The proportion is 3 : 1. Warming, in his work, "Lagoa Santa," which deals with a portion of the Brazilian vegetation, enumerates 300 trees and herbaceous or woody plants, and the larger majority of these plants have simple leaves (lanceolate, elliptic, etc.).

CHAPTER XIX

MICROSCOPIC LEAF STRUCTURE

In two previous papers* I have described the leaf structure of the sand-dune plants of Bermuda and of the strand plants of New Jersey, namely, the dune plants and the salt-marsh plants. Bibliographic notes are given in each of these papers, so it would be superfluous to repeat what I have emphasized. Since these papers were written a number of recent accounts have been published and these bear directly upon the microscopic leaf structure of plants. Woodhead,† in his "Ecology of Woodland Plants in the Neighborhood of Huddersfield," gives an excellent account of the effect of environment on structure, using 8 species as examples. The structure of each species is illustrated by figures of the microscopic structure of stems and leaves. Later, in 1910, appeared the magnificent monograph of Jean Massart entitled, "Esquisse de la Geographie Botanique de la Belgique." Considerable attention is given in this book to the leaf structure of a selected number of Belgian plants of ecologic interest. The October (1912) number of the Botanical Gazette‡ contains an illustrated study of the "Comparative Anatomy of Dune Plants," by Anna M. Starr. Only two of the plants studied by her are found in the pine-barren region of New Jersey.

The response of leaf structure to light and transpiration is found in the palisade layers, their presence on the upper and under sides of the leaves and their arrangement, so that the central part of the leaf becomes palisade throughout, if the light is intense and the transpiration excessive. When both leaf surfaces are illuminated equally, the leaf may be termed isophotic; when illuminated unequally, diphotic. Diphotic leaves which show a division into palisade and spongy parenchyma have been called by Clements diphotophylls. Isophotic leaves are of three types, viz., the staurophyll, or palisade leaf; the diplophyll, or double leaf; the spongophyll, where the rounded parenchyma cells

---

* HARSHBERGER, J. W.: The Comparative Structure of the Sand-Dune Plants of Bermuda. Proceedings, American Philosophical Society, XLVII: 97–110, with 3 plates, 1908; The Comparative Leaf Structure of the Strand Plants of New Jersey. Proceedings American Philosophical Society, XLVIII: 72–89, with 5 plates, 1909.

† Linnæan Society's Journal—Botany, XXXVII: 333–406, October, 1906.

‡ LIV: 265–305.

make up the bulk of the leaf in cross-section. Succulent leaves are those in which water storage is accomplished. The depression of the stomata, the development of a thick cuticle, the presence of a hypodermis of thick-walled cells, the presence of hairs and the formation of air-still chambers by a rolling or a folding of the leaf are all structures which assist in the regulation of transpiration.

The following is a classification of the different leaf structures of 55 species of pine-barren plants. The species are arranged systematically under each head.

Thin Cuticle.—Myrica carolinensis, Quercus ilicifolia, Q. prinoides, Q. stellata, Sassafras variifolium, Azalea viscosa (West Plain), Andromeda (Neopieris) mariana, Gaylussacia dumosa, Vaccinium vacillans, V. atrococcum, V. macrocarpon.

Thick Cuticle.—Chamaecyparis thyoides, Pinus rigida, Xerophyllum asphodeloides, Alnus rugosa, Quercus ilicifolia (East Plain), Q. marylandica, Corema Conradii, Ilex glabra, Kalmia latifolia, Chamaedaphne (Cassandra) calyculata, Dendrium (Leiophyllum) buxifolium, Epigaea repens, Arctostaphylos uva-ursi, Gaultheria procumbens.

Thick-walled Epidermis.—Schizaea pusilla, Pinus rigida, Chamaecyparis thyoides, Xerophyllum asphodeloides, Pyxidanthera barbulata.

Hypodermis Present.—Pinus rigida, Chamaecyparis thyoides, Xerophyllum asphodeloides, Castalia odorata (in patches), Corema Conradii.

Single Row of Palisade.—Pteris aquilina, Chamaecyparis thyoides, Comptonia asplenifolia, Quercus marylandica [Lakehurst], Q. prinoides, Q. stellata, Arenaria caroliniana, Lespedeza frutescens, Hudsonia ericoides, Acer carolinianum, Sassafras variifolium, Clethra alnifolia [Cedar Swamp], Azalea viscosa, Gaylussacia frondosa, Vaccinium atrococcum, Liatris graminifolia, Eupatorium verbenaefolium.

Two or More Rows of Palisade.—Sagittaria longirostra, Smilax glauca, S. laurifolia, Orontium aquaticum, Myrica carolinensis, Alnus rugosa, Quercus ilicifolia, Q. marylandica [West Plain], Q. prinoides [Shamong], Q. stellata [Shamong], Magnolia glauca, Castalia odorata, Tephrosia (Cracca) virginiana, Ilex glabra, Euphorbia ipecacuanhae, Corema Conradii, Arctostaphylos uva-ursi, Azalea viscosa, Chamaedaphne (Cassandra) calyculata, Clethra alnifolia (pine woods), Epigaea repens, Gaultheria procumbens, Gaylussacia dumosa, G. resinosa, Kalmia latifolia, Dendrium (Leiophyllum) buxifolium, Andromeda (Lyonia) ligustrina, Andromeda (Neopieris) mariana, Vaccinium corymbosum, V. macrocarpon, V. pennsylvanicum, V. vacillans, Chrysopsis falcata, C. mariana, Solidago stricta, Pyxidanthera barbulata, Melampyrum lineare.

Stomatal Guard Cells at Surface.—Pteris aquilina, Sagittaria longi-rostra, Comptonia asplenifolia, Alnus rugosa, Quercus ilicifolia, Q. marylandica, Arenaria caroliniana, Magnolia glauca, Sassafras varii-folium, Tephrosia (Cracca) virginiana, Acer carolinianum, Hudsonia ericoides, Azalea viscosa, Clethra alnifolia [Cedar Swamp], Epigaea repens, Gaultheria procumbens, Gaylussacia dumosa, G. frondosa, G. resinosa, Andromeda (Lyonia) ligustrina, Andromeda (Neopieris) mari-ana, Vaccinium atrococcum, V. corymbosum, V. macrocarpon, V. pennsylvanicum, V. vacillans, Pyxidanthera barbulata, Chrysopsis falcata, C. mariana, Eupatorium verbenaefolium, Solidago stricta.

Stomatal Guard Cells Projecting.—This refers to the projection of the guard cells beyond the general epidermal surface. Schizaea pusilla, Myrica carolinensis, Corema Conradii, Clethra alnifolia (pine woods) Melampyrum lineare.

Stomatal Guard Cells Slightly Depressed.—Smilax glauca, S. laurifolia, Castalia odorata, Lespedeza frutescens, Euphorbia ipecacuanhae, Ilex glabra, Chamaedaphne (Cassandra) calyculata, Dendrium buxifolium, Liatris graminifolia.

Stomatal Guard Cells Deeply Depressed.—Pinus rigida, Orontium aquaticum (upper), Xerophyllum asphodeloides, Arctostaphylos uva-ursi, Kalmia latifolia.

Leaves Glabrous.—Pteris aquilina, Pinus rigida, Smilax glauca, S. laurifolia, Xerophyllum asphodeloides, Alnus serrulata, Euphorbia ipe-cacuanhae, Arctostaphylos uva-ursi, Clethra alnifolia, Epigaea repens, Gaultheria procumbens, Kalmia latifolia, Andromeda (Neopieris) mari-ana, Vaccinium pennsylvanicum, V. vacillans, Pyxidanthera barbu-lata.

Leaf Hairy on Under Side.—Quercus ilicifolia, Q. marylandica, Q. prinoides, Castalia odorata, Magnolia glauca, Lespedeza frutescens, Azalea viscosa [West Plain], Gaylussacia frondosa, Vaccinium atrococ-cum, V. corymbosum.

Leaf Hairy on Both Sides.—Comptonia asplenifolia, Myrica carolinen-sis, Quercus stellata, Sassafras variifolium, Corema Conradii, Tephrosia (Cracca) virginiana, Chamaedaphne (Cassandra) calyculata, Gaylus-sacia dumosa, G. resinosa, Andromeda (Lyonia) ligustrina, Melampy-rum lineare, Chrysopsis mariana, Eupatorium verbenaefolium, Liatris graminifolium, Solidago stricta.

Epidermal Surface Papillate.—Schizaea pusilla, Orontium aquaticum, Smilax glauca, Alnus rugosa, Comptonia asplenifolia, Lespedeza frutes-cens, Tephrosia (Cracca) virginiana, Azalea viscosa, Chamaedaphne (Cassandra) calyculata, Andromeda (Neopieris) mariana.

Leathery Leaf.—Smilax laurifolia, Myrica carolinensis, Quercus ilicifolia, Q. marylandica, Q. stellata, Castalia odorata, Ilex glabra, Arctostaphylos uva-ursi, Chamaedaphne (Cassandra) calyculata, Epigaea repens, Gaultheria procumbens, Gaylussacia dumosa, G. resinosa, Kalmia latifolia, Dendrium (Leiophyllum) buxifolium, Vaccinium macrocarpon, V. pennsylvanicum.

Wiry Leaf.—Schizaea pusilla, Chamaecyparis thyoides, Pinus rigida, Xerophyllum asphodeloides, Arenaria caroliniana, Corema Conradii, Hudsonia ericoides, Pyxidanthera barbulata.

Deciduous Leaf.—Pteris aquilina, Smilax glauca, Comptonia asplenifolia, Myrica carolinensis, Alnus rugosa, Quercus ilicifolia, Q. marylandica, Q. prinoides, Q. stellata, Castalia odorata, Magnolia glauca, Tephrosia (Cracca) virginiana, Acer carolinianum, Sassafras variifolium, Azalea viscosa, Clethra alnifolia, Gaylussacia dumosa, G. frondosa, G. resinosa, Andromeda (Lyonia) ligustrina, Andromeda (Neopieris) mariana, Vaccinium atrococcum, V. corymbosum, V. pennsylvanicum, V. vacillans.

Evergreen Leaf.—Chamaecyparis thyoides, Pinus rigida, Smilax laurifolia, Xerophyllum asphodeloides, Ilex glabra, Arctostaphylos uva-ursi, Chamaedaphne (Cassandra) calyculata (?), Epigaea repens, Gaultheria procumbens, Kalmia latifolia, Dendrium buxifolium, Vaccinium macrocarpon, Pyxidanthera barbulata.

Latex Tubes.—Euphorbia ipecacuanhae.

Mucilage.—Quercus marylandica [West Plain], Castalia odorata, Lespedeza frutescens, Sassafras variifolium.

Rhomboidal Crystals (Calcium Oxalate).—Smilax glauca, Lespedeza frutescens, Andromeda (Neopieris) mariana, Vaccinium corymbosum.

Sphaerocrystals.—Orontium aquaticum, Quercus ilicifolia, Arenaria caroliniana, Acer carolinianum, Ilex glabra, Chamaedaphne (Cassandra) calyculata, Clethra alnifolia, Epigaea repens, Gaultheria procumbens, Gaylussacia resinosa, Ilex glabra, Kalmia latifolia.

Idioblasts.—Castalia odorata.

Resin Hairs.—Myrica carolinensis, Gaylussacia resinosa.

Resin Canals.—Pinus rigida.

The following classificatory heads are based on the general structure of the leaf, *i. e.*, the arrangement of the palisade and the spongy parenchyma.

Diphotophyll.—Pteris aquilina, Orontium aquaticum, Sagittaria longirostra, Smilax glauca, S. laurifolia, Comptonia asplenifolia, Myrica carolinensis, Alnus rugosa, Quercus ilicifolia, Q. marylandica, Q. prinoides, Q. stellata, Castalia odorata, Magnolia glauca, Sassafras variifolium, Lespedeza frutescens, Tephrosia (Cracca) virginiana, Corema conradii, Acer carolinianum, Ilex glabra, Hudsonia ericoides, Arctostaphylos uva-ursi, Azalea viscosa, Chamaedaphne (Cassandra) calyculata, Epigaea repens, Gaultheria procumbens, Gaylussacia dumosa, G. frondosa, G. resinosa, Dendrium buxifolium, Andromeda (Lÿonia) ligustrina, Andromeda (Neopieris) mariana, Vaccinium atrococcum, V. corymbosum, V. macrocarpon, V. pennsylvanicum, V. vacillans, Pyxidanthera barbulata, Melampyrum lineare, Chrysopsis falcata, C. mariana, Eupatorium verbenaefolium, Liatris graminifolia.

Diplophyll.—Chamaecyparis thyoides, Arenaria caroliniana, Euphorbia ipecacuanhae, Solidago stricta.

Staurophyll.—Smilax glauca (?), S. laurifolia (?), Arctostaphylos uva-ursi (incipient), Kalmia angustifolia, K. latifolia.

Spongophyll.—Pyxidanthera barbulata (?).

Sclerophyll.—Schizaea pusilla, Pinus echinata, P. rigida, Xerophyllum asphodeloides.

Roll Leaf Structure.—Corema Conradii.

Insectivorous Leaf.—Drosera rotundifolia.

DETAILED MICROSCOPIC STRUCTURE OF LEAVES

The sections of leaves and other parts, which are figured and described in this chapter, were made free hand with a razor, then stained with Bismarck brown and mounted for permanency in Canada balsam. The drawings were made by Edinger's drawing and projection apparatus, manufactured by E. Leitz, Wetzlar. A description of this instrument will be found in the trade circular of that firm and need not be recounted here. The drawings were first made in lead pencil and finished in India ink. Using the same magnification for each drawing, the sections are drawn to the same scale, a highly desirable end in making a comparison of leaf section with leaf section. In order to attain the greatest accuracy, each drawing before it was inked was compared with the corresponding section under the microscope, and the following descriptions were written with the drawings and the microscopic sections before me. The drawings of stomata were not made to scale. The source of the material is given in each instance where such data have been kept.*

Pteris aquilina [Lakehurst].—The bracken-fern has dull-green, ternately compound fronds at the summit of an erect, stout petiole (2 to 9 dm. high) from a creeping root-stock. Both sides of the leaf are glabrous with thin-walled epidermal cells of unequal size. The guard cells

---

* In the reproduction for printing, each drawing has been reduced one-half.

of the stomata on the lower side project slightly.   There is a single row of palisade cells.   A typic diphotophyll (Fig. 206).

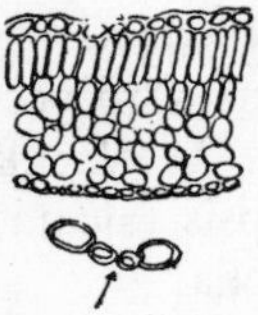

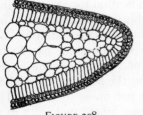

FIGURE 206
Pteris aquilina, Lake-hurst.

Schizaea pusilla [Cedar Swamp, Shamong].—The curly-grass fern is a small inconspicuous plant with distinct sterile and fertile fronds that are very slender, linear and wiry, usually curled.   As shown by the figure, the structure of the stipe of both fronds is essentially alike, that of the fertile being larger and thicker than that of the sterile.   The epidermal cells have thick outer walls that are distinctly papillate, with a slight boss, or projection, on each epidermal cell.   The stomata are confined to the lower or curved surface, and the guard cells project considerably beyond the general surface. The chlorenchyma cells have slightly thicker walls than usual, and in the leaf-trace bundle the phloem occurs on the abaxial side of the bundle, which is usually roughly elliptic in cross-section with its long axis tangential to the stele.*   A spongophyll (Fig. 207).

FIGURE 207
Schizaea pusilla, Cedar Swamp, Shamong, June 28, 1910.

Chamaecyparis thyoides.—The white-cedar has small, scale-like or awl-shaped, closely appressed, imbricated leaves often with a small gland on the back.   The epidermis is protected by a thick cuticle, and is supported by a single row of hypodermal sclerenchyma cells.   A single row of palisade cells is found on both sides of the leaf.   The spongy parenchyma is rather compact.   A diplophyll (Fig. 208).

Pinus rigida [Lakehurst].—The leaves of the pitch-pine are in fascicles of three. Their length varies from 5 to 12 centimeters, and they are dark-green in color. The thick-walled epidermal cells are covered with a cuticle, and they are reinforced by silvery, thick-walled hypodermal sclerenchyma, which is represented by two or three rows of cells that increase in number at the margin of the leaf.   The guard cells of the stomata are thick-walled and are depressed below the

FIGURE 208
Chamaecyparis thyoides, Cedar Swamp, Shamong (Chatsworth), June 28, 1910.

* SINNOTT, EDMUND W.: The Evolution of the Filicean Leaf-trace.   Annals of Botany, XXV: 171, January, 1911.

surface, until they are found on a level with the first row of hypodermal sclerenchyma cells.   The walls of the rounded, mesophyll parenchyma cells are ruminate by projections of the wall into the cell cavity.   Three

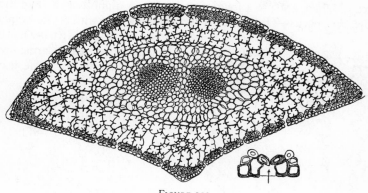

FIGURE 209
Pinus rigida, Lakehurst.

resin canals are found and there are two leaf-trace bundles surrounded by large parenchyma cells and a bundle sheath.   A leaf from a tree found on the Upper Plain is thicker transversely, the parenchyma cells are larger, and the resin canals larger and more open (Fig. 209).

Orontium aquaticum [Shamong].—The golden-club has long-petioled, entire, oblong, and nerved leaves.   The thin-walled epidermal cells of the petiole are vertically placed.   The parenchyma cells are spaced widely by large, rounded, intercellular spaces, so that a reticulum, or net, of these cells is formed with the collateral vascular bundles attached here and there to several meshes of this parenchymatous net.   A transverse section of the leaf-blade shows the upper epidermal cells strongly papillate and interspersed with stomata, the guard cells of which are deeply depressed.   There

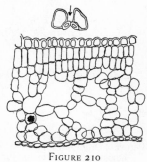

FIGURE 210
Orontium aquaticum, Cedar Swamp, Shamong, N. J., June 28, 1910.

are at least two rows of palisade cells, and the spongy parenchyma has large, irregular, intercellular spaces.   Sphaerocrystals are present in these cells.   The lower epidermal cells are large, with their outer walls smooth and not papillate.   A diphotophyll (Figs. 210 and 211).

Sagittaria longirostra [Shamong].—This species of arrow-leaf has broadly ovate-oblong, obtusish, sagittate leaves. The epidermal cells of both leaf surfaces are large and thin-walled. Stomata level with the surface are present on both leaf surfaces and opposite large intercellular spaces. There are 3 to 4 rows of palisade. A diphotophyll (Fig. 212).

Smilax glauca.—The saw-brier is a climbing vine with terete branches and somewhat four-angled, glaucous branchlets armed with scattered prickles. The leaves

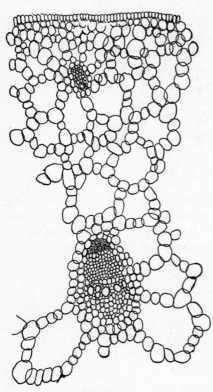

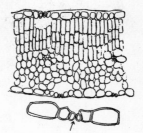

FIGURE 212
Sagittaria longirostra, Shamong, August 29, 1910.

are ovate and glaucous beneath. The upper epidermis consists of thin-walled, brick-like cells. The

FIGURE 211
Petiole of Orontium aquaticum, Cedar Swamp, Shamong (Chatsworth), June 28, 1910.

FIGURE 213
Smilax glauca, East Plain, August 31, 1910.

epidermal cells of the lower side are papillate and the guard cells of the stomata on the lower surface are depressed only slightly. There are two rows of short, rounded palisade cells and the short, transversely placed, spongy parenchyma cells are aligned with those of the palisade. Rhomboidal crystals are present. A diphotophyll simulating a staurophyll (Fig. 213).

Smilax laurifolia [Cedar Swamp, Shamong].—This woody, climbing plant has thick, evergreen, coriaceous leaves (6–12 cm. long). The regular, brick-like, upper and lower epidermal cells are covered by a thick cuticle. The stomata have their guard cells depressed slightly below the surface. There are two to three rows of palisade cells and the loose parenchyma cells are aligned with the palisade. A diphotophyll, or a potential staurophyll (Fig. 214).*

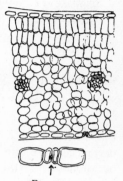

FIGURE 214

Smilax laurifolia, Cedar Swamp, Shamong, June 28, 1910.

Xerophyllum asphodeloides.—The turkey's-beard is an herb with a short stem from a thick, tuberous root-stock bearing a crown of long, needle-shaped, wiry leaves. The upper, flat surface of the leaf has small, thick-walled epidermal cells, reinforced by three to several rows of hypodermal sclerenchyma connected with each of the several bundles of the leaf. The lower curved surface of the leaf is provided with a few large, stiff, tooth-like hairs. The epidermal cells are thick-walled, supported by several rows of hypodermal sclerenchyma. The stomata, found on both surfaces, have thick-walled, depressed guard cells covered by two projecting tooth-like epidermal cells. The bundles, which are of the usual monocotyledonous, collateral type, are surrounded completely with a bundle-sheath of sclerenchyma. The mesophyll parenchyma cells, which are hexagonal or pentagonal in shape, are very compact and the walls are thicker than the walls of ordinary fundamental tissue. A sclerophyll (Fig. 215).

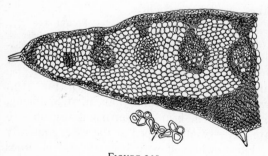

FIGURE 215

Xerophyllum asphodeloides, between Shamong and the Plains, June 28, 1910.

Comptonia asplenifolia [Lakehurst].—The sweet-fern is a shrub 3 to 6 decimeters tall, with sweet-scented, fern-like, lanceolate pinnatifid leaves that are deciduous. Both surfaces of the leaves are hairy, with straight, rather stiff, unicellular hairs. The epidermal cells are slightly

* See KEARNEY, T. H.: Contributions U. S. National Herbarium, 5: 486, f 86, 1901.

papillate and the guard cells on the lower side are at the surface.

There is a single row of palisade in the leaves from Lakehurst and two rows in the leaves from the Upper Plain, thus increasing the thickness of the leaves. In both leaves there is open loose parenchyma. A diphotophyll (Figs. 216 and 217).

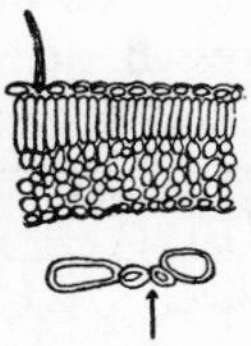

FIGURE 216
Comptonia aspleni-
folia, Cedar Swamp,
Shamong, June 28,
1910.

FIGURE 217
Comptonia asplenifolia,
West Plains, June 28,
1910.

Myrica carolinensis [Cedar Swamp, Shamong].—The bayberry is a shrub 1–2 meters tall, with oblong, entire, or crenately toothed leaves. The upper epidermal cells are thin-walled, with a cuticle and an occasional long, stiff unicellular hair. There are 2 to 3 rows of palisade cells and the spongy layer is a net of parenchyma cells. The lower epidermis has similar hairs and stomata, the guard cells of which project slightly. Occasional multicellular hairs occur in depressions of the lower epidermis. A diphotophyll (Fig. 218).

Alnus rugosa (=*A. serrulata*) [Shamong].—The smooth alder is a shrub or small tree with obovate, sharply serrate leaves, smooth or slightly pubescent beneath. The regular, brick-like, upper epidermal cells are covered with a thin cuticle. There are 3 rows of palisade cells. The spongy parenchyma is open. The guard cells of the sto-

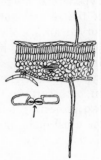

FIGURE 218
M y r i c a carolinensis,
Cedar Swamp, Sha-
mong, June 28, 1910.

mata on the lower side are at the surface. A diphotophyll (Fig. 219).

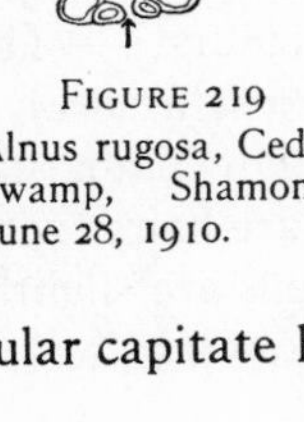

FIGURE 219
Alnus rugosa, Cedar
Swamp, Shamong,
June 28, 1910.

Quercus ilicifolia (=*Q. nana*) [Lakehurst, Shamong, Lower Plain].—The bear or scrub oak is a shrub with obovate leaves, wedge-shaped at the base, angularly 3 to 7 lobed, white, downy beneath and glabrous above. A thin cuticle is present on the upper surface of the leaf of the plants from Shamong and the Lower Plain. The epidermal cells are thin-walled, and the bent, unicellular hairs of the lower surface are clustered with four or five hairs arising from a single point of attachment with occasional multicellular capitate hairs. The under surface of the Shamong leaf is more hairy than in the Lakehurst leaf. There are two rows of palisade cells in the leaf of the Lakehurst plant and three rows in the leaf of the Shamong plant. The parenchyma is rather compact. The stomata have their guard cells at the surface. Sphaero-crystals are present in some of the cells. A dipho-tophyll (Fig. 220).

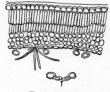

FIGURE 220
Quercus ilicifolia, East Plain, August 31, 1910.

Quercus marylandica [Lakehurst and Upper Plain].—The black-jack oak is a small tree with broadly wedge-shaped leaves, widely dilated and somewhat 3 to 5 lobed at the summit, rusty pub-escent beneath, shining above. The large cells of the upper epidermis are protected by a thick cuticle, and in the epidermal cells from leaves gathered on the Upper Plain there is found a gum-resin-like substance absent in the cells of the leaves from Lakehurst. There is a single row of true palisade cells in the Lakehurst leaves and two rows in the leaves of the specimen from the Upper Plain, and the lower epidermis displays the stomatal guard cells at the surface. The large, unicellular hairs are present on the lower surface and also branched, stalked, and capitate multicellular hairs. The spongy parenchyma is rather open. A diphotophyll (Fig. 221).

FIGURE 221
Quercus marylandica, Lakehurst.

Quercus prinoides [Lakehurst and Shamong].—This species of oak is a low, spreading bush with undulate leaves covered beneath by a close white tomentum. The upper epidermal cells, which are thin-walled, are protected by a clearly defined cuticle. There is a single row of palisade, while in the leaf from the pine-barrens at Shamong there are two rows and the spongy parenchyma is rather compact. The stomata on the under side have their guard cells at the surface. The hairs characteristic of the under side are straight, tufted and unicellular. A diphoto-phyll (Fig. 222).

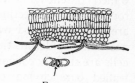

FIGURE 222
Quercus prinoides, Sha-mong, June 28, 1911.

Quercus stellata (=*Q. minor*) [Lakehurst, Shamong, Lower Plain].—The post-oak is a tree with thick, obovate, lyrate, pinnatifid leaves, sinuately cut into 5 to 7 rounded lobes, the upper of which are much larger and broader. The upper surface is rough, with hairs stellately arranged, and the lower surface is brownish-downy with curved stiff hairs in tufts. A cuticle is present. A single row of palisade is found in the leaves from Lakehurst and Lower Plain and three in the leaf from Shamong. The guard cells of the stomata on the lower side are flush with the surface. A diphotophyll (Fig. 223).

Quercus velutina [Lakehurst].—The quercitron-oak is a tree with leaves variously divided. Both surfaces of the leaf are smooth, but the pubescence is in tufts in the axils of the veins beneath. The upper epidermal, thin-walled cells are provided with a cuticle. The palisade cells are found in a single row, and the stomata have guard cells flush with the surface. A diphotophyll.

FIGURE 223
Quercus stellata, Shamong, June 28, 1911.

Arenaria caroliniana [Shamong].—The pine-barren sandwort is a densely tufted plant with closely imbricated but spreading, awl-shaped leaves. The thin-walled epidermis, which covers the surface of the approximately semicircular leaf in cross-section, is without a cuticle. Palisade cells are developed on both surfaces and the rather compact loose parenchyma with sphaerocrystals surrounds a centrally placed bundle system with strongly marked bundle sheath. Two small patches of sclerenchyma occupy the corners of the leaf. The guard cells of the stomata are at the surface on both sides of the leaf. A diplophyll (Fig. 224.).

Castalia (Nymphaea) odorata.—The leaves of the water-lily are orbicular, 0.5 to 2.2 dm. in diameter, deeply cordate at the base, with the margin entire. The under surface is a deep crimson-brown. The microscopic structure of the petiole is as follows: The epidermis consists of thin-walled parenchyma cells, the outer wall rounded and projecting. Thin-walled hairs are found scattered over the surface. Beneath the epidermis the hypodermal cortex cells are slightly collenchymatous, and thin-walled parenchyma cells are interspersed

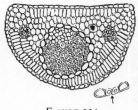

FIGURE 224
Arenaria caroliniana, Shamong, June 28, 1911.

between large, open intercellular canals into some of which star-like idioblasts project, their walls being roughened with crystalline deposits. The bundles are collateral with the phloem external and the xylem internal. A round canal accompanies invariably each bundle. Mucilage cells are present also.

The microscopic structure of the leaf blade is of interest. The upper epidermis of thin-walled cells alone shows the stomata with their guard cells depressed slightly below the surface. There are four to five rows of palisade cells and a wide area of spongy parenchyma with extremely large intercellular air chambers, crossed by chains, or bridges, of parenchyma cells. Into these air spaces project large, thick-walled, star-shaped idioblasts, which penetrate also to the upper epidermal cells between those constituting the palisade. The collateral bundles with xylem

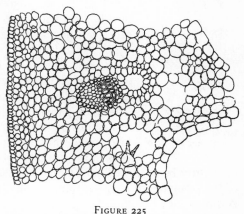

FIGURE 225

Castalia (Nymphaea) odorata, Cedar Swamp, Shamong, June 28, 1910.

uppermost are tied to the lower epidermis by patches of sclerenchyma.

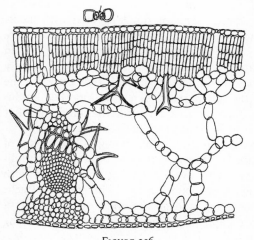

FIGURE 226

Castalia (Nymphaea) odorata, Cedar Swamp, Shamong, June 28, 1910.

Peculiar flattened epidermal hairs are found on the under surface. A diphotophyll (Figs. 225 and 226).

Magnolia glauca (= *M. virginiana*) [Shamong and Lower Plain].—The sweet-bay is a tree with broadly lanceolate to elliptic obtuse leaves, 8 to 15 cm. long, glaucous beneath. The small upper and lower epidermal cells are without cuticle. Long, straight, 3- to 4-celled, silky hairs are found on the lower side. The guard cells are level with the surface. There are two rows of palisade cells and the spongy parenchyma is somewhat compact. A diphotophyll (Fig. 227).

Sassafras variifolium [Lakehurst].—The sassafras is a low tree in the pine-barrens with ovate, entire, 1- to 3-lobed, pubescent, mucilaginous leaves. The upper epidermis, which is cuticularized, is provided sparingly with short, stiff, unicellular hairs. There is a single row of open palisade tissue with large, rounded, lysigenous cavities* filled with mucilage.

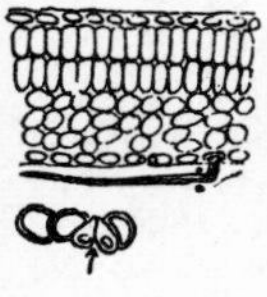

FIGURE 227

Magnolia glauca, Cedar Swamp, Shamong, June 28, 1910.

The loose parenchyma cells are flat and irregular in shape. The lower epidermis of thin-walled cells shows a number of straight, or slightly bent, unicellular hairs. The guard cells of the stomata are at the surface of the under side. A diphotophyll (Fig. 228).

FIGURE 228

Sassafras officinale, Lakehurst.

Drosera rotundifolia.—The sundew is a plant of such unusual biologic interest that an enlarged drawing of one of its leaf tentacles is added. The stalk of the tentacle is supplied with a few spiral tracheae. A plexus of spiral tracheids in the enlarged terminal portion of the tentacle is characteristic. This plexus is covered with modified epidermal and subepidermal cells which secrete a glairy fluid to which small insects adhere. When stimulated by insects, the tentacles curve inwardly over the central part of the leaf, thus trapping the insect, the soft parts of which are absorbed gradually by the digestive action of the secretions that are formed for that purpose (Fig. 229).

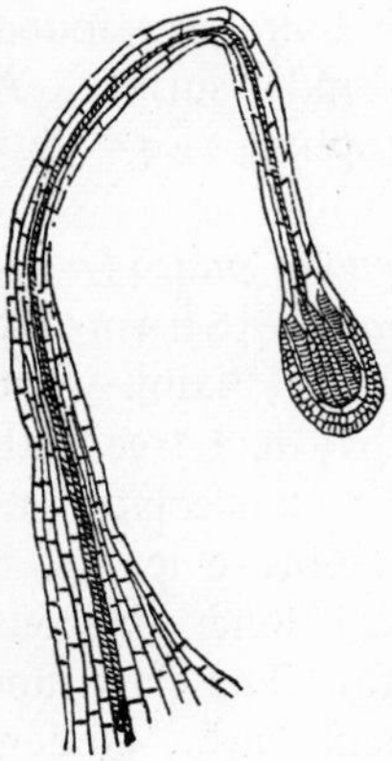

FIGURE 229

Tentacle of Drosera rotundifolia.

Lespedeza frutescens [Shamong].—The stems of this plant (1.5 to 7 dm. tall) have compound trifoliate leaves with oval to oblong firm leaflets with finely appressed pubescence. The epidermal cells are thin-walled, and on the under surface some of them are papillate. There is a single row of long palisade cells. Rhomboidal crystals of calcium oxalate are present in some of the cells. Some of the spongy parenchyma cells apparently have a mucilaginous material which stains deeply with Bismarck brown. The guard cells are depressed

* In Solereder's Systematic Anatomy of Dicotyledons, 11: 703 (1908), these are referred to in the order as "enlarged sac-like cells."

slightly.   Straight hairs occur on the under leaf surface.   A diphoto-phyll (Fig. 230).

Tephrosia (Cracca) virginiana [Lakehurst, Upper Plain and Lower Plain].—The goat's-rue is a per-ennial plant with an erect stem, 3 to 6 dm. high, covered with silky villous hairs when young.   The yellowish-white flowers, marked with purple, are in dense racemes.   The leaves are compound, and there are 17 to 19 linear-oblong leaflets.   The upper epi-dermis consists of thin-walled, arched cells raised as papillae in some places, in others replaced by the stiff and slightly curved, unicellular hairs.   There are 2 rows of palisade cells in the leaves from plants gathered on the Upper Plain and the pine-barrens at Lakehurst, and three in the leaves from a plant growing on the

FIGURE 230
Lespedeza frutes-cens, Shamong, Au-gust 29, 1910.

FIGURE 231
Tephrosia virginiana, Lakehurst.

Lower Plain.   The loose parenchyma consists of rounded cells with large intercellular spaces.   The lower epidermal cells are thin-walled, and the guard cells of the stomata on both the upper and lower sides are flush with the surface.   A typic diphoto-phyll (Fig. 231).

Euphorbia ipecacuanhae [Shamong and Lower Plain].—This spurge has leaves that vary extremely in form and color.   The shape varies from obovate, or oblong, to narrowly linear and glabrous.   The color varies from green to deep-red.   The epidermal cells on both the upper and lower sides of the leaf are thin-walled.   The guard cells are depressed on both surfaces, and a single row of palisade cells is found on the upper and lower sides, the lower being more prominent in the leaf from Shamong, with spongy parenchyma between.   Latex tubes are found in all cross-sections.   A diplophyll (Fig. 232).

FIGURE 232
Euphorbia ipecacu-anhae, East Plain, August 31, 1910.

Corema Conradii [Lower Plain].—The broom-crow-berry has nearly verticillate, linear, heath-like leaves. In cross-section these leaves show a typic roll struc-ture, with both margins rolled together until only a narrow, hair-protected slit is left.   The whole ex-posed outer surface is covered with a thick cuticle, beyond which project multicellular capitate hairs. The chambered inner surface displays two kinds of hairs—multicellular, capitate hairs and straight, unicellular hairs, which are found most numerously along the two inturned edges of the leaf, thus guarding the slit-like opening between the leaf margins.   The stomata are found on the lower, or inner, surface of the leaf and the guard cells project beyond the general surface.   Beneath the upper epidermal cells are large, looped cells filled with a material suggesting a gum-resin.   This material stains deeply with Bismarck brown.   The palisade cells arearranged in 2 to 3 layers and the spongy parenchyma is open with large, irregular, intercellular spaces.   A diphotophyll (Fig. 233).

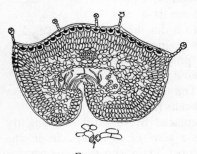

FIGURE 233
Corema Conradii, East Plain, August 20, 1911.

Ilex glabra [Shamong, between Shamong and the Plains, Lower Plain].—The inkberry is a shrub 6 to 9 decimeters tall, with oblong, smooth leaves.   A study of sections of leaves gathered from plants in the above localities enables one to note the influence of environment on leaf structure. The upper and lower epidermal cells of leaves from all three localities are protected by a thick cuticle.   The stomata, confined to the lower surface, have their guard cells depressed the thickness of the cuticle. Sphaerocrystals are present in the spongy parenchyma of the leaves from the three localities.

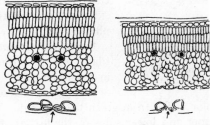

FIGURE 234
Ilex glabra between Shamong and the Plains, June 28, 1910.

FIGURE 235
Ilex glabra, Cedar Swamp, Shamong, June 28, 1910.

The leaf from the white-cedar swamp at Shamong displays three rows of palisade cells and a very open spongy parenchyma.   This leaf is thinner than the other two.   The leaves from the Lower Plain and from east of Shamong have four rows of palisade cells and a more compact loose parenchyma. A diphotophyll (Figs. 234 and 235).

FIGURE 236
Acer carolinianum, Cedar Swamp, June 28, 1910.

Acer carolinianum [Cedar swamps, Shamong and edge of Upper Plain].—The Carolinian red maple has small, obovate leaves, with three short lobes.   The upper epidermal cells are large, with the upper wall arched slightly.   The lower epidermal cells are thin-walled and papillate.

The stomata, confined to the lower surface, have superficial, obliquely-placed guard cells.   There is a single row of palisade cells and the spongy parenchyma is semi-compact.   A diphotophyll (Fig. 236).

Hudsonia ericoides [Upper Plain].—This plant is a bushy, heath-like, small shrub, with slender, awl-shaped leaves which vary in outline in different parts of the serial cross-sections.   The epidermal cells are thin-walled, with the stomata at the surface.   There is a single row of palisade cells and in front of each stoma there is an intercellular chamber developed in the otherwise rather compact, spongy parenchyma.   The bundle is placed centrally.   A few straight hairs are scattered over the surface.   The leaf is a diphotophyll.   The stem section is circular, and the epidermis of thin-walled cells develops numerous long, silky, uni-cellular hairs interspersed with multicellular, stalked, capitate hairs. The bundle system is central and is surrounded by a cortex of thin-walled parenchyma cells (Figs. 237 and 238).

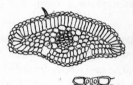

FIGURE 238
Leaf of Hudsonia ericoides, West Plain, June 28, 1910.

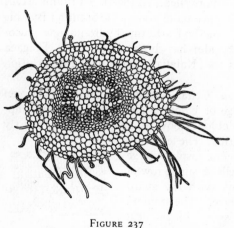

FIGURE 237
Stem of Hudsonia ericoides, West Plain, June 28, 1910.

Arctostaphylos uva-ursi [Shamong, Upper and Lower Plains].—The bearberry is a trailing plant with thick, evergreen, obovate or spatulate, entire, smooth leaves. Both the upper and the lower epidermal cells are covered with a thick, yellowish cuticle with stomata on the lower surface only.   The large, thick-walled guard cells are depressed an amount equal to the thickness of the cuticle.   There are five to six rows of palisade cells and the loose parenchyma is open with many of the cells directed vertically, so that the leaf becomes incipiently a staurophyll.   The leaves of the pine-barren plant collected at Shamong are thinner by one row of palisade cells than are those leaves gathered in the Upper and Lower Plains (Fig. 239). The leaves of the bearberry contain a glucoside, arbutin.   It has been discovered by a new method of microscopic analysis, known as the microsublimation method, that when the powdered leaves of Arctostaphylos are heated in a watch-glass and the sublimable principles collected on microscopic slides, the arbutin is split up into hydroquinone, which is deposited on the slides in several crystalline forms, characteristic of the plant. Weevers considers that arbutin, as a glucoside, is a reserve food and is stored in the leaves, and when the new leaves are formed in the spring, it is split by a suitable enzyme into sugar and hydroquinone. The sugar is used up, and the hydroquinone remains behind and combines with more sugar, so that the autumn leaves once more contain arbutin.* By this method arbutin (hydroquinone glucoside) has been detected in Epigaea repens, Gaultheria procumbens and Kalmia angustifolia, a description of which plants follows.

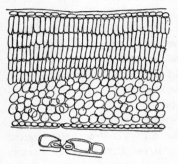

FIGURE 239
Arctostaphylos uva-ursi, West Plain, June 28, 1910.

Azalea (Rhododendron) viscosa [Cedar swamps, Shamong, between Shamong and Upper Plain, edge of Upper Plain].—The white swamp-honeysuckle is a shrub with oblong-obovate, smooth leaves, bristly along the margins and the midrib. The leaf from the white-cedar swamp at Shamong displays large thin-walled upper epidermal cells, a single row of palisade and a lower epidermis, with the guard cells of the stomata at the surface. The stomata are found only on the lower side. In the leaf from a plant collected between Shamong and the Lower Plain the thickness of the leaf has increased. A few of the upper epidermal cells are papillate, the palisade cells are longer, and the spongy parenchyma shows a larger number of layers and is more compact. In the leaf of a plant which grew at the edge of a white-cedar swamp along the Upper Plain the number of palisade layers is three and the spongy parenchyma is open. Glandular capitate hairs are found sparingly on the lower epidermis. A diphotophyll (Fig. 240).

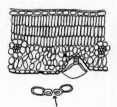

FIGURE 240
Azalea viscosa, West Plain, edge of Cedar Swamp, August 31, 1910.

* MOLISCH, HANS: Mikrochemie der Pflanze, 1913: 165–166; HAAS, PAUL, and HILL, T. G.: An Introduction to the Chemistry of Plant Products, 1913: 171; TUNMANN, O.: Ueber Folia uvae-ursi und den mikrochem. Nachweis des Arbutins. Pharm. Zentralb., 1906, XLVII, No. 46; TUNMANN, O.: Pflanzen Mikrochemie, 1913: 355.

Chamaedaphne (Cassandra) calyculata.—The leather-leaf is a low shrub with coriaceous evergreen leaves. A thick cuticle covers the small, thin-walled upper and lower epidermal cells. There are two to four rows of palisade cells. The spongy parenchyma is rather compact, with occasional sphaero-crystals. The stomata, found on the lower side of the leaf, have their guard cells depressed slightly and protected by tooth-like projections of the adjoining epidermal cells. Saucer-shaped, peltate hairs give to the lower leaf surface a rusty, scurfy appearance. Such hairs are distributed sparingly over the upper surface. A diphotophyll (Fig. 241).

FIGURE 241
Cassandra calycu-lata, Cedar Swamp, Shamong, June 28, 1910.

Clethra alnifolia [White-cedar Swamp, Shamong, pine-woods between Shamong and the Plains].—The sweet pepper-bush is a shrub 1 to 3 meters tall. The leaves are wedge-obovate, sharply serrate, and prominently veined. The leaf of the white-cedar swamp plant examined has large, rounded, thin-walled upper epidermal cells without stomata and large lower epidermal cells with a single row of palisade cells and open loose parenchyma. The guard cells are at the surface. Sphaerocrystals are present in some of the palisade cells. The leaf of the pine-barren plant has the same large, thin-walled upper and lower epidermal cells, but the guard cells of the stomata on the lower surface project slightly. There are two rows of palisade cells and the spongy parenchyma cells are more compact. Sphaerocrystals are present. A diphotophyll (Fig. 242).

FIGURE 242
Clethra alnifolia, be-tween Shamong and the Plains, June 28, 1910.

Epigaea repens [Upper Plain and Lower Plain].—The trailing-arbutus is a prostrate plant with evergreen, elliptic, leathery leaves on short petioles. The upper epidermis of large cells is protected by a thick yellowish cuticle pierced by scattered stomata, the guard cells of which are flush with the surface. The lower epidermal cells are without a cuticle. If we contrast the leaves from the pine-barren plants growing

FIGURE 243
Epigaea repens (Sun), East Plain, August 31, 1910.

FIGURE 244
Epigaea repens (Shade), Pine-barrens, March 25, 1910.

in the shade and in the sun, we find the shade leaf thinner, with more open spongy parenchyma, while the sun leaf has a cuticle and more compact spongy parenchyma. There are four rows of palisade cells in the leaves from the Lower and the Upper Plains, and the loose parenchyma consists of large, rounded cells. Sphaerocrystals are present. A diphotophyll (Figs. 243 and 244).

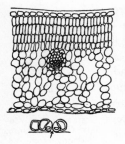

FIGURE 245

Gaultheria procumbens, East Plain, August 31, 1910.

Gaultheria procumbens [Lakehurst].—The teaberry is a perennial herb with leathery, evergreen, obovate, or oval leaves. The cuticle is thick on the upper surface, and the upper epidermal cells are thin-walled. There are three to four rows of palisade cells and some of the rounded loose parenchyma cells have sphaerocrystals. The stomata, which are confined to the lower surface, have their guard cells flush with the thin cuticle of that side. The glucoside arbutin is present (see ante). A diphotophyll (Fig. 245).

Gaylussacia dumosa [Lakehurst].—The dwarf huckleberry is a low, bushy perennial, 2 to 15 decimeters tall. Its tardily deciduous leaves are obovate-oblong, thick and shining when old, hairy on both sides. The epidermal cells are thin-walled, the upper with a thin cuticle. There are two rows of palisade cells, the cells of the lower row being slightly divergent. The stomata, as far as the sections available for study show, are confined to

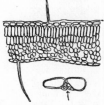

FIGURE 246

Gaylussacia dumosa, Lakehurst.

the lower side of the leaf and the guard cells are level with the surface. The hairs on both sides are stiff, unicellular, and bent slightly. A typic diphotophyll (Fig. 246).

FIGURE 247

Gaylussacia frondosa, West Plain, edge of Cedar Swamp, August 31, 1910.

FIGURE 248

Gaylussacia frondosa, East Plain, August 31, 1910.

Gaylussacia frondosa [Lakehurst, Upper and Lower Plains].—The dangleberry is a bush with obovate-oblong leaves, finely pubescent and glaucous beneath. The upper and lower epidermal cells are large, the lower being arched slightly. The hairs on the under surface are straight and several-celled. There is a single row of palisade cells in the leaves from Lakehurst and the Upper Plain, and two rows in the leaf of the plant growing on the Lower Plain. The spongy parenchyma consists of rounded cells separated by wide intercellular spaces. A diphotophyll (Figs. 247 and 248).

Gaylussacia resinosa (= *G. baccata*) [Shamong, Lower Plain, between Shamong and the Upper Plain].—The black huckleberry is a much-branched bush, 0.3 to 1 meter tall, with oval, oblong-ovate, or oblong leaves, thickly covered at first with clammy, resinous globules. The upper and lower epidermal cells are large and thin-walled, bearing in depressions multicellular capitate hairs that are resiniferous on both the upper and lower sides of the leaf. The number

FIGURE 249
Gaylussacia resinosa, Shamong, June 28, 1910.

of palisade layers varies from 1 to 2, perhaps 3, and the stomata, confined to the lower surface of the leaf, have their guard cells flush with the surface. Sphaerocrystals are present in leaves from some localities. A diphotophyll (Fig. 249).

Kalmia angustifolia [Shamong, Lower Plain].—The sheep-laurel is an evergreen shrub, 1 meter high, with opposite to verticillate (in threes), narrowly oblong, obtuse leaves, glabrate underneath. The cells of the lower and of the upper epidermis are protected by a thin cuticle. The upper epidermal cells are larger and more arched than the lower, and the upper wall is thickened considerably. The stomata are confined to the lower surface and the guard cells are depressed only slightly

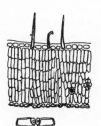

FIGURE 250
Kalmia angustifolia, East Plain, August 31, 1910.

below the surface. As the leaf is a typic staurophyll, there are 4 to 5 rows of palisade cells in the leaf from Shamong and 6 rows in the leaf from a twig gathered on the Lower Plain. Sphaerocrystals are abundant in the palisade cells. Long, erect, unicellular hairs are present on the upper side of the leaf from the Lower Plain (Fig. 250).

Kalmia latifolia [West Plain].—The laurel has alternate, evergreen, leathery leaves. The cells of the upper and lower epidermis are protected by a thick cuticle. The stomata, confined to the lower surface of the leaf, have their thick-walled guard cells as thick as the cuticle. There are 3 to 4 well-defined rows of palisade cells, and the cells of the spongy

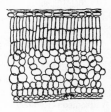

FIGURE 251
Kalmia latifolia (Sun), Pine-barrens, March 25, 1910.

parenchyma are aligned with those of the palisade. Sphaerocrystals are present. Sections made from leaves of plants from the pine-barrens (sun) and from the plains agree practically in all details, except that the spongy parenchyma of the pine-barren leaf is meager. A staurophyll (Figs. 251 and 252).

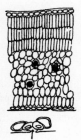

FIGURE 252
Kalmia latifolia, East Plain, August 31, 1910.

Dendrium (Leiophyllum) buxifolium [Lower Plain]. —The sand-myrtle is a shrub, 1 to 9 decimeters tall, with oval, or oblong, glabrous, leathery leaves. The upper and lower epidermal cells are protected by a thick cuticle. The guard cells of the stomata, found only on the lower leaf-side, are depressed to the depth of the protective cuticle. There are 3 to 4 rows of palisade cells and the spongy parenchyma cells are separated by large, intercellular spaces. A diphotophyll (Fig. 253).

Andromeda (Lyonia, Xolisma) ligustrina [Shamong and between Shamong and the Upper Plain].—The male-berry is a bush, 0.5 to 3 meters tall, with obovate to lanceolate, oblong, serrulate, or entire leaves. The upper and lower epidermal cells, provided with multicellular hairs, are thin-walled. The stomata, found only on the lower leaf-side, have their guard cells at the surface. There are a number of rows of palisade and a rather compact, spongy parenchyma of irregular cells. A diphotophyll (Fig. 254).

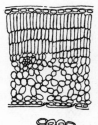

FIGURE 253
Dendrium (Leiophyllum) buxifolium, East Plain, August 31, 1910.

FIGURE 254
Andromeda (Lyonia) ligustrina, between Shamong and the Plains, June 28, 1910.

Andromeda (Lyonia, Neopieris) mariana. — The stagger-bush is a deciduous shrub with oblong to oval leaves, 3.5 to 8 centimeters long. The upper epidermis of thin-walled cells is protected by a thin cuticle. There are two rows of palisade cells. Rhomboidal crystals of calcium oxalate are present. The lower epidermal cells are slightly papillate and the guard cells of the stomata are at the surface. A diphotophyll.

Vaccinium atrococcum [Lower Plain].—The high black blueberry is a shrub, 1 to 4 meters tall, with entire leaves, downy or woolly underneath, even when old. At flowering time the leaves are hardly unfolded.

The upper epidermal cells are protected by a cuticle. The lower epidermal cells are smaller in size and some are modified into two kinds of hairs, viz., long, straight, unicellular hairs and capitate ones. The stomata are confined to the lower surface and have their guard cells on the epidermal level. There is a single row of palisade cells. A diphotophyll (Fig. 255).

FIGURE 255
Vaccinium atrococcum, East Plain, August 31, 1910.

Vaccinium corymbosum [Edge of White-cedar Swamp, Upper Plain].—The high swamp-blueberry is a shrub 1 to 4 meters tall. The leaves are half grown at flowering time, which is later than the preceding species. The ovate to elliptic, lanceolate, smooth leaves are slightly pubescent beneath. The epidermal cells of the upper and lower sides of the leaf are thin-walled. Stomata with their guard cells level with the surface are found only on the lower side. There are 2 rows of palisade cells and a very open spongy parenchyma. A diphotophyll (Fig. 256).

FIGURE 256
Vaccinium corymbosum, West Plain, edge of Cedar Swamp, August 31, 1910.

Vaccinium macrocarpon.—The cranberry plant is a trailing one, with oblong-elliptic, blunt, evergreen, smooth, leathery leaves, 6 to 17 millimeters long, 2 to 8 millimeters broad. The large upper and lower epidermal cells are protected by a thin cuticle. There are two rows of short palisade cells with large chloroplasts. The spongy parenchyma is open and of rounded cells. The stomata with their guard cells at the surface are confined to the lower epidermis. A diphotophyll (Fig. 257).

FIGURE 257
Vaccinium macrocarpon.

FIGURE 258
Vaccinium pennsylvanicum, West Plain, June 28, 1910.

Vaccinium pennsylvanicum [Shamong, Upper Plain].—The early sweet blueberry is a dwarf plant with green, warty, glabrous stems. The leaves are lanceolate, or oblong, distinctly serrulate, bright green, smooth and shiny on both sides, hairy on the midrib above and below. The epidermal cells are large and thin-walled. The stomata are confined to the lower epidermis with their guard cells level with the surface. There are 2 rows of palisade cells and the spongy parenchyma is rather compact in the leaf of the plant from the Upper Plain, more open in the leaf from the pine-barrens at Shamong.   A diphotophyll (Fig. 258).

Vaccinium vacillans [Shamong and Lower Plain].—The late low blueberry is a small shrub, 3 to 9 decimeters tall. The leaves are glabrous, obovate or oval, very pale, or dull, glaucous, at least underneath. The upper epidermis of thin-walled cells is covered with a thin cuticle. The lower epidermis of smaller, thin-walled cells is pierced by stomata, the guard cells of which are level with the surface. There are two rows of

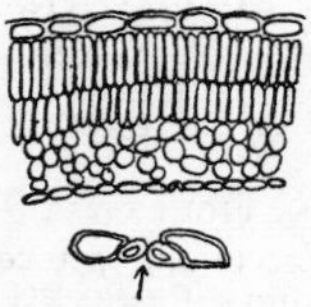

FIGURE 259
Vaccinium vacillans, East Plain, August 31, 1910.

palisade cells in the leaves from Shamong and the Lower Plain. A diphotophyll (Fig. 259).

Pyxidanthera barbulata.—The flowering-moss is a prostrate, creeping plant, with narrowly oblanceolate, awl-pointed leaves, 3 to 8 millimeters long, and of two colors, brownish and green. The epidermal cells of both lower and upper leaf surfaces have a very thick outer wall. The stomata are confined to the flat surface, while the curved surface is without stomata, and here the several rows of palisade cells are located. The guard cells of the stomata are at the surface. A diphotophyll, approaching a spongophyll (Fig. 260).

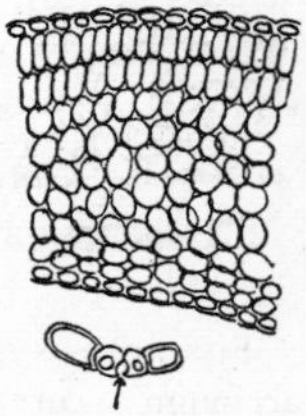

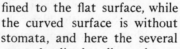

FIGURE 260
Pyxidanthera barbulata (Shade), March 25, 1910.

Melampyrum lineare (= *M. americanum*).—The cow-wheat is an annual plant with opposite, linear-lanceolate leaves. The upper and lower epidermal cells are large, the upper exceptionally so, some of them being modified into short, canine-tooth-like hairs. The stomata have large projecting guard cells.

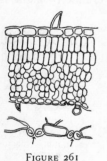

FIGURE 261
Melampyrum lineare. East Plain, August 31, 1910.

There are two rows of palisade cells and the spongy parenchyma cells are large.   A diphotophyll (Fig. 261).

Chrysopsis falcata [Lakehurst].—The sickle-leaved, golden aster is a very woolly perennial herb, 1 to 3 decimeters tall, with linear leaves. The cuticle is developed faintly and the epidermal cells of the upper and lower sides are large and thin-walled. There

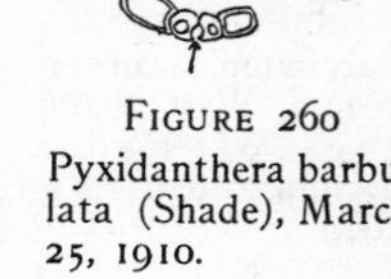

FIGURE 262
Chrysopsis falcata, Lakehurst.

are 3 rows of palisade cells and the stomata developed on the lower side of the leaf have their guard cells level with the general surface. A diphotophyll (Fig. 262).

Chrysopsis mariana [Shamong].—The golden aster is a perennial plant, silky, with long, weak hairs. The oblong leaves have thin-walled epidermal cells and the stomata on the lower and upper surfaces have slightly protruding guard cells. There are two rows of palisade cells, more pronounced in the leaf from the Lower Plain. A diphotophyll (Fig. 263).

Eupatorium verbenaefolium [Shamong].—This plant is roughish pubescent. The leaves are ovate-oblong to ovate-lanceolate, slightly three-nerved and coarsely toothed. The epidermal cells are large and thin-walled, bearing tooth-like multicellular, short, rigid hairs on both sides. A diphotophyll.

Liatris graminifolia [Shamong].—This species of blazing-star has linear, one-nerved leaves, borne on a stem 3 to 9 decimeters tall. The upper and lower large epidermal cells are without a cuticle, and both surfaces display stomata with slightly depressed guard cells. There is a single row of palisade cells and the loose paren-chyma cells are rounded. Vesicular hairs are found in slight pits, or depressions on both the upper and lower epidermal surfaces. A dipho-tophyll.

FIGURE 263
Chrysopsis mariana, East Plain, August 31, 1910.

Solidago stricta [Shamong].—The wand-like goldenrod is a smooth perennial with small, entire, appressed, lanceolate-oblong, thickish

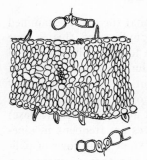

leaves. The epidermal cells on both sides are large and rounded with stomata on both sur-faces, the guard cells slightly projecting. The palisade parenchyma occurs on both sides in a loose, somewhat irregular arrangement of the cells. The spongy parenchyma occupies a central position. Short canine-tooth-like hairs are found on both surfaces. A dipho-tophyll (Fig. 264).

SYNOPSIS OF LEAF STRUCTURE

FIGURE 264
Solidago stricta, Shamong, August 29, 1911.

An analysis of the leaf structures of Bermuda strand plants, of New Jersey strand plants, of New Jersey salt marsh plants, as well as those of the pine-barren region, brings out some interesting facts about the relative abundance of the different types of leaf structure (Fig. 265).

Such a comparison is important, because it enables us to point in part to this or that structure as peculiar to plants growing under different environmental conditions. The plant names are omitted, and only a simple enumeration is attempted.

ENUMERATION OF LEAF STRUCTURE

Special Leaf Structure	17 Bermuda Dune Plants	20 New Jersey Dune Plants	11 New Jersey Salt-Marsh Plants	55 Plants of Pine-Barren		
				Pine Forest	White-Cedar Swamp	Plains
1. Cuticle Present	3	3	1	18	4	2
2. Thick Epidermis	8	4	5	3	2	..
3. Hypodermis Present	..	1	2	2	2	1
4. Single Row Palisade	9	13	7	11	6	..
5. Two or more Palisade Rows	8	7	4	24	9	2
6. Guard Cells Depressed	10	8	5	10	4	..
7. Leaves Hairy	5	7	3	17	7	2
8. Leaf Surface Papillate	..	1	..	5	5	..
9. Leathery Leaf	..	5	2	13	4	..
10. Wiry Leaf	..	2	3	5	2	1
11. Latex Present	1	1	..	1	..	..
12. Crystals Present	2	4	..	9	3	..
13. Idioblasts	..	1	..	..	1	..
14. Diphotophyll	5	10	4	29	14	1
15. Diplophyll	9	3	2	3	1	..
16. Staurophyll	4	2	1	4	1	..
17. Spongophyll	2	2	2	1	..	..
18. Succulent Leaf	5	3	6	..	..	..

It will be noted in the accompanying table that the plants examined from the six different localities are diphotophylls, and this is true especially of the plants that grow in the pine forest. The diplophyll structure, which is occasioned probably by illumination on both leaf surfaces, is a prominent feature of the dune plants of Bermuda and New Jersey, as is also the staurophyll. The spongophyll is found in dune plants and salt-marsh plants, while it is absent almost entirely (1 case) in the plants of the pine-barrens. Hairy leaves are frequent in pine-barren plants, as also those with a cuticle. Two or more rows of palisade cells are more common in pine-barren plants than a single row, while in the dune plants a single row of cells is more frequent than a double row, or over. More dune plants have a thick epidermis than do pine-barren plants. The papillate leaf surface is a feature of the pine-barren plants and leaf succulency is a dune and salt-marsh characteristic. Enough has been said to show that the table is instructive and enables us to contrast the influence of the different ecologic factors upon the structure of the plants concerned (Fig. 265).

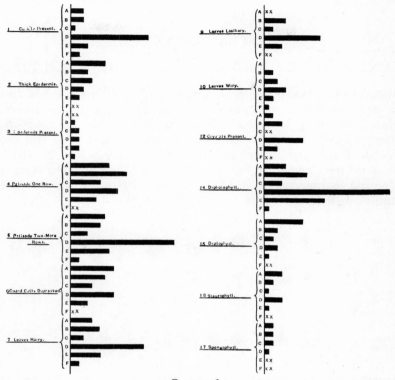

FIGURE 265

Comparative frequency of various leaf structures of plants of (A) Bermuda dunes; (B) New Jersey dunes; (C) New Jersey salt marsh; (D) New Jersey pines; (E) New Jersey cedar swamp; (F) New Jersey plains; XX indicates absence of structure.

CHAPTER XX

CONE AND SEED PRODUCTION OF THE PITCH PINE

August Renvall* has presented, in a monograph published at Helsingfors in 1912, the periodic phenomena of the reproduction of the Scotch-pine, Pinus silvestris, at the polar limit of trees in Finland. He considers, after a valuable historic introduction, the formation of staminate and ovulate flowers, working out statistically the frequency data, the sterility of the trees, and other important details by biometry. The years of cone and seed production also are discussed by Renvall, who uses the statistic method.

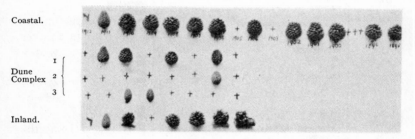

FIGURE 266

Arrangement of pitch-pine cones to show the years of cone production and by the open cones the years of seed discharge. Note that the coastal pine cones are larger than the inland.

Following out the suggestions of this important paper I undertook, during the latter part of the summer of 1912, to apply somewhat similar methods to the study of the pitch pine, but I have omitted the purely mathematic method and I have emphasized that phase of the study which appeals directly to the forester, who is concerned with the reproduction of the species. Several trees were taken in each locality and a single branch or two on each tree were chosen upon which to make the estimation of cone production (Fig. 266). The localities are given in each enumeration.

* RENVALL, AUGUST: Die periodischer Erscheinungen der Reproduktion der Kiefer an der polaren Waldgrenze, Fennia 29, No. 4, 1912, pages I-XII, and 154, with colored map.

BELMAR PARK, N. J.—The first tree studied had a circumference measurement of 76 decimeters, or an approximate diameter of 28 decimeters. A lower, south-bending limb was taken. The numbers after the year indicate the number of cones found on the portion of the branch corresponding to that date.

First Limb.—1912=1; 1911=4; 1910=3; 1909=1; 1908=2; 1907 =3; 1906=4; 1905=1 broken off; 1904=2 one broken off; 1903=2 broken off; 1902=1; 1901=1; 1900=1; 1899=0; 1898=0; 1897=0; 1896=1 detached and hanging by the bark.

Second Limb.—1912=0; 1911=2; 1910=4; 1909=0; 1908=2; 1907=2 one broken off; 1906=0; 1905=4; 1904=0; 1903=1 broken off; 1902=2; 1901=2; 1900=0; 1899=0; 1898=1 broken off; 1897= 0; 1896=0; 1895=2.

The second tree, which was from twenty-five to thirty years old, was chosen in the same locality. 1912=1; 1911=4; 1910=2; 1901=1 dried up; 1908=0; 1907=0; 1906=2; 1905=2; 1904=2 one broken off; 1903=0.

The third tree was twenty-five to thirty years old and grew in the same locality. 1912=2; 1911=6 two in one cluster, four in another cluster separated from each other; 1910=2; 1909=0; 1908=1; 1907= 1; 1906=2; 1905=0.

The fourth tree grew in the same locality, but in a place more exposed to the wind and sun. 1912=2; 1911=0; 1910=3; 1909=0; 1908=0; 1907=2; 1906=3; 1905=0.

The fifth tree grew in the same locality, but was exposed to the northeast winds from the nearby ocean. 1912=2; 1911=0; 1910=2; 1909 =0; 1908=1; 1907=1; 1906=1; 1905=1; 1904=0; 1903=0.

It will be seen from the above enumeration that from 1905 to 1912 practically every year was marked by the production of cones on one of the five trees. The year 1905 was perhaps the worst year for cone production, as only one tree out of five, as evidenced by a single branch, produced cones in that year. Each year from 1900 to 1905 on the first tree was marked by the production of cones. Then the years 1897, 1898, 1899 were bad cone years and 1896 a good cone year. Practically all of the cones on these five trees were open, showing that the winged seeds had been discharged during the late fall of the year of their maturity. To determine whether the liberated seeds germinated, or were in a viable condition, which is an important consideration to the forester, the expedient was adopted of cutting down young trees with a saw. The annual rings were counted and the diameter of the tree measured. The years in which viable seeds were produced was ascertained by subtracting the number of annual rings from the year of study, 1912.   Many factors are influential in the production of seeds and cones and these are spread over two years, hence the conditions of the year preceding the actual formation of cones and seeds may have been influential.   With this explanation the statement given above as to the year in which the ripe cone and ripe seeds were formed is not misleading.   Ten trees were thus measured at Belmar Park with the following results:

TREE	NUMBER OF RINGS	DIAMETER	YEAR OF GERMINATION
1.................	18	9.0 cm.	1894
2.................	14	8.0 "	1898
3.................	14	5.5 "	1898
4.................	14	5.5 "	1898
5.................	10	3.4 "	1902
6.................	10	2.5 "	1902
7.................	8	2.2 "	1904
8.................	8	2.6 "	1904
9.................	7	2.0 "	1905
10................	4	1.8 "	1908

It will be seen, as far as the estimation of the 10 young trees indicates, that there were viable pine seeds produced, as evidenced by those which germinated in the years 1894, 1898, 1902, 1904, 1905, and 1908, although, as previously shown, the pitch-pine trees produced seed every year (Fig. 268), but the conditions suitable for germination may have been absent or the seed may not have been fertilized, and other suitable factors may have been absent during the years unaccounted for in the above table

The second locality was chosen in the dune complex area at Spring Lake, where the bases of the trees were surrounded with the marram-grass, Ammophila arenaria, and clumps of the wax-myrtle, Myrica carolinensis.

The first tree was about 2 meters (6 feet) tall, of a spreading, wind-swept form, with 18 rings, and grew near the inner edge of the dune complex.   The diameter of the tree was 70 centimeters.   The count of the cones was made on a branch of the lee side.   1912=1; 1911=2; 1910=2 tightly closed; 1909=0; 1908=1 tightly closed; 1907=0; 1906=1; 1905=0.

The second tree was a low-spreading specimen near the outer edge of the dune complex at Spring Lake.   The first branch showed the following: 1912=1; 1911=2; 1910=2; 1909=1.   The lower, cone-bearing branch of this tree, well protected by the dense crown of the tree, showed: 1912=0; 1911=0; 1910=0; 1909=0; 1908=0; 1907=0; 1906=2 tightly closed.

The third tree grew at the extreme outer edge of the dune complex. It was a tree 1 meter tall with yellowish-brown leaves. The whole tree had only 2 cones both unopened, and by a count of annual rings, it was 9 years old. The whole appearance of the tree betokened stress of growth conditions.

First Limb.—1912=0; 1911=0; 1910=0; 1909=1 tightly closed.
Second Limb.—1912=0; 1911=0; 1910=1 tightly closed.

The fourth tree, over 3 meters tall, grew in the thicket back of the dune complex. 1912=1; 1911=0; 1910=0; 1909=2 green; 1908=0; 1907=1 tightly closed.

The fifth tree in this locality grew at the outer edge of the thicket and was 10 feet tall. A branch on the protected side was studied. 1912=5 in one whorl, 3 in another whorl lower down; 1911=3; 1910=0; 1909=2 tightly closed; 1908=1 tightly closed. Two other trees were cut down in this thicket, one showed 9 annual rings and, therefore, began its growth in 1903, reaching a stem diameter of 7 centimeters; the other showed 13 rings and began its growth in 1899 (Fig. 266).

A study of these dune trees show that the cones were not produced every year, and even in those years when cones were developed their production amounted to nothing in the regeneration of the forest, as long as they remained tightly closed. It shows that the wind-swept trees at the outer part of the dune complex are at the edge of their natural range as long as the present conditions control, because these trees rarely produce cones, and those that are formed, never open to let their seeds fall out. The advance of the forest into the dune complex as the pioneer trees prepare the way for those that follow, therefore, depends on the seeds which are produced on trees that grow some distance back from the outer edge of the tree line of the dune complex.* These winged seeds are carried to the tension line between the dunes and the forest by the currents of wind that blow. Not every year witnesses the start of a new tree. In the area studied new trees began their growth in the years 1894, 1899, and 1903.

To determine the cone production under normal conditions of forest growth a locality was chosen at Spring Lake, about 1 kilometer (½ mile) inland. The pine wood opposite to the "Plantation" in Spring Lake was chosen, where there was a uniform stand of tall pine trees with

---

* *Cf.* observations by ROBERT F. GRIGGS on the encroachment of the Sitka spruce on the tundra at the northern limit of forest on Kodiak Island, Alaska, Bull. Torr. Bot. Club, July, 1914.

at least forty years of growth. The heart-wood covered 6 annual rings. A number of young trees were cut down at the southern edge of the woods. The following data were obtained:

Tree	Number of Rings	Diameter in Centi-meters	Year of Germination
1..............	14	6.0	1898
2..............	12	2.5 Suppressed.	1900
3..............	12	3.5	1900
4..............	10	4.8	1902
5..............	7	1.8	1905
6..............	5	1.3	1907

The largest young trees, averaging from fifteen to twenty years, grew on the west, or lee side, of the taller trees. The younger trees form a fringe to the woods, because they start best in well-lighted situations, as the pitch pine is intolerant of shade. Tree No. 2 shows this in the small diameter, as compared with a tree fourteen years old growing in a sunny position. Nearly all of the young trees beneath the larger ones were similarly suppressed. There is very little typic pine-barren undergrowth in this particular grove, because the taller pine trees probably grew up in an open field forty to fifty years ago.

A study was made of a tall pitch-pine tree some distance west of the "Plantation." The circumference of this tree was 9.7 decimeters, which would make its diameter about 3.2 decimeters. 1912=2; 1911=4 green unopened; 1910=2; 1909=3; 1908=0; 1907=2; 1906=3; 1905=2; 1904=2; 1903=0; 1902=0; 1901=1.

Another locality was chosen 6 kilometers (4 miles) inland, i. e., 3 kilometers (2 miles) west of Point Pleasant. It was found that there was no uniformity in the production of cones on different branches. A tree twenty-six years old was found to be 7.6 meters tall. The diameter, 6 decimeters above the ground, was found to be 1.2 decimeters. Rapid growth was made in the first fifteen years, and in the later eleven years of growth the rings were small and difficult to see. The year 1911 was an exceptionally good year in this locality, as indicated by green cones on nearly all of the pine trees for the growth of that year, but, as emphasized previously, we must presume that pollination was successful the year before. The tree studied was cut down in order to get at the cones (Fig. 266).

First Limb (topmost).—1912=2; 1911=1; 1910=2; 1909=0; 1908=0; 1907=1; 1906=1; 1905=0.

Second Limb.—1912 = 1; 1911 = 2; 1910 = 1; 1909 = 0; 1908 = 3; 1907 = 1; 1906 = 2; 1905 = 1 with the cone partially imbedded in the bark. A number (3) of small trees were cut down.

NUMBER OF TREE	NUMBER OF RINGS	HEIGHT IN METERS	DIAMETER IN CENTIMETERS	YEAR OF GERMINATION
1	18	1.82	3.3	1894
2	15	1.52	2.5	1897
3	10	1.21	2.8	1902

A study of a tree in the heart of the pine-barrens on the Speedwell Road outside of Shamong (Chatsworth), N. J., on November 13, 1912, resulted in the following: 1912 = 0; 1911 = 3; 1910 = 0; 1909 = 1; 1908 = 1; 1907 = 2; 1906 = 3; 1905 = 2; 1904 = 3; 1903 = 2; 1902 = 3. On another smaller tree cones were found for the year 1912. It will be noted that in the heart of the pine-barrens the production of cones that open finally to discharge their seeds is the normal course of events. If we take the photograph of the museum preparation illustrating the production of pine cones under various environmental conditions, we find that the pine cones of the coastal trees are larger than those produced in the interior, as far as the observations of the writer go. I would account tentatively for this difference in size as due to a larger amount of moisture in the air of the coastal pine-barrens, as contrasted with the drier air inland. The conditions of the interior, however, are such as to produce smaller cones (Fig. 266).

The following conclusions have been reached from such a statistic study of the seed and cone production of the pitch-pine tree in the New Jersey pine-barrens. The regeneration of the pitch pine, *Pinus rigida*, depends on the following factors which are operative over two years:

First: The formation of viable seeds. This is conditioned on:
(*a*) Proper pollination.
(*b*) Certain fertilization.
(*c*) Proper development of the embryo.
(*d*) Absence of seed-eating insects and rodents.
Second: On the opening of the pine cones at the end of the second year, when they reach maturity. This normally takes place in the pine-barrens of New Jersey in the Indian summer days of November.
Third: On a suitable seed-bed.
Fourth: On climatic conditions suitable for germination.
Fifth: On plenty of light.

Sixth: On freedom from too much competition.

Seventh: Absence of fire for a few years after germination.

The absolute success of the establishment of a pitch-pine tree is determined best by an actual study of young trees in all stages of growth toward maturity.

LAW OF CONE AND SEED PRODUCTION IN THE PITCH PINE IN THE PINE-BARREN REGION OF NEW JERSEY

Cone production in the pine-barren region of New Jersey (coastal and inland) is an annual event, followed by the opening of the cones in the late fall of the year after their inception, or later, and the discharge of viable, winged seeds. The omission of this event is a local phenomenon and not of general occurrence in any one year. Natural annual regeneration is assured if the factors described above are favorable.

VIVIPARY IN THE ACORNS OF QUERCUS MARYLANDICA

On September 5, 1912, at Spring Lake, New Jersey, an oak tree of this species was found with a viviparous acorn. This condition was induced probably by the heavy rains, followed by mists and fogs, that had prevailed for a few days previously. An examination of this acorn showed that the embryo had swollen sufficiently to crack open the acorn covering, which had started to glaze. This splitting of the acorn shell revealed the presence of a small lenticular acorn at the base and inside the margin of the cupule. This vivipary of the acorns has not been studied exhaustively in America, but in Europe it has been studied in Quercus robur. Guppy* states that this is exhibited, not only in the occasional germination of the fruits on the tree, but in the actual stages of growth of the seed within its shell before maturity is reached. The steady growth of the seed on the tree long after the pericarp, or shell, has begun to dry is noteworthy. The vital connection with the parent plant is maintained by the attachment of the base of the fruit to its cupule. When the acorn begins to brown this attachment to the cupule begins to loosen, the result evidently of the drying of the pericarp, or shell. In the case of the viviparous acorn, it falls to the ground and usually dies, but it must happen frequently in moist mild weather that it continues the growth commenced on the tree, and if covered by protecting leaves, it may survive to the next spring and grow into an oak tree. Guppy suggests in connection with this viviparous habit, that if acorns are taken before they enter the rest period, that is while the peri-

* GUPPY, H. B.: Studies in Seeds and Fruits, 310 and 432.

carp is still green but the embryo is mature, they can be induced to keep up an uninterrupted growth without entering the rest period. This suggestion was tested with a number of acorns of species of oak, viz., Quercus alba, Q. marylandica, Q. prinus. Acorns of these three species were planted in a box filled with sphagnum. The planting was done on September 12, 1912, and the results recorded on October, 22 1912. Two series of acorns were sown. One set had a portion of the shell removed, exposing the embryo. The other set was planted with an uninjured shell covering. There was a slight advantage in the rate of germination of the cut acorns, as contrasted with the uncut. Practically all of the green acorns of the chestnut oak, Quercus prinus, the white oak, Q. alba, and the black-jack oak, Q. mary-landica, germinated (Fig. 267).

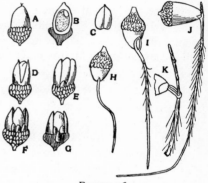

Sachs* and other botanists have maintained that, even under the most favorable conditions of vegetation, dormant periods occur in the course of the life of the plant. Under circumstances when the plant would be in a condition to grow most vigor-ously, because it is provided with reserve materials, water and oxy-gen are at its disposal and a suf-ficiently high temperature might be expected to call forth the in-ternal activities, yet every ex-

FIGURE 267

Stages in vivipary of Quercus marylandica.

A, Acorn of Quercus marylandica.
B, Acorn of Quercus marylandica dis-sected.
C, Embryo of Quercus marylandica.
D, First stage of vivi-pary in acorn of Quercus mary-landica.

E, F, G, Later stages of vivipary.
H, I, Stages in germi-nation of green acorns of Quercus marylandica.
J, Stage in germination of green acorn of Quercus alba.
K, Details of such germination.

ternally perceptible vital motion nevertheless ceases, and it is only after some months of rest that the growth commences anew, and this frequently under circumstances which appear far less favorable— especially at a conspicuously lower temperature. The experiments with acorns detailed above and experiments performed by Guppy† with the seeds of Iris pseudacorus, Vicia sepium, Arenaria peploides and Quercus robur show that a rest period is not essential for the germina-tion of acorns and other seeds, but that by taking immature acorns, whose embryo has not ceased to grow, and planting them the period of

* SACHS, JULIUS: The Physiology of Plants: 350.
† GUPPY, H. B.: Studies in Seeds and Fruits: 421.

growth is maintained without cessation, or a rest period, and the result is the elongation and growth of the embryo into a young seedling plant, as fully demonstrated in the figures. The germinative capacity of so-called unripe seeds does not seem to have been appreciated by foresters and gardeners, who layer their tree seeds in boxes of sand kept slightly moist and stored in a cool place over the winter. The acorns can be planted while green and carried over the winter in the greenhouse in the actively growing condition and in the spring, they can be planted in the open (Fig. 267).

CHAPTER XXI

NOTES ON A FEW INSECT GALLS OF THE PINE-BARRENS

Modern ecology has concerned itself with the habitat relationships of plants, and no more interesting field presents itself than the association of insects and plants of such a character that a gall, or cecidium, is the product.* The cecidologist is an ecologist, or biologist, who has chosen a narrow field of study, that of the galls. There are, he tells us, various kinds of cecidia caused by specific organisms, such as insects, nematodes, fungi and bacteria. We are concerned in this monograph only with insect galls. An insect gall is an abnormal growth of plant tissue produced by an insect which inserts an egg into the plant tissue from which hatches out a larva which is the exciting cause of the gall formation. Adler has shown that there was no foundation for supposing that the gall mother injected any irritating secretion whatever, and Beyerinck proved that the fluid ejected by the gall-fly is without taste or smell, and absolutely unirritating if injected under the skin. Both of these authors show that it is not in the gall-mother, but in the larva, that we must seek for the cause of the gall growth; and that it is the nature of the salivary secretion, and the manner of feeding of the larva, peculiarities inherited by each species which give the characteristic structure to each specific kind of gall. Mel. T. Cook† has shown that "the morphologic character of the gall depends upon the genus of the insect producing it, rather than upon the plant on which it is produced." The necessity for the continuance of the excitation during the whole period of gall growth is shown by its cessation when the larva has been destroyed by parasites.

* Consult BEUTENMÜLLER, WILLIAM: Catalogue of Gall Producing Insects found within Fifty Miles of New York City. Bull. Amer. Mus. Nat. Hist., IV: 245–278; ADLER, H.: Alternating Generations: A Biological Study of Oak Galls and Gall Flies, 1894; CONNOLD, EDWARD T.: British Vegetable Galls, 1902; BEUTENMÜLLER, WILLIAM: Insect Galls of the Vicinity of New York. Guide Leaflet, No. 16, Amer. Mus. Journ., IV, No. 4, October, 1904.

† COOK, MEL. T.: Galls and Insects Producing Them. Ohio Naturalist, II: 263–278, 1901; Cecidology in America. Bot. Gaz., 49: 219–222, March, 1910; MAYR, GUSTAV L.: Mittel-Europaischer Eichen-Gallen in Wort und Bild, 1907; KÜSTER, DR. ERNST: Die Gallen der Pflanzen, 1911.

The association of the insect and host plant is of ecologic interest. The fact that a specific larva can produce a specific gall of a definite shape, size, color, and longevity is one of the most remarkable in the whole range of biology. That out of elements so diverse in nature as the living cells of the host plants and the larval body of an insect, can result a gall of great beauty at times and always the same in morphologic structure, if produced on a given plant by a given insect, is remarkable.

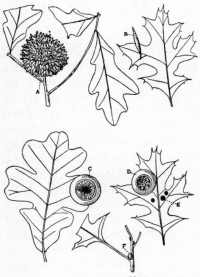

FIGURE 268

Insect galls on pine-barren trees.

A, Oak seed gall, Andricus seminator Harr.
B, Amphibolips coelebs O. S.
C, Holcaspis centricola O. S.
D, Oak Apple, Amphibolips confluentus Harr.
E, Cecidomyia pisum, Fitch.
F, Coccid.

A few insect galls were collected about 3 kilometers (2 miles) west of Point Pleasant in September, 1912. These galls with one exception were identified by Professor P. P. Calvert, of the University of Pennsylvania, and one of them by Dr. William Beutenmüller, of the American Museum of Natural History (Fig. 268).

On the white oak, Quercus alba, was found a large gall, the oak-seed gall, due to Andricus seminator Harris. This gall, which is found on the twigs of the white oak in June, is composed of a woolly substance, and is irregularly rounded, about 4 centimeters in diameter. Inside are numerous seed-like bodies adhering round the twig and very much resembling seeds, hence the common name. The gall is pure white, or white tinged with red, but by September it assumes a rusty-brown shade. Later it breaks to pieces and drops off the twig. The insect is hymenopterous, belonging to the CYNIPIDAE (Fig. 268A).

The scarlet oak, Quercus coccinea, had a number of different galls upon it. The oak, or May-apple, the largest of these galls, is due to an hymenopterous insect, Amphibolips confluentus Harris, which attacks the oak leaves. The gall is large, globular and smooth outside. The interior is filled with a spongy substance surrounding a central, hard, woody kernel containing the larval cell. The gall is about 2.5 centimeters in diameter and when fresh is pale-green, soft and succulent with whitish contents. The shell becomes light-brown in September and brittle. Another uncommon gall, which lives attached to a vein on the underside of the scarlet oak leaf, is the oak spindle gall, Amphibolips coelebs Osten Sacken (Fig. 268, B). This gall is elongated, spindle-shaped, at first soft and green, later becoming umber-brown and brittle. The central kernel is held in place by radiating fibers. The scarlet oak specimen yielded a small, spheric gall, determined by Dr. William Beutenmüller to be caused by Cecidomyia pisum Fitch, a dipterous insect. This little gall distributed plentifully over the leaf surface which is roughened by it, is deep reddish-brown in color, and woody. It is about the size of a morning-glory seed. A coccid was found also attached to the twigs in the axils of the leaves.

On the leaves of the post oak, Quercus stellata, was found a large, spheric gall due to Holcaspis centricola Osten Sacken, a hymenopterous insect. This gall is about 2 centimeters in diameter, and in September is of a light yellow-brown color. The shell is smooth and brittle. Internally there are radiating fibers attaching a central body to the inner wall of the gall (Fig. 268, C).

Four of the above gall insects, viz., Andricus seminator, Amphibolips coelebs, A. confluentus, Holcaspis centricola, are hymenopterous, belonging to the family CYNIPIDAE, the "gall-wasps," or "gall-flies." The ovipositor of the adult insect is coiled partly within the abdomen, which is dilated and enlarged posteriorly, closely joined to the thorax, but not sessile. The life-cycle is often complicated, and parthenogenesis is of frequent occurrence. In some species the walls, perhaps, have been eliminated, while in others there is an alternation of generations, one having both sexes normally present, while in the other the females only occur. Cecidomyia pisum is a dipterous insect, included in the family CECIDOMYIIDAE, the "gall-gnats," or "gall-midges," which are fragile in appearance and slow in flight. The larvae are small, legless grubs, bluntly pointed at both ends, often with a chitinous process, known as a breast-bone, on the under side, near the anterior end.* A complete list of the gall insects of New Jersey is given in the volume on insects, Annual Report New Jersey State Museum, 1909. The interested botanist is referred to that publication for additional details about the pine-barren gall insects.

* Consult Report of the New Jersey State Museum. The Insects of New Jersey, 1909: 595–604; 725–734.

CHAPTER XXII

PINE-BARREN PLANTS FROM AN EVOLUTIONARY VIEWPOINT

Our study of the vegetation of the pine-barrens of New Jersey would not be complete without reference to the broader principles which underlie all such ecologic studies, namely, questions of evolution. Many of the evolutionary problems in Europe and America have been approached by a study of cultivated plants. While cultivated plants are suitable for the investigation of many phases of evolution, yet they are not suited so perfectly for this purpose as the wild plants which exist in virgin plant formations. For this reason the pine-barren vegetation of New Jersey, which exists practically in undisturbed condition in its original regional surroundings, affords unusual opportunities for research upon plants in a state of nature. De Vries has emphasized the importance of the experimental treatment of the species question, but in the impossibility of studying all species experimentally the ecologist and taxonomist, with different ends in view, can contribute much material which will be of lasting value. The taxonomist is prone to make new species, because his object is to describe a plant so that it may be identified readily, and this, from the very nature of the case, demands a more or less artificial classification. The species of the taxonomist, as L. Cockayne* has proved abundantly, are frequently ideas merely and not living entities. The ecologist, on the other hand, emphasizes the influence of the environment in modifying the form of the species, and he finds that the descriptions of the taxonomist do not always fit the plant, even with an elastic interpretation of the specific characteristics of the plant. With him a specific name is a means, rather than an end, because in his description of vegetation the scientific names are useful as a means of describing intelligibly the associated plants of a particular plant formation and the epharmonic structures of these plants. He recognizes the necessity of the substitution of elementary species, as the raw material for the evolutionary process, rather than the Linnaean historic or taxonomic species.

---

* COCKAYNE, L.: Observations concerning Evolution, derived from Ecological Studies in New Zealand. Trans. New Zealand Instit., XLIV: 1—50, 1912.

ELEMENTARY SPECIES.—Elementary species are physiologic conceptions. They are forms whose characters, whether large or small, are heritable, and thus they become of phylogenetic importance. Among the pine-barren plants no one plant possesses more interest from the standpoint of elementary species, than Euphorbia ipecacuanhae,* in which Professor de Vries took the greatest interest, when the writer showed him the plant some years ago in its natural habitat in New Jersey. This taxonomic species consists of distinct forms that merit specific rank as elementary species. The

FIGURE 269

The spurge, Euphorbia ipecacuanhae, illustrating in its various forms the origin of elementary species.

leaves in the different types range in shape from linear-lanceolate to broadly elliptic, and in color from green to deep reddish-brown, and these leaf and color characters are heritable (Fig. 269).

FIGURE 270

The cowbane, Oxypolis rigidior, in its several varieties which are probably elementary species. The lower figure represents the variety longifolia.

The cowbane, Oxypolis rigidior, and its variety longifolia, probably represent two elementary species. The variety has narrow leaf segments, rarely over 4 to 5 mm. wide. The margin of the segments are generally entire, but the narrowest are sometimes lobed (Fig. 270).

The writer has been interested for some time in Vaccinium corymbosum and V. atrococcum. Both are tall blueberries growing in wet thickets. The first shrub flowers in May, when the leaves are partly expanded and the fruit is black with a bloom. Vaccinium atrococcum flowers earlier than the preceding, and its leaf development follows usually the opening of its flowers. The fruit is black without bloom. Two other

* DE VRIES, HUGO: The Mutation Theory [Engl. transl.], II: 605, 1910.

taxonomic species are distinguished, viz., Vaccinium virgatum Ait. and V. caesariense Mackenzie. According to Mackenzie, there are apparently three forms of tall blueberry native to the coastal plain of New Jersey, viz.: (1) a form with finely serrate leaves (virgatum), somewhat pubescent below, and restricted to the pine-barrens; (2) a form with entire leaves, partially pubescent below, particularly the veins (corymbosum); and (3) an entire-leaved, absolutely glabrous form (caesariense). Perhaps we should think of all these four forms as elementary species, viz., Vaccinium atrococcum, caesariense, corymbosum, virgatum.

The small-flowered dogbane, Apocynum medium Greene, as a taxonomic species, includes probably two growth forms which have been segregated from it, namely, A. Milleri Britton and A. urceolifer G. S. Miller. These are probably all elementary species.

FIGURE 271

Different forms of branches from pine trees growing in different localities. Left, pine forest at Belmar; Second, branch from a protected tree, sea dune thicket; Third, branch from a small tree in dune complex; Right, branch from exposed side of the same tree from which No. 3 branch came.

VARIATION.—Darwin emphasized the study of individual differences of organisms in his discussion of variation. He considered that they formed the material upon which natural selection could act, and hence they accumulate. De Vries and others deny that these fluctuating characters are accumulated indefinitely, and as far as the writer's study of plants enables him to form an opinion, he agrees with the position of de Vries. We can, by statistic methods, fix the limits of such fluctuations, as they oscillate around an average.

A number of interesting cases of fluctuating variability occur among the pine-barren plants of New Jersey. The pitch pine, Pinus rigida, shows a great diversity of tree forms, which have been described in a former section. It is difficult to determine which is the normal form of the tree, although in fact no one has any difficulty in pointing out a pitch-pine tree when he sees one. Such fluctuations show how difficult it is to give a description of a tree which will be comprehensive enough to include all of the different growth forms (Fig. 271).

The common arrowhead, Sagittaria longirostra, of the pine-barren region, shows leaves that vary from narrowly linear to broadly hastate, though the narrow type of leaf is more frequent (Fig. 272). Panicum Commonsonianum is a plentiful grass throughout the pine-barrens, first described by Ashe. Related to it is another slightly different grass, distinguished by Nash as the variety Addisonii. Between these two

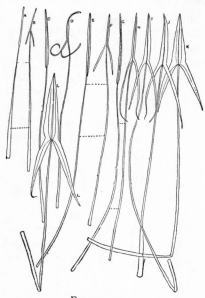

FIGURE 272

Leaf variation of Sagittaria longirostra.

A–B, Leaves from a single plant collected by S. S. Van Pelt at Tom's River, July 19, 1906. Narrow leaf an early juvenile form at the outside of the rosette.

C, Leaf of plant collected by Bayard Long at Lakehurst, August 8, 1908.

D, Leaf of plant collected by J. H. Grove at New Egypt, August 14, 1907.

E–F, Leaves of a single plant collected by Bayard Long at Dover Forge, September 9, 1907. Narrow leaf from outer part of rosette an early juvenile form.

G–H, Leaves of a single plant collected by S. S. Van Pelt, near Job's Bridge, between Hammonton and Batsto, August 6, 1905.

I, Leaf of a plant from Parkdale collected by S. S. Van Pelt, August 17, 1905.

J, Leaf from Lakehurst plant collected August 8, 1908, by Bayard Long.

K, Leaf of a Pasadena plant collected by Bayard Long on July 24, 1908.

L, Leaf of a plant collected by Bayard Long at Lakehurst, August 8, 1908.

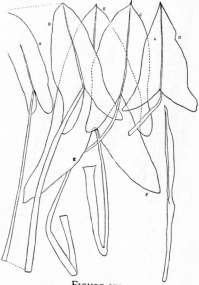

FIGURE 273

Leaf variation of Peltandra virginica.

A–B, Leaves of a single plant as denoted by connecting dotted line, collected by Bayard Long at Manahawkin on September 16, 1910.

C, Leaf of a plant collected by J. H. Grove at New Egypt, June 22, 1906.

D, Leaf of a plant collected by Alexander Mac Elwee at Forked River, June 22, 1895.

E, Leaf of a plant collected by Bayard Long along Hospitality Branch at Folsom on June 10, 1911.

F, Spathe of the plant.

forms are a great number of intergradations. The green arrow arum, Peltandra virginica (Fig. 273), shows also a great variety of leaf forms,

FIGURE 274

Leaf variation of pickerel weed, Pontederia cordata.

A–B, Leaves from a plant collected between Tuckerton and Atsion by C. F, Saunders and W. N. Clute, July 3 to 6, 1899.
C, Leaf from pickerel weed found at Toms River, June 2, 1895, by Alexander Mac Elwee.
D, A whole plant collected by A. N. Leeds at Mays Landing, August 20, 1893.
E, Leaf from a plant collected at the mouth of Hamilton Creek, August 21, 1910, by Bayard Long.

as also the pickerel weed, Pontederia cordata, which has a series of leaf forms from broad to narrow, the narrow forms being distinguished as the variety angustifolia by Pursh (Fig. 274).

The number of flowers varies considerably in the turk's-cap lily, Lilium superbum, according to locality. Usually only a single flower is found on plants growing in the pine-barren swamps, while in richer swamps of West Jersey it bears a large cluster of flowers, as many as fifteen to twenty. The colic-root, Aletris farinosa, exists in a form with short leaves and more nearly spheric flowers found near the plains. Smilax tamnifolia (Fig. 275), Quercus falcata (Fig. 276), show the greatest diversity of leaf forms, which are cases of fluctuating variability, and in a less pro-

FIGURE 275
Variation in the leaves of Smilax tamnifolia.

nounced way we note the same in the leaves of Quercus alba, Q. ilicifolia, Q. marylandica, Q. stellata, and Q. velutina. (See Figs. 277 and 278.) The forms of the spatterdock (Nymphaea) in the pine region of New Jersey vary considerably. Stone believes that Nymphaea rubrodisca is referable to N. variegata; as also those referred to N. microphylla, which is connected with N. variegata by a full series of intermediates. The flower color of Lupinus perennis shows great fluctuations from deep-purple to white. As intermediate colors exist between these two colors, we must classify these color variations as of the fluctuating sort. The meadow-beauty, Rhexia virginica, shows variations in flower color to deep magenta, but never the pale pink of R. mariana.

FIGURE 276
Variation in the leaves of Spanish oak, Quercus falcata (= Q. triloba).

MUTATIONS.—The origin of new species in the sense of de Vries by mutation can be proved only by experiment under which the heredity of the new character may be proved. In a state of nature, therefore, we can infer only that such sports or mutants have arisen. The new forms are distinct from each other and do not show intermediate types, such as we find in variations of the fluctuating type. In surveying the plants of the pine-barrens it seems that the following cases are mutations.

Pogonia ophioglossoides is a bog orchid, usually with a single, rose-colored flower. Occasionally mutants are found with white flowers. The water lily, Castalia odorata, has beautiful, creamy white flowers. Sometimes a pink color suffuses the whole flower. Some authors call

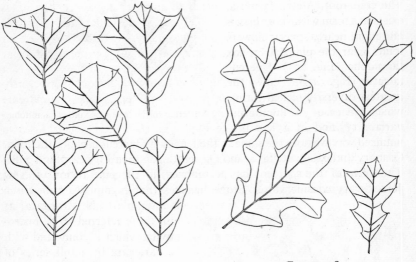

FIGURE 277
Leaf variations of oak, Quercus marylandica.

FIGURE 278
Leaf variations of Quercus stellata and Q. ilicifolia.

this the variety rosea. Polygala cruciata has flowers that are normally the color of red clover, that is, rose-colored, but sometimes as a mutation we find plants with greenish-purple flowers, and this statement is true for Polygala viridescens, with rose-colored flowers. Occasional forms have green flowers. These two color forms were described originally as different species. The calico-bush, Kalmia latifolia, has masses of flowers of deep pink, but white-flowered bushes are not infrequent. The bird-foot violet, Viola lineariloba, has large, lilac-purple flowers. As a mutant of this species, the true Viola pedata, we have flowers in which the two posterior lateral petals are a dark velvety purple. The broom crowberry of the plains, Corema Conradii, exists in two color forms, deep green and reddish-brown. These forms are both found in adjacent patches in the bright sunlight and are probably mutants. As a mutant of the widely distributed red-maple, Acer rubrum, we have the variety tridens of Wood, or Acer carolinianum of Britton. The leaves of the mutants are small, with three short lobes, the middle lobe being triangular.

EPHARMONY.—Epharmony is the harmony between the structure of a plant and the external factors. Vesque* says: "Tous les organes de la plante peuvent s'adopter au mileu incerte ou animé qui les entoure, mais à des degres divers, et c'est précisément sur cette inegalité que repose la subordination des caractères; mais au mileu de tous ces organes il y en a dont la nature depend 'uniquement de l'adaptation,' savoir la structure anatomique des organes vegetatifs en tant qu'elle est en relation directe avec l'air, le sol et l'eau, c'est ce qui je propose d'appeler l'epharmonie."

PLASTICITY OF SPECIES.—Nothing has been brought out more clearly by ecologic studies in the pine-barrens of New Jersey than the extreme plasticity of many species and structures, and their response to a change of environment. This fact is so evident that the idea of normal loses its value. For example, the black-jack oak, Quercus marylandica, may exist in the pine-barrens as a tree and as a shrub. The pitch pine, Pinus rigida, we have seen has a great variety of growth forms and its plasticity is manifested also in the size of its cones.

RESPONSE TO ECOLOGIC FACTORS.—Warming, in his "Oecology of Plants," 1909: 1 and 81, has summed up our knowledge about these matters, so that only a few examples are concerned.

(a) SOIL

Pyxidanthera barbulata, as we have seen, exists in three forms which have been conditioned on the character of the upper layers of the forest soil. The wandering stems of one form of this plant are due to soil conditions. Two other forms of this plant occur in the pine-barren region of New Jersey, as forms due to the influence of the substratum as previously noted.

(b) LIGHT

Quite a number of plants show two distinct forms which grow in the shade and in the light. The pitcher-plant, Sarracenia purpurea, has green ascidia when growing in deep, shady white-cedar swamps. In open sunny places it is generally a deep shade of reddish-brown. The

* Ann. Sc. Nat., ser. 6, XIII (1882), page 9.

leaves of the laurel are elliptic and leathery in the sun-plant. The leaves from a bush growing in the shade are larger, thinner and longer. The pyxie, Pyxidanthera barbulata, has a sun form and a shade form.

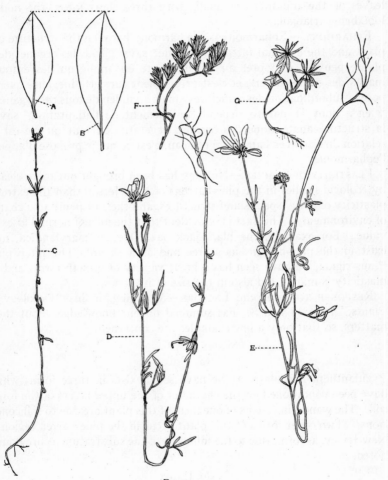

FIGURE 279

Influence of sun and shade on pine-barren plants.

A, Leaf of laurel, Kalmia latifolia, from plant growing in bright sun and dry soil.

B, Leaf of laurel, Kalmia latifolia, from plant growing in the shade.

C, Bidens trichosperma growing in deep shade in Cedar Bog, two miles west of Pt. Pleasant, N. J., September 9, 1912.

D, Bidens trichosperma growing in medium shade in Cedar Bog.

E, Bidens trichosperma growing in bright sunny places.

F, Pyxie, Pyxidanthera barbulata, growing in bright sun and dry sandy soil.

G, Pyxie, Pyxidanthera barbulata, growing in dense shade and leaf mould.

The former is denser and more cushion-like with smaller leaves, the latter is more open, less compact in growth with larger leaves (Fig. 279).

(c) WIND

The wind-shearing action on trees is noticeable, especially in the pitch-pine trees, Pinus rigida.   In the coastal pine-barrens, the action of the wind in producing one-sided trees and partially prostrate trees is worthy of note. The action of the wind on the vegetation of the plains has been described. (Figs. 31, 32, 33, 34, 35, 36, 37.)

(d) WATER

Quite a few pine-barren plants have forms that are due to the action of water. Most plants of Sagittaria graminea have leaves that are lanceolate, but in wholly submerged plants the leaf-blades are partly, or entirely absent, being represented by linear phyllodia. The water club-rush, Scirpus subterminalis, has in its usual submerged forms long, narrow leaves, streaming in the direction of the current of water. In ponds where the water has been drained off it often grows upright, with much shorter and stiffer stems. Eriocaulon septangulare, in its submerged form, has well-developed leaves, about as long as the scape. Other leaves are only half the length of the scape, while plants on the edge

FIGURE 280
Leaf forms of several species of pond weeds, Potamogeton.

A, Potamogeton confervoides with narrow linear submerged leaves, collected by Bayard Long at Bamber, May 8, 1907.

B, Potamogeton epihydrus with (a) narrow linear submerged leaves and (b) broad floating leaves; collected by Bayard Long at Manahawkin, September 1, 1907.

C, Potamogeton Oakesianus with broad floating leaves, collected by Bayard Long at Folsom along Hospitality branch of Great Egg Harbor River, July 27, 1909.

of a pond or swamp are often 75 to 100 millimeters in height, with leaves 25 millimeters long, according to Stone's statement (324). The bayonet-rush, Juncus militaris, has curious, submerged, thread-like leaves that arise from the root-stock and grow in the water-like masses of waving hair. The low water-milfoil, Myriophyllum humile, exists in the form of small plants creeping in the mud (humile) or in a submerged form (capillacea), and those with an emersed spike (natans). Several species of Potamogeton and Utricularia show the influence of an aquatic environment on the form and structure of the plant body (Figs. 280 and 281).

CONVERGENT EPHARMONY.—It is evident that many growth forms among the pine-barren plants may be referred with confidence to the influence of the stimuli of various external factors. Some of these forms are mere fluctuations, or oscillations, about a norm, others are hereditary and remain constant. Cockayne in his study of New Zealand vegetation enunciates a principle applicable also to the New Jersey pine-barren vegetation, that, "*It is a fact of the greatest significance that identical forms are found side by side amongst species belonging to unrelated families.*" A few illustrations of this principle will be mentioned, as they have come within the notice of the writer, as cases of convergent epharmony.

(a) ANNUAL WIRY STEMS, REPEATEDLY BRANCHED

Anychia canadensis is an annual herb of the family Caryophyllaceae. Crotonopsis linearis is an annual herb of the family Euphorbiaceae. In growth forms they are much alike. The first plant has a very slender, filiform, usually erect stem, which is forked repeatedly with opposite, small, elliptic, oval, or sometimes oblanceolate leaves. The second plant has wiry stems, repeatedly forked, and bearing small oblong-ovate to linear-lanceolate leaves. It can be distinguished in the vegetative condition from the first by its silvery-scurfy appearance. Both plants are of about an equal height.

(b) ANNUAL HERBS WITH ARTICULATE STEM AND SUBULATE LEAVES

Three herbs approach each other in growth forms. They are Polygonella articulata (Polygonaceae), Sarothra gentianoides (Hypericaceae) and Bartonia virginica (Gentianaceae). The first plant is an annual herb with slender, jointed, wiry, erect stem diffusely spreading, or branched with linear or linear-subulate leaves. The plant is found in barren, sandy places in the coastal pine-barrens. Sarothra gentianoides is an erect annual fastigiately branched, and with wiry branches. The leaves are minute and subulate. It also grows in sandy soil. Bartonia virginica is an erect, stiff annual with filiform, wiry stem beset with leaves that are reduced to subulate scales appressed to the stem. It is found in moist, sandy soil.

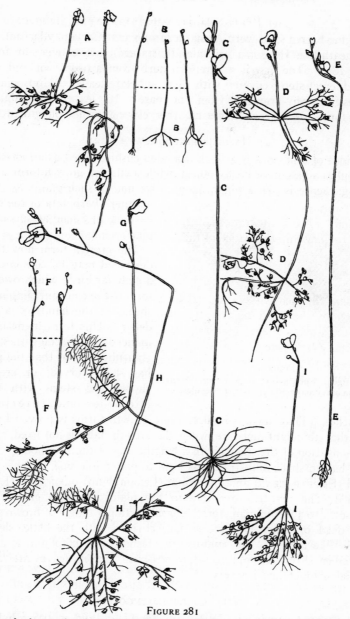

FIGURE 281

Details of submerged and aerial parts of several species of Utricularia. Drawings from specimens in the Herbarium of the Philadelphia Botanical Club, where data will be found.
A, Utricularia clandestina. B, U. cleistogama. C, U. cornuta. D, U. inflata. E, U. juncea. F, U. subulata. G, U. intermedia. H, U. fibrosa. I, U. purpurea.

(c) PERENNIAL HEATH-LIKE PLANTS

The pine-barren sand-wort, Arenaria caroliniana (Caryophyllaceae), and the heath-like Hudsonia ericoides (Cistaceae), are convergent forms (Fig. 282). The first is a perennial herb from a deep root and with tufted, erect stems covered with rigid, subulate leaves. Hudsonia ericoides is bushy-branched from the base. Its leaves are subulate. Both forms are suggestive of each other, especially the young plants.

(d) HASSOCK-LIKE GROWTH FORMS

The idea of a hassock is a thick, rounded cushion, used as a foot-stool. It implies a besom, or bushy object, while a cushion growth form in the ecologic sense is like a pin-cushion, as we find in such plants as Pyxidanthera barbulata of the pine-barrens, or Silene acaulis of alpine regions.

FIGURE 282
Convergent Epharmony in Sandwort, Arenaria caroliniana and Hudsonia ericoides.

The growth form now to be described may be compared to a globular bush, which owes its form to the constant clipping by shears in the hands of a gardener. This form in nature is induced in a dry soil with strong transpiration, so that the plant has short curved, or crooked shoots and stems with short internodes and interlaced branches. That dryness induces these growth forms is proved by the fact that abundant moisture, when the growth habit is not fixed, causes an elongation of the shoots with a lengthening of the internodes. Two pine-barren shrubs illustrate this form to perfection, viz., Corema Conradii (EMPETRACEAE) and Dendrium (Leiophyllum) buxifolium (ERICACEAE). The first shrub native to the plains exists in great round masses, 1 to 3 feet in diameter, the basal portion a tangle of brown stems and dead branches, but the surface covered with the little, slender, needle-like leaves. The sand-myrtle, Dendrium buxifolium, is a low, evergreen shrub resembling a low box-bush, but its leaves are larger, although tough and leathery.

(e) PROSTRATE SHOOTS

Many species growing in dry, warm, sandy soil and in other habitats have prostrate shoots. A few examples from different families may be cited from the pine-barren region of New Jersey. The following are of interest: Epigaea repens (Fig. 193), Gaultheria procumbens (Fig. 188), Arctostaphylos uva-ursi (ERICACEAE), (Fig. 197), Euphorbia ipecacuanhae (EUPHORBIACEAE), Galactia regularis (LEGUMINOSAE) (Fig. 180).

(f) LIANES

The climbing plants of the New Jersey pine-barrens belong prominently to five genera of four families of flowering plants. In the climbing habit they approach each other, although of diverse morphologic structure. They are species of Smilax (SMILACACEAE), Dioscorea villosa (DIOSCOREACEAE), Vitis aestivalis, Ampelopsis (Psedera) quinquefolia (VITACEAE), and Rhus radicans (ANACARDIACEAE). Species of Ipomoea, Convolvulus, Cuscuta (CONVOLVULACEAE), have twining stems, while two leguminous plants, Clitoria mariana * with twining habit, and Apios tuberosa, with climbing habit, are herbs to be included in this category.

(g) PLANTS WITH AERENCHYMA

Five aquatic plants, belonging to four different families of flowering plants, possess aerenchyma. They are Hypericum adpressum (HYPERICACEAE) found at May's Landing at the edge of the pine-barrens, Decodon verticillatus (LYTHRACEAE), Rhexia virginica and R. aristosa (MELASTOMACEAE) and Ludvigia sphaerocarpa (ONAGRACEAE). Aerenchyma is a tissue-like cork, which has its own phellogen, but it consists of thin-walled, non-suberized cells between which are large air-containing, intercellular spaces. The tissue shows itself as a white, spongy covering. In the five plants above mentioned, it is found around the bases of the stems and as in a number of other helophytes, or swamp plants in other parts of the world.

(h) CLEISTOGAMY

A number of plants native to the pine-barren region are characterized by the formation of underground cleistogamous flowers which produce normally a larger number of good seeds than the conspicuous flowers in an aerial position. Such cleistogenes are Amphicarpon Purshii (GRAMINACEAE), Viola sagittata, V. emarginata, V. primulifolia, V. lanceolata (VIOLACEAE) (Fig. 283) and Utricularia cleistogama (LENTIBULARIACEAE) (Fig. 281 B).

In the cleistogamous flowers of the violets, see figures (Fig. 283)

* Very frequently erect, not twining.

adapted from Church,* the pollen grains are fewer in number than in normal flowers and the ovules are about half the normal number. The anthers lie against the ovary wall, and the distance from the uppermost pollen grains to the end of the rudimentary style is extremely small (about 5 or 6 diameters of the pollen grain). The pollen grains germinate within the pollen sac and the wall of the anther box is ruptured and the pollen tubes grow across to the style which they penetrate, thus accomplishing the act of close pollination.

The list of converging growth forms which have been described in categories (*a*) to (*h*) inclusive is not exhaustive, because we have omitted a description of a number of distinct kinds, such as the rosette form.

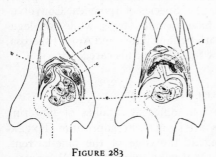

FIGURE 283
Cleistogamous flowers of violet.

a, Inturned sepals.
b, Covering inturned petals.
c, Stamen and anther with pollen.
d, Stigmatic region.
e, Ovules.
f, Growth of pollen from anther across to stigma (Cleistogamous pollination).

The above account, however, will serve as an introduction to a very attractive field of ecologic investigation.

PERSISTENT JUVENILE FORMS

Cockayne† states that about 200 species of New Zealand vascular plants belonging to 37 families show a more or less well-marked distinction between the juvenile and adult stages of development while, perhaps, in 100 species the differences are very great indeed. In the absence of exact data for the pine-barrens of New Jersey, and to draw attention to this important phase of evolutionary investigation, a single example of the persistence of a juvenile form may be cited. The red cedar, Juniperus virginiana, in the early stages of growth has awl-shaped, divergent leaves, which in most trees are replaced by closely appressed scale leaves. In a few trees which I have seen in the coastal plain of New Jersey the juvenile state persists until the tree reaches a considerable size. Herbarium specimens of these persistent juvenile forms have been preserved. One way in which these persistent juvenile forms differ from the adult forms is in the bright green color of the foliage, as contrasted with the darker greens of the tree, with closely appressed scale leaves (Fig. 284).

* CHURCH, A. H.: Types of Floral Mechanism, Part I, page 98 (1908).

† COCKAYNE, L.: Observations concerning Evolution derived from Ecological Studies in New Zealand. Trans. New Zealand Institute, XLIV, 1911.

HYBRIDIZATION

A number of natural hybrids have been collected in the pine-barren region of New Jersey. The white fringed orchid, Blephariglottis blephariglottis, is concerned as one of the parents in the formation of two hybrids. Blephariglottis bicolor Rafinesque is B. blephariglottis × B. ciliaris, and has been found at Bamber by Bayard Long on August 25, 1909, according to Witmer Stone. B. Canbyi Ames is a hybrid of B. blephariglottis and B. cristata, and has been collected a number of times in the pine-barrens.

A hybrid between Quercus phellos and Q. ilicifolia was found by J. E. Peters at May's Landing, while east of Farmingdale, at the northern edge of the pine-barren region, the writer found Quercus heterophylla, the Bartram oak, which has been proved by the experiments of MacDougal * to be a hybrid of Quercus phellos and Quercus rubra as parental forms.

The fifth case of a natural hybrid that has come to my notice is one produced in a natural cross between Eupatorium resinosum and E. perfoliatum discovered by a former student of mine, Bayard Long of the Philadelphia Botanical Club.

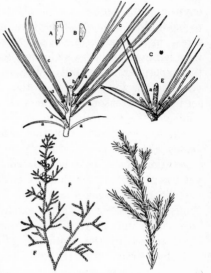

FIGURE 284

Juvenile and adult leaf forms in pine and juniper.

A, Winged seed of Pinus rigida.
B, Winged seed of Pinus echinata.
C, Seed of white cedar Chamaecyparis thyoides.
D, Piece of a sprout from a fire-injured pitch pine, showing (a) reverted juvenile leaves; (b) sheathing leaves; (c) secondary needle leaves.
F, Adult shoot of red cedar, Juniperus virginiana with berries.
G, Young shoot of red cedar, Juniperus virginiana, showing persistence of juvenile needle-like leaves.

CONCLUSION

The above evolutionary studies have been treated in a general way and from the observational standpoint. The field has not been exhausted and much important work remains to be done in the field

* MACDOUGAL, D. T.: Hybridization of Wild Plants. The Botanical Gazette, 43: 53, January, 1907.

under experimental conditions. The above account is, therefore, a short introduction to a subject of considerable botanic interest. It provides an outline of what may be accomplished by intensive methods of research work, spread over a period of years. Preferably the investigation should be conducted in the field under controlled conditions, rather than in the laboratory where even with the best surroundings the environmental factors are purely artificial. An ideal system would be to have plots of land, an hectare or two in extent, located in the pine forest, in the plains, in a cedar swamp, and in a deciduous swamp, surrounded by a barbed-wire fence 2 to 3 meters in height, so as to protect the experimental ground. Within the inclosures the instruments devoted to the investigation of the climatic and other factors could be set up with a reasonable hope of their remaining undisturbed during the course of the research work.

The foregoing survey of the vegetation of the pine-barrens of New Jersey, although not exhaustive, presents some interesting facts about a vegetation which, although related to that of the southern coastal plain, yet has many features that are entirely distinct. We have seen in the first place that in no other region of North America does the pitch pine, Pinus rigida, cover such an extended area of country as a dominant tree. The pine-barrens of the southern states, as we have seen, are characterized by the dominance of other pine species, notably the long leaf-pine, Pinus palustris, the loblolly-pine, Pinus taeda, the yellow-pine, Pinus echinata, and the slash-pine, Pinus caribaea. These are all dominant in the area of country representing their natural range, and on soils which are different for each of the pine species above mentioned.

Secondly, the pine-barren vegetation of New Jersey is an old vegetation in a long occupancy of the territory in which it is found. The age of this forest is attested by its survival through the changes of several geologic epochs antedating the glacial period, when the northern part of the State was covered with ice.

Thirdly, the pine-barren vegetation has maintained itself against the encroachment of other types of vegetation. This is attested by the difficulty that introduced plants have in gaining headway against the exclusive closed formation of pine-barren plants, and even today, after long occupancy of the region by man, introduced weeds play a relatively unimportant part in the flora.

Fourthly, nowhere in the northern United States, east of the mountains, is there such a large geographic area which exists in such a wild and pristine state. This is due undoubtedly to the rapid settlement of the richer and more desirable lands of the western states, which have yielded larger financial returns with a smaller outlay of capital. The pine-barren soils, although suitable for the growth of many vegetable crops, yet yield their returns only after an expenditure of money and labor which until recent years has not been considered commensurate with the returns. With the settlement of the western states, attention will be turned to the poorer sandy soils of these northern pine-barrens, which yield fair returns if properly treated. Already the effect of the settlement of Italians and Russian Jews in the southern part of the territory is felt in the removal of the pine forests and in the cultivation of the soil. As early as 1888 the following statement was made: "The remainder of the region is being rapidly improved and brought under cultivation. Twenty-five per cent. of the pine region is cleared in Camden County, 30 per cent. in Gloucester, and 63 per cent. in Salem. Perhaps the most advancing portion of the region is Cumberland County, where 36 per cent. is now cleared. It is true that a good part of this is the older settled country west of the Cohansey, but large inroads have been made in the eastern portion within 25 years, particularly in the vicinity of Vineland." To this statement might be added one, that at Woodbine, large areas have been cleared by the Baron de Hirsch estate, and rapid inroads have been made in the pine forest along the railroad between Newfield and May's Landing.

Fifthly, the future development of the region, according to the writer's acquaintance with it will be along four lines. Agriculture and horticulture will remain always the most important industries of the pine region. The preservation of the pine forest in large tracts will be necessary to preserve a proper balance between forested areas and agricultural areas. At least 30 per cent. of the region should be kept permanently in the form of state forests, owned and managed by the State in perpetuity. Manufacturing interests will be centered in the larger towns, where the factory employee will be within reach of his own home ground, where supplementary supplies of food can be raised in his own kitchen-garden.

Lastly, the region, having a salubrious climate, will naturally attract the health seeker and sufficient stretches of pine-forest should be preserved for the benefit of this class of the community. The future development of the region into one of unusual prosperity is only a matter of time. Its close proximity to the vast populations that center about New York City and Philadelphia, will make ultimately of the region a valuable agricultural country. The State of New Jersey has begun to have a far-sighted policy in the development of its roads and its inland water ways.   In addition to these improvements the State should take over large blocks of forest for the future use and pleasure of its people.   Its streams should be kept from pollution and stocked with fish.   The larger forest tracts should be stocked with game and forest paths constructed, so that the largest number of people, healthy and sick, wealthy and poor, could derive benefit from the life-giving air of the pines.

INDEX

An almost complete list of pine-barren plants, 548 species out of 555 native ones, will be found on pages 196–208. As these names are arranged systematically, they are not included in the index. Additional lists of plants will be found on pages 83, 127, 139, 145, 172, 180, 186, 189, 190.

ABAMA americana, roots and stems of, 225
Abbe, Cleveland, cited, 194
Acer carolinianum, 305; figure of microscopic leaf structure, 274, microscopic leaf structure, 274
  rubrum, 112, 133, 134, 137, 140; mutation of, 305
  tridens, 305
Acorn vivipary, 292
Additional pond plants, 144
  swamp plants, 138
Adler, H., cited, 295
Aerenchyma, 311
Age of larger pines of Lower Plain, 153
  of pine sprouts on plains, 153
  of tree sprouts of plains, 154
Agriculture, future, 315
Aletris farinosa, forms of, 303
Algoid leaves, 253
Allen, Grant, cited, 188
Alnus rugosa, figure of microscopic leaf section, 268; microscopic leaf structure, 268
Alpine meadows, study of, 175
Amentiferous trees, flowering period of, 211
American and European plant formations compared, 168
Ampelopsis quinquefolia as liane, 311
Amphibolips coelebs, the oak spindle gall, 297
  confluentus, gall insect on scarlet oak, 296
Amphicarpon Purshii as a cleistogene, 311
Analyses of white-cedar trees, 119
Andricus seminator, gall on white oak, 296, 297
Andromeda ligustrina, figure of microscopic leaf structure, 280; microscopic leaf structure, 280
  mariana, figures of, 237; root system of, 237; microscopic leaf structure, 280
Andropogon scoparius, roots and stems of, 224, 225
Animals of pine-barrens, 13
Annual herbs in marshes, 191; of pine-barren formation, 187

Anthropochores, 106
Anthropophile element, 106
Anthropophytes, 106
Apios tuberosa with twining habit, 311
Apocynum medium as a species, 300
  Milleri as a species, 300
  urceolifer as a species, 300
Apophytes, 106
Apparatus used in soil research, 36
Application of fertilizer to soils of plains, 165
Aquatic plants in cedar swamps, 190
Arbutin as a reserve food, 276
Archaeophytes, 107
Architectural form of pine trees, 53
Arctostaphylos heath, 168
Arctostaphylos uva-ursi, 65, 68, 311; figure of, 241; figure of microscopic leaf structure, 276; microscopic leaf structure, 275; photograph of, 157; root system of, 239
Area of pine-barrens, 178
Arenaria caroliniana, 85; figure of, 310; figure of microscopic leaf structure, 270; microscopic leaf structure, 270; heath-like form, 310
  peploides var. robusta, 5
Artificial ponds, 144
Asclepias amplexicaulis, figure of, 234; root system of, 234
  obtusifolia, root system of, 234
Aspects of vegetation, 180
  seasonal, of white-cedar swamps, 127
Association coefficient, 175
Associations in cedar swamp stream, 125
Aster nemoralis, 5
  radula, 5
Atlantic Highlands, 3
Autumn aspects, 80; of cedar swamps, 128
  leaf colors, 195, 209
Average rate of growth of white-cedar, 119
Avivivect seeds, 93
Azalea viscosa, 112, 134, 135, 137; figure of microscopic leaf structure, 276; microscopic leaf structure, 276

317

BACCHARIS halimifolia, 95
Bailey, L. H., book by, 252
Balsams, 28
Baptisia tinctoria, figure of, 233; root system of, 233
Bark for drugs, 28
    of pitch-pine, 98
Barren grounds of Canada, 9
Barren Hill, 9
Barrens, application of term, 9
Bartonia virginica, 308
Basket form of pitch-pines, 152
Bassett, William F., cited, 172
Bastard toad-flax, root system of, 230
Bayberry, microscopic leaf structure, 268
    root system of, 228
Bay-like leaves, 255
Beacon Hill area, 3; formation, 1
Bearberry, 65, 68; microscopic leaf structure, 275; photograph of, 157; root system of, 239
Bear-oak, microscopic leaf structure, 268; root system of, 230
Beauchamp, William M., cited, 10
Bermuda dune plants, 285
Beutenmüller, William, cited, 295, 296
Beyerinck, M. W., quoted as to galls, 295
Bidens trichosperma, sun and shade forms, 306
Biologic spectrum, 186
    types, 186; of marsh formations, 190; of pine-barren plants, 185
Bird-distributed seeds, 93
Bird-foot violet, mutations, 304
Birkenheide, 86, 89
Black-jack oak, 89, 101; microscopic leaf structure, 269; root system of, 230
Blackman, Leah, quoted, 14
Black oak, 102
Bladder leaves, 257
Bladderwort, figures of, 309
Blakeslee, Albert F., cited, 53
Blephariglottis blephariglottis × B. ciliaris, 313
    blephariglottis × B. cristata, 313
    Canbyi, a hybrid, 313
Block Island, 5
Blueberry, 85, 133, 134; culture, 238; root system of, 238
Bluff forest at Somers Point, 91
Bog asphodel, roots and stems of, 225
Bog formation, 48
Bog plants, additional list of, 126
Bogs, 26
    of northern New Jersey, 115
Boulder plants, 219
Boxing pine trees for turpentine, 22, 23
Bracken, 133; roots and stems of, 220
Branch swamps, 136
Briggs, L. J., cited, 45, 47

Broom-crowberry, 151, 304; microscopic leaf structure, 273
Brown, Harry P., cited, 209
    Stewardson, cited, 156
Bulbous shoots, 218
Bush fruits, 173

CAATINGAS, 145
Caleche of Mexico, 166
Calico-bush, mutations, 304
Calopogon pulchellus, roots and stems of, 227
Calvert, P. P., cited, 296
Campos of Brazil, 145
Cannon, W. A., cited, 217, 220; his classification of roots, 247
Canopy of pine forest, 60
Cape Cod, 5; plants of, 5
Cape May deposits, 5
Capillary rise of water, 42
Carex hormathodes, 5
    silicea, 5
    sterilis, 5
    trisperma var. Billingsii, 5
Carolinian types, 5
Cassandra calyculata, 143; in savanna, 149; root system of, 239
Cassia chamaecrista, root system of, 233
Castalia odorata, 144; figure of microscopic leaf structure, 271; figure of microscopic petiole structure, 271; flowering and fruiting of, 210; microscopic leaf structure, 270; mutations, 304; photograph of, 123; root system of, 232
Caudex multiceps, 218
Caudices, 217
Cause of dwarf trees of Coremal, 159
    of galls, 295
Cecidology, 295
Cecidomyia pisum, 297
Cedar (red) leaf forms, 253
Cedar swamp formation, 48, 114, 189; blending with deciduous swamp, 134
Cedar swamp at Lakehurst, 114; at Mays Landing, illustration of, 116; between Whitings and Bamber, illustration of, 121; blending with pine forest, 112; synecology of, 120; yield of, 20
Census of plants in lots at Belmar, 83
    of vegetation, 175
Cereal crops, 173
Chamaecyparis thyoides, 112; base of, 118; figure of microscopic section of leaf, 264; microscopic leaf structure, 264; photograph of, 126
Chamaedaphne calyculata, figure of microscopic leaf structure, 277; microscopic leaf structure, 277; root system of, 239
Chamaephytes, 185
Chamberlain, T. C., cited, 1

Charcoal, 20
Chemic analysis of soils, 34
Chimaphila maculata, figure of, 235; root system of, 236
Chrysopsis falcata, figure of microscopic leaf structure, 282; microscopic leaf structure, 282
  mariana, figure of, 243; figure of microscopic leaf structure, 283; microscopic leaf structure, 283; root system of, 242
Church, A. H., cited, 312
Cinnamon-fern, 80, 134, 135
Circinate ptyxis, 253
Cladonia rangiferina, 152
Clammy azalea, 134
Clasping leaves, 257
Classification of alien plants, 106; of leaf structures, 260; by root systems, 246; of root systems by Cannon, 220; of shoots by Hess, 217, 218, 219
Cleistogamy, 311; in violet, figures of, 312
Clements, F. E., cited, 175, 259
Clethra alnifolia, 112, 137; figure of, 235; figure of microscopic leaf structure, 277; microscopic leaf structure, 277; root system of, 235
Climbing plants in marshes, 191
Clitoria mariana, with twining stems, 311
Coastal oak forest, 104
  pine-barrens, 82
Cockayne, L., cited, 298, 308, 312
Coefficient of association, 175; of genera, 176, 181
Cole, Rex Vicat, cited, 53
Collection of turpentine, 22
Colors of flowers in white-cedar swamps, 190; of marsh plants, 92; tabulated, 188
Colors of leaves in autumn, 195, 209
Colors of pine-barren vegetation, 77
Columbia formation, 1
Comandra umbellata, root system of, 230
Commencement of cambial activity in pitch-pine, 209, 210
Comparison of American and European plant formations, 168
  of quadrats, 180
Competition of root systems, 251
Compound palmate leaves, 256; pinnate leaves, 256; trifoliate leaves, 256
Comptonia asplenifolia, 67, 85; figure of, 228; figures of leaf sections, 268; microscopic leaf structure, 267; root system of, 228
Conard, Henry S., cited, 210
Conclusion, 313
Conditions in cedar swamps, 115
Cone production, law of, 292; on different trees, 287; of pitch-pine, 286
Cones of pitch-pine, figure of, 286

Connold, Edward T., cited, 295
Convergent epharmony, 308
Convolvulus with twining stem, 311
Cook, Mel T., cited, 295
Corema Conradii, 5, 6, 143, 151, 155, 169; botanic characters of, 156; figure of microscopic leaf structure, 274; forms of, 304; growth forms, 310; microscopic leaf structure, 273; photograph of, 152, 157, 158; on Nantucket, 155; time of flowering, 156
Coremal, 151, 158, 169; cause of dwarf trees, 159; measurement of species, 159; of New Jersey, 168; on Nantucket, 165; size of plants, 159; soils treated experimentally, 162; synecology of, 154; synopsis concerning, 168; theories concerning, 159
Corn culture in soil from plains, 162, 163, 165; in top-soil, 169
Coville, F. V., cited, 238
Cowbane as an elementary species, 299
Cow-lily, root system of, 231
Cow-wheat, root system of, 242
Cranberry, 25
  bog, 25; formation, 48
  culture, 25
  sorting house, 26
  storage house, 26
Cranmer, George, cited, 160
Crested shoots, 218
Crotonopsis linearis, 308
Cryptophytes, 185
Crystals, rhomboidal, 262
Cultivated plants of the pine-barrens, 172
  trees of the pine-barrens, 172
  vines, 173
Culture of corn in plains soils, photograph of, 162, 163, 165; in top-soil, 169
  of cranberry, 25
  of peas in soil from plains, 162; photograph of, 163; in top-soil, 169
  of sunflowers in soils from plains, 164, 167
  of wheat in soils from plains, 164
Cup-and-gutter method of turpentine gathering, 22
Curves showing percolation of water, 39
Cuscuta with twining stems, 311
Cuticle thick in leaves, 260
  thin in leaves, 260
Cypripedium acaule, roots and stems of, 226, 227

Dams, 144
Dandelion leaf form, 255
Darwin's discussion of variation, 300
Date of pollination of pitch-pine, 209
Davis, Charles A., cited, 117

Deciduous-leaved species, 188
  leaves, 255, 262
  swamp at Belmar, photograph of, 135; swamp and cedar swamp blending, 134; swamp formation, 48, 133; photograph of, 132; transition to, 132; of transition region, 136
  trees in cedar swamps, 189; in marshes, 192
  vegetation of Delaware Valley, 9
Decodon verticillatus with aerenchyma, 311
Defebaugh, J. E., cited, 19
Delgar, Colonel E., of Denmark, work of, 161
Demarcation between plains and pine forest, 158
Dendrium buxifolium, 71, 72; figure of, 236; figure of microscopic leaf structure, 280; growth forms, 310; microscopic leaf structure, 280; root system of, 236
Denmark, reforestation of, 161
Density, floral, 176
Detailed root and stem study, 219 et seq.
Development of pitch-pine seeds, 209
De Vries, Hugo, cited, 298, 299; views, 300
Diageic plants, 185
Diagram of savanna, 147; showing succession, 49
Diameter of pitch-pine trees, 288, 290, 291
Dioscorea villosa as liane, 311
Diphotic leaves, 259
Diphotophyll, 259, 262
Diplophyll, 259, 263
Diseases of white cedar, 129
Dispersal of coastal plain plants, 6
Dissected submerged leaves, 256
Distribution of roots, 245
  of transition plants, 100
Districts, floral, 15
Dodder leaf form, 257
Dormancy of plants, 195
Drosera rotundifolia, figure of leaf tentacle, 272; microscopic leaf structure, 272
Drought, relation of plants to, 249
Drude, Dr. O., cited, 190
Drug plants, 27, 28
Dulichium spathaceum, roots and stems of, 224
Dumb-watch, 17
Dwarf chestnut-oak, root system of, 230
  pine trees of plains, 167
  trees of Coremal, cause of, 159; of Hempstead Plains, 171

Early glacial time, 6
  saw mills, 18
  settlements in West Jersey, 13
East Plains, 151
Edinger's drawing apparatus, 263
Eichenheide, 86, 168

Elatine americana, 5
Elementary species, 298, 299
Elevations in plains, 151
Engler, Prof. Adolf, 63
Enumeration of pine-barren plant families, 182, 183
Epharmony, 305; convergent, 308
Ephemerophytes, 107
Epidermis, papillate, 261; thick-walled, in leaves, 260
Epigaea repens, 311; figure of, 239; figures of microscopic leaf structure, 277; microscopic leaf structure, 277; root systems of, 237
Epigeic plants, 185
Epökophytes, 107
Ergasiolipophytes, 106
Ergasiophygophytes, 107
Ergasiophytes, 106
Eriocaulon septangulare, 5; floating and submerged forms, 307
Eriophorum virginicum in savanna, 148
Eupatorium resinosum × E. perfoliatum, 313
  verbenaefolium, microscopic leaf structure, 283; root system of, 242, 244
Euphorbia ipecacuanhae, figure of, 234, 299; figure of microscopic leaf structure, 273; as an elementary species, 299; microscopic leaf structure, 273; root system of, 234
European and American plant formations compared, 168
Evergreen leathery leaves, 256
  leaves, 262
  species of pine-barrens, 188
  trees in cedar swamps, 189; in marshes, 191
Evolutionary viewpoint, 298
Exotic place names, 11
Experimental inclosures, 314
  treatment of Coremal soils, 162
Experiments with soils, 35

Facies of pine-barren formation, 56; of pine-barren vegetation, 57
Factors influencing regeneration of pitch-pine, 291, 292
Fall of pine-barren streams, 8
False asphodel, roots and stems of, 225
Families of pine-barren plants, 182, 183
Fascicled leaves, 253
Fernald, M. L., cited, 5
Ferns in cedar swamps, 189; in marsh formations, 190; of pine-barren formation, 187
Fertility of soils, theories, 32
Fertilization of egg-cell in pitch-pine, 209
  of pine, 291

Fertilizer, application to plains soils, 165
Field controlled experiments, 314
Fire-denuded pitch-pines, 51, 52
Fire, influence in formation of plains, 169
    strips, 100
Fires on plains, 159
Flat pine-barren facies, 65
Floating leaves, 256
Floral density, 176; districts, 15; procession, 193
Flower colors, numbers of species with, 189; of lupine, 303; of marsh plants, 192; summarized, 192; tabulated, 188; of white-cedar swamp plants, 190
    crops, 173
Flowering and fruiting of Castalia odorata, 210; of pine-barren plants, 210; of amentiferous trees, 211; of Pyxidanthera barbulata, 210 moss, 152; root system of, 240 periods, 193; of pine-barren plants, 196 et seq.
Flowers, 29; numbers of different months, 212, 213; procession of, 213, 214, 215
Fluctuating variations, 300
Fluctuations of plants, 300
Fly-trap leaves, 257
Fodder crops, 173
Forest canopy, 60
Formation of bogs, 117; of plains, 151; of pond plants, 143; of savannas, 145
Formations of America and Europe compared, 168; phytogeographic, 48
Forms of leaves, 252
Frequency of individuals, 176
Fritsch, F. E., cited, 216
Frost, first killing, 194; last killing, 194
Fruit of Pyxidanthera barbulata, 210
Fruit trees of pine-barrens, 172
Fruiting period of plants, 193, 196 et seq.
Fruits, cultivated bush, 173; small, 173
Fungi on white-cedar, 131
Fungous diseases of white-cedar, 129, 130, 131
Future of pine-barren region, 315

GALACTIA regularis, 311; figure of, 233; root system of, 233
Gall-gnats, 297
Gall-midges, 297
Galls, cause of, 295; figures of, 296; of insects, 295
Game preserve, 316
Garden crops, 174
Gates, Frank C., cited, 239
Gathering of flowers, 29; of greens, 29
Gaultheria procumbens, 311; figure of, 237; figure of microscopic leaf structure, 278; microscopic leaf structure, 278; root system of, 236

Gaylussacia baccata (=resinosa), microscopic leaf structure, 279
    dumosa, figure of microscopic leaf structure, 278; microscopic structure of leaf, 278
    dumosa var. Bigelowiana, 5
    frondosa, 69, 112, 135; figures of microscopic leaf structures, 278; microscopic leaf structure, 278
    resinosa, 69, 73, 101; figure of, 240; figure of microscopic leaf structure, 279; microscopic leaf structure, 279; root system of, 238
General observations on pine-barren vegetation, 175
    remarks on leaf forms, 258
Generic coefficient, 176, 181
Geographic location of pine-barrens, 8; of plains, 169
    place names, 9
Geography of pine-barren region, 8
Geologic map of pine-barren region, 3
Geophytes, 185
German plant formations, 85
Germination of unripe acorns, 293
Gifford, John, cited, 11
Glabrous leaves, 261
Glacial period, 1
Gnarled trees of plains, 156
Goat's-rue, root system of, 232
Goebel, K., cited, 252
Golden aster, root system of, 242
Golden-club, 147; leaf structure, 265
Graebner, Dr. Paul, cited, 48, 85, 166, 167
Grand period of growth, 194
Graph of leaf structure, 285; representing water-retentiveness, 44; showing rate of percolation, 39
Grass form of leaves, 253
Grasses in cedar swamps, 189; in marsh formations, 190
Great Egg Harbor River, 7
Greens, 29
Gridiron method, 176
Griggs, Robert F., cited, 289
Groom, Percy, cited, 249
Growing season, 194
Growth in pitch-pine stem, 210
    of white cedars, average rate of, 119
Guard cells deeply depressed, 261; projecting, 261; present at surface, 261; slightly depressed, 261
Guppy, H. B., cited, 292, 293

HAAS, PAUL, and Hill, T. G., cited, 276
Habenaria blephariglottis, 5; roots and stems of, 227
Haberlandt, Prof. G., cited, 53
Hall, A. D., cited, 32

Hall, William L., cited, 19
Hapaxanthic species, 217
Harper, Roland M., cited, 108, 144, 170, 182; photograph by, 140
Harshberger, J. W., cited, 4, 9, 15, 17, 25, 48, 129, 161, 181, 259
Hassock-like growths, 310
Hawley, L. F., cited, 23
Heat of soils, 250
Heather leaf forms, 253
Heathland, 166; of Nantucket, 168, 170; photograph of, in Nantucket, 170; of New Jersey, 170; physiognomy of, 167
Heath-like plants, 310
Helophytes, 185
Hemicryptophytes, 185
Hempstead Plains, 108, 170
Henkel, Alice, cited, 29
Herbs, annual, 187; as drugs, 28; with articulate stem, 308; perennial, 187; with subulate leaves, 308
Herty, Charles H., cited, 22
Hess, Eugen, cited, 217
Hieracium venosum, figure of, 242; root system of, 242
High pine-barren facies, 60
Hillman, Sarah C., cited, 25
Holcaspis centricola on Quercus stellata, 297
Holly, 90, 91
Holm, Theo., cited, 226
Hopkins, Albert A., cited, 24
Horticulture, future, 315
Huckleberry picking, 27; root system of, 238
Hudsonia ericoides, 5, 75, 112, 310; figure of, 231, 310; figures of microscopic leaf structure, 275; microscopic leaf structure, 275; root system of, 231
    tomentosa, figure of, 231; root system of, 230
Huntington, Annie Oakes, cited, 53
Hybridization, 313
Hydrophytes, 185
Hypericum adpressum with aerenchyma, 311
    densiflorum, 140; figure of, 232; root system of, 231
Hypodermis in leaves, 260

Idioblasts, 262; in leaves of water-lily, 271
Ilex glabra, 69, 133; figures of microscopic leaf structure, 274; microscopic leaf structure, 274; root system of, 234
    opaca, 90, 91, 94, 95, 96
Impervious subsoil of the plains, 169
Importance of plant families, 183, 184
Inaccessibility of pine-barrens, 12
Indian names of places, 10
    trails, 13

Individual frequency, 176
Industries of pine-barrens, 17
Influence of geography on peoples and industries, 12; of water, 307; of wind, 307
Inkberry, microscopic leaf structure, 274; root system of, 234
Insect-catching plants in cedar swamps, 189
Insect galls, 295
Insectivorous leaf, 263
Inter-glacial time, 6
Introduced trees, 106
    weeds, 105
Investigation of pine-barren soils, 31
Ipomoea with twining stems, 311
Iris leaf form, 254
Iron-stained savanna pools, 149
Isophotic leaves, 259
Italians in South Jersey, 315

Jaccard, Paul, cited, 175
Jepson, W. L., cited, 53
Juncus militaris, submerged form, 307
    pelocarpus, 5
Juniperus, juvenile form of, 312
    virginiana, 90, 91, 94, 95, 96; wind-swept, 87
Juvenile forms of Juniperus, figure of, 313; of New Zealand plants, 312
    reverted leaves of pine, 313

Kalmia angustifolia, 68, 112, 149; figure of microscopic leaf structure, 279; microscopic leaf structure, 279; root system of, 237, 238
    latifolia, 70, 77, 113, 134; figure of, 239; figure of microscopic leaf structure, 279, 280; figures of sun and shade leaves, 306; microscopic leaf structure, 279; mutations of, 304; root system of, 237
Kearney, T. H., cited, 267
Kiefern-Heide, 85, 86, 168
Klebahn, H., cited, 167
Kraus, G., cited, 248
Krout, Dr. A. F. K., cited, 115
Küster, Dr. Ernst, cited, 295

Lafayette epoch, 1
Lagoa Santa, 258
Lalang vegetation, 145
Lanceolate leaves, 254
Latex tubes, 262
Laurel, 77, 133, 134; microscopic leaf structure, 279; root system of, 237
Law of cone and seed production in pitch-pine, 292
Layers in pine forest, 61, 62, 63
    of soil, 245, 246

Leaf colors in autumn, 195, 209
  crops, 173
  forms, 252; in Denmark, 258; in
    Lagoa Santa, South America, 258
  layer, 64
  leathery, 262
  mould layer, 245, 246
  structures, graph of, 285; microscopic,
    259; synopsis of, 283; table of, 284
Leather-leaf, 149; root system of, 239
Leathery leaves, 255, 262
Leaves as drugs, 28
  deciduous, 262; glabrous, 261; hairy
    on both sides, 261; hairy on
    under side, 261
Lee, Francis Bazley, quoted, 14
Leiophyllum buxifolium, figure of, 236;
  microscopic leaf structure, 280; root sys-
  tem of, 236
Lespedeza frutescens, figure of microscopic
  leaf structure, 273; microscopic leaf
  structure, 272
Leucothoë racemosa in savanna, 149
Level pine-barrens, 61
Lianes, 311
Liatris graminifolia, figure of, 244; micro-
  scopic leaf structure, 283; root system of,
  244
Life-giving air of pines, 316
Light and regeneration of pitch-pine, 291
Light forms of Sarracenia purpurea, 305
Limodorum tuberosum, roots and stems of,
  227
Linear leaves, 253; not grass-like, 254
Linnaean species, 298
Liquidambar styraciflua, 75
List of 548 pine-barren plants, 196–208
Llanos of Venezuela, 145
Loblolly pine-barrens, 9
Local plant names, 16
Log school house at Speedwell, 19
Long, Bayard, cited, 193, 210, 313
Long Island, 5; pine-barrens, 107, 109;
  plants, 5
Long-leaf pine-barrens, 9
Lophiola aurea in savanna, 149
Low or wet pine-barrens, 75
Lower Plains, 151; photograph of, 152, 160,
  161
Lubbock, Sir John, quoted, 252
Ludvigia sphaerocarpa with aerenchyma,
  311
Lumbering, 17
Lupinus perennis, 73, 82; color of flowers,
  303; figure of, 233; root system of, 233
Lycopodium leaf forms, 253

MacDougal, D. T., cited, 313
Magnolia glauca, 134, 135; illustration of,
  120; figure of microscopic leaf structure,
  272; microscopic leaf structure, 271

Maltby, Robert D., cited, 172
Manasquan River, 6
Map showing geologic history, 3
Marsh formation, 48
Martha's Vineyard, 5
Massart, Jean, cited, 259
Maturity of pitch-pine cones, 209
Maxwell, Hu., cited, 19
Mayr, Gustav L., cited, 295
McCourt, W. E., cited, 117
Meadow-beauty, colors of flowers, 303
Meadow Brook, L. I., valley of, 171
Measurement of rhizome of Pteris, 221, 222
  of species of the Coremal, 159
Mechanic analysis of soils, 33
Megaphanerophytes, 185
Melampyrum americanum (= lineare), root
    system of, 242
  lineare, figure of, 242; figure
    of microscopic leaf struc-
    ture, 282; microscopic leaf
    structure, 282; root sys-
    tem of, 242
Mesophanerophytes, 185
Method of pine-barren investigation, 57
  of squares, 176
Microphanerophytes, 185
Microscopic leaf structure, 259; structure
  of leaves, detailed, 263
Microsublimation method, 276
Milk-pea, root system of, 233
Minerals in soils, 34
Miocene, 1, 2
Mixed forest at Browns Mills Junction, 99;
  at Somers Point, 93
  pine-oak formation, 48; succession,
    103
Moccasin-flower, roots and stems of, 226,
  227
Moisture equivalent, 47
Moisture holding capacities of soils, 45
Molisch, Hans, cited, 276
Monotropa leaf form, 257
Months of growth, 194, 195
  of preparation, 195
Mucilage, 262
Mullica River, 7
Multicipital root, 218
Mutation theory, 299
Mutations, 303
Myrica carolinensis, 88, 94, 95, 96, 149;
  figure of microscopic leaf structure, 268;
  microscopic leaf structure, 268; root sys-
  tem of, 228
Myriophyllum humile, 308; var. natans,
    308; var. capillacea, 308
  tenellum, 5

Nabalus leaf form, 255
Names, local plant, 16; of early English
    settlers, 14; of places, 10

Nanophanerophytes, 185
Nantucket, 5; Coremal on, 165; heathland, 168; heathland, photograph of, 170; oceanic climate of, 166
Narthecium americanum, roots and stems of, 225
Needle leaves, 253
Neocene, 1
Neophytes, 107
Newfoundland, 5; flora, 5
New Jersey dune plants, 285; plant formations, 85; salt marsh plants, 285
New Zealand plants with persistent juvenile forms, 312; vegetation, 308
Nichols, George E., cited, 108
Nitrogen in soils, 35
Notes on distribution of transition plants, 100
Nova Scotia plants, 5
Number of pine-barren plants, excluding pines, 180; of pine-barren plants including pines, 180
Numbers of flowers in different months, 212, 213
Numerical sequence of flower development, 212
Nymphaea (Nuphar) advena, root system of, 231
    microphylla, 303
    rubrodisca as a species, 303
    variegata, 231, 303
Nyssa sylvatica, 75, 112; wind-swept, 87

OAK-BOTTOM formation, 104
    coppice formation, 48; succession, 102
    heath, 168
Oceanic climate of Nantucket, 166, 169, 170; of New Jersey, 170
Oekiophytes, 107
Oleoresins, 28
Oliver, F. W., cited, 176
Opposite leaves, 257, 258
Orchids in cedar swamps, 190
Origin of designation " pine-barrens," 8
Original size of pitch-pine, 17
Orontium aquaticum, 147; figure of microscopic leaf structure, 265; figure of microscopic section of petiole, 266; microscopic leaf structure, 265
Ortstein, 166, 167
Ortstein-Kiefer, 167
Ortsteintöpfe, 167
Osmunda cinnamomea, 80, 135
Ostenfeld, C. H., cited, 216, 217
Oxypolis rigidior, figure of, 299
    rigidior longifolia, an elementary species, 299

PALISADE leaf, 259
    of two or more rows, 260; of single row of cells, 260

Palmately compound leaves, 256
Panicum Addisonii as a variety, 301
    Commonsianum as a species, 301
    depauperatum, roots and stems of, 224, 225
    implicatum, 5
    virgatum, 90, 94, 95, 96
Parasites in cedar swamps, 190
Parasitic plants, 187; in marshes, 191
Parmelee, C. W., cited, 117
Patanas of Ceylon, 145
Paulsen, Ove, cited, 186
Peas, culture of, in plains soil, 162, 163; in top-soil, 169
Peat, 26; formation of, 117
Peltandra virginica, figures of leaves, 301
Peneplain, Pre-Pensauken, 1
Pennypacker, J. L., quoted, 28
Pensauken Sound, 170; submergence, 3
Percentages of leaf forms, 258
Percolation of water, curves showing, 39; rate of, 38
Perennial herbs in cedar swamps, 190; in marshes, 191; of pine-barrens, 187; with sedge- or grass-like leaves in marshes, 191
Periods of flowering and fruiting, 196 et seq.
Periwig shoots of Pyxidanthera, 241
Periwigs, 218
Persistent juvenile forms, 312
Pervious soils, 250
Petersen, H. E., work of, 216
Phanerophytes, 185
Phosphorus in soils, 35
Physiognomy of heathland, 167; of savannas, 146
Physiography, 1
Phytogeographic formations, 48
    Survey of North America quoted, 2
Phytophenology of pine-barren vegetation, 193
Pickerel-weed leaf forms, 302
Picking huckleberries, 27
Pinchot, Gifford, cited, 126, 153, 160
Pine-barren formation, 48, 49
    vegetation as a closed formation, 314; old, 314
Pine-barrens, area of, 178; at northern limit, 96; kinds of, 314; types of, 9
Pine forest at Mays Landing, 82; at edge of salt marsh, 94; in Lebanon Reserve, 81; interior at Somers Point, 93; on old fields, 89
Pine grove in Shark River Bay, 103
Pine knots, 20
Pines, xerophytism of, 249
Pinnate compound leaves, 25
Pinus caribaea, 9
    echinata, 9, 50, 67
    palustris, 9

Pinus rigida, 9, 50, 69, 75, 91, 95, 96, 112, 133, 135, 137; and its fluctuations, 300; figures of branches, 300; figure of microscopic leaf section, 265; low trees in savanna, 152; microscopic leaf structure, 264; roots of, 223, 224
taeda, 9
Pitch, 23
Pitchered leaves, 257
Pitcher-plant, photograph of, 122; in savanna, 148
Pitch-pine, 50, 75, 91, 135; and its fluctuations, 300; bark of, 98; cone production of, 286; original size of, 17; roots of, 223, 224; seed, figure of, 313; seed, production of, 286; trunk of, 98
Place names, 9; of exotic origin, 11
Plains, 151; age of trees of, 153; formation, 48, 151; trees of, 152
Plant names, local, 16
Plants, additional, of white-cedar swamp, 126; as weeds, 105; found in quadrats, 180, 181; systematic list of 548 pine-barren, 196–208
Plasticity of species, 305
Pleistocene, 1
Plow-sole, 166
Pogonia ophioglossoides, mutations of, 304; roots and stems of, 226
Poison-oak, 137
Pollen-tube formation in pitch-pine, 209
Pollination of pine, 209, 291
Polygala cruciata, mutations, 304
Polygonella articulata, 308
Pond formation, 48, 143
plants, additional, 144
Ponds, 143; artificial, 144
Pontederia cordata, figure of, 302; figures of leaf variations, 302
Post-glacial time, 6
Post-Miocene, 1, 6
Post-oak, microscopic leaf structure, 269
Post-Pensauken uplift, 2, 5
Potamogeton confervoides, figure of, 307
epihydrus, 307
figures of, 307
Oakesianus, 5; figure of, 307
Potash in soils, 35
Precipitation, 248
Preparation months, 195
Pre-Pensauken peneplain, 1, 3
Procession of flowers, 193; of flower development, 213, 214, 215
Propagation of shoots, 216
Prostrate shoots, 310
Protozoans of soils, 161
Proximity of centers of population, 315
Pteris aquilina, 69, 112, 133; microscopic drawing of leaf, 264; microscopic leaf structure, 263; root and stems of, 200

Pyrus arbutifolia var. atropurpurea, 5
Pyxidanthera barbulata, 74, 76, 152, 310; figure of, 241; in flower, 75; in fruit, 75; figures of sun and shade plants, 306; flowering and fruiting of, 210; figure of microscopic leaf structure, 282; microscopic leaf structure, 282; root system of, 240; sun and shade plants, 305
Pyxie, root system of, 240

Quadrat method, 175
1, north side of Como Lake, 177
2, one mile east of Lakehurst, 178
3, one-half mile east of Sumner, 179
Quadrats, 60; comparison of, 180; method of, 176
Quercus alba, 91, 95, 96, 133, 135; leaf forms, 303
falcata, figures of leaf forms, 303
heterophylla, a hybrid, 313
ilicifolia, 65, 68, 69, 109, 112; figure of, 229; leaf forms, 303; microscopic leaf studies, 268; root system of, 230
marylandica, 63, 67, 69, 84, 89, 101, 112; figures of, 228; figures of leaves, 304; figure of microscopic leaf structure, 269; leaf forms, 303; microscopic leaf structure, 269; root system of, 230
phellos × Q. ilicifolia, 313
phellos × Q. rubra, 313
prinoides, 66, 67, 69, 109; figure of microscopic leaf structure, 262; microscopic leaf structure, 269; root system of, 230
prinus, 135
stellata, 63, 67, 69, 90, 96; figures of leaves, 304; figure of microscopic leaf structure, 270; leaf forms, 303; microscopic leaf structure, 269; photograph of, 156
velutina, 64, 67, 91, 102; leaf forms, 303; microscopic leaf structure, 270

Radial shoots, 219
Radical leaves, 254
Rainfall, 248; and pine-barren plants, 249
Rate of growth of white-cedars, 119
of percolation of water, 39
Rattlesnake poison, cure for, 28
Rattlesnake-weed, root system of, 242
Raunkiaer, Ch., cited, 185
Red cedar, 90, 91
gravel layer, 245, 246
maple, 133, 134
Redfield, John H., cited, 6

Reed marsh formation, 138
Reed, William G., cited, 194
Reforestation of Denmark, 161
Regeneration of fire-denuded pitch-pines, 51, 52
Reindeer lichen, 152
Relative importance of plant families, 183, 184
Renvall, August, cited, 286
Resin canals, 262
  hairs, 262
Response to factors, 305; to light, 305; to soil, 305
Retention of water, 43
Rhexia aristosa with aerenchyma, 311
  virginica, flower colors of, 303; with aerenchyma, 311
Rhizomes, 219; as drugs, 28
Rhomboidal crystals, 262
Rhus copallina, figure of, 235; root system of, 235
  radicans as liane, 311
  vernix, 137
Rhynchospora fusca, 5
Rikli, Dr. M., cited, 106
Rings in pitch-pine trees, 288, 290, 291
River bank formation, 48
  plants, 144
Rivers of pine-barren region, 6, 7
Roads through pine forest, 14, 74; through pine forest in Lebanon Reserve, 81
Rocky Hill, 3
Roll leaf, 263
Root classification by W. A. Cannon, 247
  competition, 251
  crops, 173
  distribution, remarks on, 245
  systems, classification of plants by, 246
  systems of desert plants, 220
Roots and stems, detailed study of, 219
  as drugs, 27; and rainfall, 249
Rootstock, 219
Rosin, 23, 24
Rubble plants, 219
Run-off in pine forest, 66
Russian Jews in South Jersey, 315

SABLE Island plants, 5
Sachs, Julius, cited, 293
Sagittaria graminea, 307
  longirostra, figures of leaves, 301; figure of microscopic leaf structure, 266; microscopic leaf structure, 266; variation of, 301
Sagittate leaves, 255
Salisbury, R. D., cited, 1
Salt marsh at Somers Point, 96
  marsh to pine forest near Ocean Gate, 95; at Somers Point, 94

Salubrious climate of pine-barren region, 315
Sand dunes of Nantucket invaded by pitch-pines, photograph, 171
Sand layer, 245, 246
Sand-myrtle, root system of, 236
Sarothra gentianoides, 308
Sarracenia purpurea, 63; in savanna, 148; photograph of, 122
Sassafras variifolium, 69, 73, 135; figure of, 230; figure of microscopic leaf structure, 272; microscopic leaf structure, 272; root system of, 230
Saunders, C. F., cited, 143
Savanna along Wading River, photographs of, 145, 146
  diagram of, 147; formation, 48, 145; physiognomy of, 146; terraces, 147
Saw-brier, roots and stems of, 226
Saw-mill on Darby Creek, 18, 19
Saw-mills, early, 18
Schematic section of cedar swamp to pine-barrens, 110
Schizaea pusilla, 5; illustration of, 118; microscopic leaf structure, 264; microscopic section of leaf, 264
Scirpus subterminalis, 5; floating submerged, 307
Sclerophyll, 263
Scrape of turpentine, 22
Scrub-oak formation, 48
Sea dunes to pine thicket, 90
  islands of coast, 93
Season of growth, 194
Seasonal aspects of white-cedar swamps, 127
  study of plants, 193
Sedentary species, 217
Sedges in cedar swamps, 189; of marsh formations, 191
Sedgy slough in plains, 161
Seed bed, suitable, 291
  development in pitch-pine, 209
  production, law of, 292; of pitch-pine, 286
Sericocarpus asteroides, figure of, 243; root system of, 242
  conyzoides, root system of, 242
Serpentine barrens, 9
Settlements in West Jersey, 13
Settlers, names of, 14
Shantz, H. L., cited, 45, 47
Shear, C. L., cited, 26
Sheep-laurel, microscopic leaf structure, 279; root system of, 237
Sherff, Earl E., cited, 217
Shoemaker's wax, 24
Shoots, classification of, by Hess, 217, 218, 219
  propagation of, 216; structure of, 216

Shreve, Forrest, cited, 2
Shrubs in cedar swamps, 189
in marshes, 191; of pine-barren formation, 187; with baccate fruits in marshes, 191; with capsular fruits in marshes, 191; with drupaceous fruits in marshes, 191
Silene acaulis, 310
Sinnott, Edmund W., cited, 264
Size of plants of Coremal, 159
Skunk cabbage, 135
Slash pine-barrens, 9
Slough in plains, 161
Sluggish run-off in pine-forest, 66
Small fruits, 173
Smilax as liane, 311
glauca, figure of microscopic leaf structure, 266; microscopic structure of leaf, 266; roots and stems of, 226
laurifolia, figure of microscopic leaf structure, 267; microscopic leaf structure, 267
rotundifolia, 135
tamnifolia, figure of, 303
Snow scene in pine forest, 81
Soil fertility, theories concerning, 32
layers, 245, 246; numerals, 34; protozoans, 161; stations, 37, 38
Soils, chemic analysis of, 34; experiments with, 35; mechanic analysis of, 33; of Coremal treated experimentally, 162; pervious character of, 250; research on, 31
Solereder, Hans, cited, 272
Solidago stricta, figure of microscopic leaf structure, 283; microscopic leaf structure, 283
uniligulata, 5
Sorting house for cranberries, 26
Sour gum, 75
Sparganium americanum, roots and stems of, 224
Spatterdock, root system of, 231
Spatulate leaves, 254
Species, elementary, 299; Linnaean, 298; plasticity of, 305; taxonomic, 298, 299; ranging from Florida to Massachusetts, 5
Spectrum, biologic, 186
Spergularia rubra, 5
Sphaerocrystals, 262
Sphagnum, 115; uses of, 26
Spongophyll, 259, 263
Spot-bound species, 217
Spotted wintergreen, root system of, 236
Spreading shoots, 218
Spring aspect, 79
Squares, method of, 176
Staminate flower development in pitch-pine, 209
Stand of pitch-pine in Lebanon Reserve, 62
Starr, Anna M., cited, 259

Stations for soils, 37, 38
Staurophyll, 259, 263
Stem analyses of white-cedar, 119
Stereoscopic photograph of pine forest, 61
Still for turpentine, 24
Stone, Witmer, cited, 5, 15, 143, 156, 193, 313
Storage house for cranberries, 26
Stories of low pine-barrens, 75, 76, 77
Stream bank formation, 139
Streams of plains, 151
Strips for fire protection, 100
Structure of leaves, classification, 260; of shoots, 216
Stunted trees of plains, 152
Submerged dissected leaves, 256
Subsoil, impervious, of plains, 169
Subulate leaves, 253
Succession, diagram showing, 49; of pine-barren formations, 49
Successional formations, 100
Sucker, 219
Summer aspect of pine-barrens, 80
Sundew, microscopic leaf structure of, 272
Sunflower culture in soils from plains, 164
Swamp plants, additional, 138
Sweet-bay, 134
Sweet-fern, 85; microscopic leaf structure, 267; root system of, 228
Sweet-gum, 75
Sweet-magnolia, illustration of, 120
Sweet-pepper bush, root system of, 235
Symplocarpus foetidus, 135
Synecology of cedar swamps, 120; of Coremal, 154
Synopsis of leaf structure, 283
Systematic list of 548 pine-barren plants, 196–208

Table of leaf structure, 284; showing plants in flower at different dates, 212
Talbot formation of Maryland, 5
Tansley, A. G., cited, 176, 216
Tar, 23, 24
Taxonomic species, 298, 299
Taylor, Norman, cited, 2, 4, 118, 194
Teaberry, microscopic leaf structure, 278
Tension line between pine forest and cedar-swamp, 109
Tephrosia virginiana, 72; figure of, 232; figure of microscopic leaf structure, 273; microscopic leaf structure, 273; roots, 232
Terraces of savanna, 147
Theories concerning Coremal, 159
Theory of mutation, 299
Therophytes, 185
Thomas, Gabriel, cited, 22
Tiresome-weed, 17
Tofieldia racemosa, roots and stems of, 225

Tolley, Howard R., cited, 194
Toms River, 7; photograph of, 140
Torches of pine splints, 20
Toxic secretions by roots of plants of Core-mal, 160
Trailing arbutus, microscopic leaf structure, 277; root system of, 237
Trails of Indians, 13
Transitional pine-barrens, 87
Treat, Mrs. Mary, cited, 29, 78, 81, 105, 129
Tree sprouts, age of, 154
Trees, architectural form, 53, 54, 55, 56; cultivated for fruit, 172; cultivated in the pine-barrens, 172; in cedar swamps, 189; in marshes, 192; introduced, 106; of pine-barren formation, 186; of plains, 152; with aggregate fruits in marshes, 192; with capsular fruits in marshes, 192; with drupaceous fruits in marshes, 192; with pomaceous fruits in marshes, 192; with winged fruits in marshes, 192
Trifoliate compound leaves, 256
Trunk of pitch-pine, 98
Tubeuf, Prof. C. von, 63
Tundra, 9
Tunmann, O., cited, 276
Turf-forming shoots, 219
Turkey's-beard, 77; microscopic leaf structure, 267; roots and stems of, 225
Turk's-cap lily, number of flowers, 303
Turpentine industry, 21
    still, 24
Types, biologic, 185, 186
Typha leaf form, 254

UNDERSHRUBS of pine-barrens, 187; in marshes, 191
Upper plains, 151; photograph of, 156
Uses of pitch-pine, 19; of sphagnum, 26; of white-cedar, 20
Utricularia clandestina, 5; figure of, 309
    cleistogama as cleistogene, 311; figure of, 309
    cornuta, figure of, 309
    fibrosa, figure of, 309
    figures of, 309
    inflata, figure of, 309
    intermedia, figure of, 309
    juncea, figure of, 309
    purpurea, figure of, 309
    subulata, figure of, 309

VACCINIUM atrococcum, 137; as a species, 299; figure of microscopic leaf structure, 281; micro-scopic leaf structure, 280
    caesariense as a species, 300
    corymbosum, 133, 134; as a species, 299; figure of micro-scopic leaf structure, 281; mi-croscopic leaf structure, 281

Vaccinium macrocarpon, 5; figure of mi-croscopic leaf structure, 281; microscopic leaf structure, 281
    pennsylvanicum, 69, 85, 101, 112; figure of, 240; figure of microscopic leaf structure, 281; microscopic leaf struc-ture, 281; root system of, 238
    vacillans, figure of, 241; figure of microscopic leaf structure, 282; microscopic leaf struc-ture, 282; root system of, 239
    virgatum, as a species, 300
Vahl, Martin, cited, 185
Valley of Meadow Brook, 171
Variable leaf forms, 256
Variation, 300
Vegetation of Hempstead Plains, Long Island, 171; of pine-barrens, general, 175; of New Zealand, 308
Vegetative propagation of shoots, 216
Vermeule, C. Clarkson, cited, 8, 12, 178
Verticillate leaves, 258
Vesque, J., cited, 305
Vines, cultivated, 173
Viola emarginata as a cleistogene, 311
    lanceolata as a cleistogene, 311
    lineariloba, 304
    pedata, 304
    primulifolia, 311
    sagittata as a cleistogene, 311
Violet leaf form, 255
Vitis aestivalis as a liane, 311
Viviparous acorns, figure of, 293; of Quercus marylandica, 292
Vries, H. de (see De Vries)

WADING RIVER, 7; savannas, 146
Waldheiden, 86
Wallace, Dr. Daniel L., soil analyses by, 34
Wandering shoots, 219
    species with epiterranean run-ners, 217; with subterranean shoots, 217
Warming, Eugen, cited, 216, 239, 252, 258, 305
Water, influence on plants, 307; retention, 43
Watering Place Pond, 143; photographs of, 143, 144
Water-lily, 144; flowering and fruiting of, 210; root system of, 232
Waxberry, 88, 149
Wax-ends, 24
Weather as a business risk in farming, 194
Weeds of pine-barrens, 105
Weevers, Th., cited, 276
West Plains, 151
Western pine-barren limit, 97

Wheat cultures in soils of plains, photograph of, 164
Whippoorwill-shoe, 17
Whitbeck, R. H., cited, 17
White cedar, 134; rate of growth of, 119; stem analyses of, 119
White-cedar swamp, colors of flowers in, 190; formation, 189; photograph of, 127
White cedar, uses of, 20
White fringed orchid, roots and stems of, 227, 228
White oak, 95, 133
White-topped aster, root system of, 242
Wicomico of Maryland, 2
Wild animals of pine-barrens, 13
  character of pine-barren country, 314
  indigo, root system of, 233
Wilting coefficient of soils, 45
Wind influence, 307
Wind-swept trees, 87

Winter aspect, 78; of cedar swamps, 129
Wintergreen, root system of, 236
Wiry leaves, 262
Wiry-stemmed plants, 308
Witches'-broom on pitch-pine, 60
Woodhead, T. W., cited, 259
Woody plants in marshes, 191

XEROPHYLLUM asphodeloides, 77; figure of microscopic leaf structure, 267; microscopic structure of leaf, 267; roots and stems of, 226
Xerophytic pines, 249

YAPP, R. H., cited, 216, 217, 251
Years of cone production in pitch-pines, 287, 288, 289, 290, 291
Yellow pine, 50, 67
  pine-barrens, 9
Yucca leaf form, 254

A CATALOGUE OF SELECTED DOVER BOOKS
IN ALL FIELDS OF INTEREST

A CATALOGUE OF SELECTED DOVER BOOKS
IN ALL FIELDS OF INTEREST

WHAT IS SCIENCE?, *N. Campbell*
The role of experiment and measurement, the function of mathematics, the nature of scientific laws, the difference between laws and theories, the limitations of science, and many similarly provocative topics are treated clearly and without technicalities by an eminent scientist. "Still an excellent introduction to scientific philosophy," H. Margenau in *Physics Today*. "A first-rate primer . . . deserves a wide audience," *Scientific American*. 192pp. 5⅜ x 8.
60043-2 Paperbound $1.25

THE NATURE OF LIGHT AND COLOUR IN THE OPEN AIR, *M. Minnaert*
Why are shadows sometimes blue, sometimes green, or other colors depending on the light and surroundings? What causes mirages? Why do multiple suns and moons appear in the sky? Professor Minnaert explains these unusual phenomena and hundreds of others in simple, easy-to-understand terms based on optical laws and the properties of light and color. No mathematics is required but artists, scientists, students, and everyone fascinated by these "tricks" of nature will find thousands of useful and amazing pieces of information. Hundreds of observational experiments are suggested which require no special equipment. 200 illustrations; 42 photos. xvi + 362pp. 5⅜ x 8.
20196-1 Paperbound $2.75

THE STRANGE STORY OF THE QUANTUM, AN ACCOUNT FOR THE GENERAL READER OF THE GROWTH OF IDEAS UNDERLYING OUR PRESENT ATOMIC KNOWLEDGE, *B. Hoffmann*
Presents lucidly and expertly, with barest amount of mathematics, the problems and theories which led to modern quantum physics. Dr. Hoffmann begins with the closing years of the 19th century, when certain trifling discrepancies were noticed, and with illuminating analogies and examples takes you through the brilliant concepts of Planck, Einstein, Pauli, Broglie, Bohr, Schroedinger, Heisenberg, Dirac, Sommerfeld, Feynman, etc. This edition includes a new, long postscript carrying the story through 1958. "Of the books attempting an account of the history and contents of our modern atomic physics which have come to my attention, this is the best," H. Margenau, Yale University, in *American Journal of Physics*. 32 tables and line illustrations. Index. 275pp. 5⅜ x 8.
20518-5 Paperbound $2.00

GREAT IDEAS OF MODERN MATHEMATICS: THEIR NATURE AND USE, *Jagjit Singh*
Reader with only high school math will understand main mathematical ideas of modern physics, astronomy, genetics, psychology, evolution, etc. better than many who use them as tools, but comprehend little of their basic structure. Author uses his wide knowledge of non-mathematical fields in brilliant exposition of differential equations, matrices, group theory, logic, statistics, problems of mathematical foundations, imaginary numbers, vectors, etc. Original publication. 2 appendixes. 2 indexes. 65 ills. 322pp. 5⅜ x 8.
20587-8 Paperbound $2.50

THE MUSIC OF THE SPHERES: THE MATERIAL UNIVERSE — FROM ATOM TO QUASAR, SIMPLY EXPLAINED, *Guy Murchie*
Vast compendium of fact, modern concept and theory, observed and calculated data, historical background guides intelligent layman through the material universe. Brilliant exposition of earth's construction, explanations for moon's craters, atmospheric components of Venus and Mars (with data from recent fly-by's), sun spots, sequences of star birth and death, neighboring galaxies, contributions of Galileo, Tycho Brahe, Kepler, etc.; and (Vol. 2) construction of the atom (describing newly discovered sigma and xi subatomic particles), theories of sound, color and light, space and time, including relativity theory, quantum theory, wave theory, probability theory, work of Newton, Maxwell, Faraday, Einstein, de Broglie, etc. "Best presentation yet offered to the intelligent general reader," *Saturday Review*. Revised (1967). Index. 319 illustrations by the author. Total of xx + 644pp. 5⅜ x 8½.
21809-0, 21810-4 Two volume set, paperbound $5.00

FOUR LECTURES ON RELATIVITY AND SPACE, *Charles Proteus Steinmetz*
Lecture series, given by great mathematician and electrical engineer, generally considered one of the best popular-level expositions of special and general relativity theories and related questions. Steinmetz translates complex mathematical reasoning into language accessible to laymen through analogy, example and comparison. Among topics covered are relativity of motion, location, time; of mass; acceleration; 4-dimensional time-space; geometry of the gravitational field; curvature and bending of space; non-Euclidean geometry. Index. 40 illustrations. x + 142pp. 5⅜ x 8½.
61771-8 Paperbound $1.50

HOW TO KNOW THE WILD FLOWERS, *Mrs. William Starr Dana*
Classic nature book that has introduced thousands to wonders of American wild flowers. Color-season principle of organization is easy to use, even by those with no botanical training, and the genial, refreshing discussions of history, folklore, uses of over 1,000 native and escape flowers, foliage plants are informative as well as fun to read. Over 170 full-page plates, collected from several editions, may be colored in to make permanent records of finds. Revised to conform with 1950 edition of Gray's Manual of Botany. xlii + 438pp. 5⅜ x 8½.
20332-8 Paperbound $2.50

MANUAL OF THE TREES OF NORTH AMERICA, *Charles Sprague Sargent*
Still unsurpassed as most comprehensive, reliable study of North American tree characteristics, precise locations and distribution. By dean of American dendrologists. Every tree native to U.S., Canada, Alaska; 185 genera, 717 species, described in detail—leaves, flowers, fruit, winterbuds, bark, wood, growth habits, etc. plus discussion of varieties and local variants, immaturity variations. Over 100 keys, including unusual 11-page analytical key to genera, aid in identification. 783 clear illustrations of flowers, fruit, leaves. An unmatched permanent reference work for all nature lovers. Second enlarged (1926) edition. Synopsis of families. Analytical key to genera. Glossary of technical terms. Index. 783 illustrations, 1 map. Total of 982pp. 5⅜ x 8.
20277-1, 20278-X Two volume set, paperbound $6.00

IT'S FUN TO MAKE THINGS FROM SCRAP MATERIALS,
Evelyn Glantz Hershoff
What use are empty spools, tin cans, bottle tops? What can be made from rubber bands, clothes pins, paper clips, and buttons? This book provides simply worded instructions and large diagrams showing you how to make cookie cutters, toy trucks, paper turkeys, Halloween masks, telephone sets, aprons, linoleum block- and spatter prints — in all 399 projects! Many are easy enough for young children to figure out for themselves; some challenging enough to entertain adults; all are remarkably ingenious ways to make things from materials that cost pennies or less! Formerly "Scrap Fun for Everyone." Index. 214 illustrations. 373pp. 5⅜ x 8½.   21251-3 Paperbound $2.00

SYMBOLIC LOGIC and THE GAME OF LOGIC, *Lewis Carroll*
"Symbolic Logic" is not concerned with modern symbolic logic, but is instead a collection of over 380 problems posed with charm and imagination, using the syllogism and a fascinating diagrammatic method of drawing conclusions. In "The Game of Logic" Carroll's whimsical imagination devises a logical game played with 2 diagrams and counters (included) to manipulate hundreds of tricky syllogisms. The final section, "Hit or Miss" is a lagniappe of 101 additional puzzles in the delightful Carroll manner. Until this reprint edition, both of these books were rarities costing up to $15 each. Symbolic Logic: Index. xxxi + 199pp. The Game of Logic: 96pp. 2 vols. bound as one. 5⅜ x 8.
20492-8 Paperbound $2.50

MATHEMATICAL PUZZLES OF SAM LOYD, PART I
selected and edited by M. Gardner
Choice puzzles by the greatest American puzzle creator and innovator. Selected from his famous collection, "Cyclopedia of Puzzles," they retain the unique style and historical flavor of the originals. There are posers based on arithmetic, algebra, probability, game theory, route tracing, topology, counter and sliding block, operations research, geometrical dissection. Includes the famous "14-15" puzzle which was a national craze, and his "Horse of a Different Color" which sold millions of copies. 117 of his most ingenious puzzles in all. 120 line drawings and diagrams. Solutions. Selected references. xx + 167pp. 5⅜ x 8.
20498-7 Paperbound $1.35

STRING FIGURES AND HOW TO MAKE THEM, *Caroline Furness Jayne*
107 string figures plus variations selected from the best primitive and modern examples developed by Navajo, Apache, pygmies of Africa, Eskimo, in Europe, Australia, China, etc. The most readily understandable, easy-to-follow book in English on perennially popular recreation. Crystal-clear exposition; step-by-step diagrams. Everyone from kindergarten children to adults looking for unusual diversion will be endlessly amused. Index. Bibliography. Introduction by A. C. Haddon. 17 full-page plates, 960 illustrations. xxiii + 401pp. 5⅜ x 8½.
20152-X Paperbound $2.50

PAPER FOLDING FOR BEGINNERS, *W. D. Murray and F. J. Rigney*
A delightful introduction to the varied and entertaining Japanese art of origami (paper folding), with a full, crystal-clear text that anticipates every difficulty; over 275 clearly labeled diagrams of all important stages in creation. You get results at each stage, since complex figures are logically developed from simpler ones. 43 different pieces are explained: sailboats, frogs, roosters, etc. 6 photographic plates. 279 diagrams. 95pp. 5⅝ x 8⅜.
20713-7 Paperbound $1.00

PRINCIPLES OF ART HISTORY,
H. Wölfflin
Analyzing such terms as "baroque," "classic," "neoclassic," "primitive," "picturesque," and 164 different works by artists like Botticelli, van Cleve, Dürer, Hobbema, Holbein, Hals, Rembrandt, Titian, Brueghel, Vermeer, and many others, the author establishes the classifications of art history and style on a firm, concrete basis. This classic of art criticism shows what really occurred between the 14th-century primitives and the sophistication of the 18th century in terms of basic attitudes and philosophies. "A remarkable lesson in the art of seeing," *Sat. Rev. of Literature.* Translated from the 7th German edition. 150 illustrations. 254pp. 6⅛ x 9¼. 20276-3 Paperbound $2.50

PRIMITIVE ART,
Franz Boas
This authoritative and exhaustive work by a great American anthropologist covers the entire gamut of primitive art. Pottery, leatherwork, metal work, stone work, wood, basketry, are treated in detail. Theories of primitive art, historical depth in art history, technical virtuosity, unconscious levels of patterning, symbolism, styles, literature, music, dance, etc. A must book for the interested layman, the anthropologist, artist, handicrafter (hundreds of unusual motifs), and the historian. Over 900 illustrations (50 ceramic vessels, 12 totem poles, etc.). 376pp. 5⅜ x 8. 20025-6 Paperbound $2.50

THE GENTLEMAN AND CABINET MAKER'S DIRECTOR,
Thomas Chippendale
A reprint of the 1762 catalogue of furniture designs that went on to influence generations of English and Colonial and Early Republic American furniture makers. The 200 plates, most of them full-page sized, show Chippendale's designs for French (Louis XV), Gothic, and Chinese-manner chairs, sofas, canopy and dome beds, cornices, chamber organs, cabinets, shaving tables, commodes, picture frames, frets, candle stands, chimney pieces, decorations, etc. The drawings are all elegant and highly detailed; many include construction diagrams and elevations. A supplement of 24 photographs shows surviving pieces of original and Chippendale-style pieces of furniture. Brief biography of Chippendale by N. I. Bienenstock, editor of *Furniture World.* Reproduced from the 1762 edition. 200 plates, plus 19 photographic plates. vi + 249pp. 9⅛ x 12¼. 21601-2 Paperbound $4.00

AMERICAN ANTIQUE FURNITURE: A BOOK FOR AMATEURS,
Edgar G. Miller, Jr.
Standard introduction and practical guide to identification of valuable American antique furniture. 2115 illustrations, mostly photographs taken by the author in 148 private homes, are arranged in chronological order in extensive chapters on chairs, sofas, chests, desks, bedsteads, mirrors, tables, clocks, and other articles. Focus is on furniture accessible to the collector, including simpler pieces and a larger than usual coverage of Empire style. Introductory chapters identify structural elements, characteristics of various styles, how to avoid fakes, etc. "We are frequently asked to name some book on American furniture that will meet the requirements of the novice collector, the beginning dealer, and . . . the general public. . . . We believe Mr. Miller's two volumes more completely satisfy this specification than any other work," *Antiques.* Appendix. Index. Total of vi + 1106pp. 7⅞ x 10¾. 21599-7, 21600-4 Two volume set, paperbound $10.00

THE BAD CHILD'S BOOK OF BEASTS, MORE BEASTS FOR WORSE CHILDREN, and A MORAL ALPHABET, *H. Belloc*
Hardly and anthology of humorous verse has appeared in the last 50 years without at least a couple of these famous nonsense verses. But one must see the entire volumes — with all the delightful original illustrations by Sir Basil Blackwood — to appreciate fully Belloc's charming and witty verses that play so subacidly on the platitudes of life and morals that beset his day — and ours. A great humor classic. Three books in one. Total of 157pp. 5⅜ x 8.
20749-8 Paperbound $1.25

THE DEVIL'S DICTIONARY, *Ambrose Bierce*
Sardonic and irreverent barbs puncturing the pomposities and absurdities of American politics, business, religion, literature, and arts, by the country's greatest satirist in the classic tradition. Epigrammatic as Shaw, piercing as Swift, American as Mark Twain, Will Rogers, and Fred Allen, Bierce will always remain the favorite of a small coterie of enthusiasts, and of writers and speakers whom he supplies with "some of the most gorgeous witticisms of the English language" (H. L. Mencken). Over 1000 entries in alphabetical order. 144pp. 5⅜ x 8.
20487-1 Paperbound $1.25

THE COMPLETE NONSENSE OF EDWARD LEAR.
This is the only complete edition of this master of gentle madness available at a popular price. *A Book of Nonsense, Nonsense Songs, More Nonsense Songs and Stories* in their entirety with all the old favorites that have delighted children and adults for years. The Dong With A Luminous Nose, The Jumblies, The Owl and the Pussycat, and hundreds of other bits of wonderful nonsense. 214 limericks, 3 sets of Nonsense Botany, 5 Nonsense Alphabets, 546 drawings by Lear himself, and much more. 320pp. 5⅜ x 8. 20167-8 Paperbound $1.75

THE WIT AND HUMOR OF OSCAR WILDE, *ed. by Alvin Redman*
Wilde at his most brilliant, in 1000 epigrams exposing weaknesses and hypocrisies of "civilized" society. Divided into 49 categories—sin, wealth, women, America, etc.—to aid writers, speakers. Includes excerpts from his trials, books, plays, criticism. Formerly "The Epigrams of Oscar Wilde." Introduction by Vyvyan Holland, Wilde's only living son. Introductory essay by editor. 260pp. 5⅜ x 8.
20602-5 Paperbound $1.50

A CHILD'S PRIMER OF NATURAL HISTORY, *Oliver Herford*
Scarcely an anthology of whimsy and humor has appeared in the last 50 years without a contribution from Oliver Herford. Yet the works from which these examples are drawn have been almost impossible to obtain! Here at last are Herford's improbable definitions of a menagerie of familiar and weird animals, each verse illustrated by the author's own drawings. 24 drawings in 2 colors; 24 additional drawings. vii + 95pp. 6½ x 6. 21647-0 Paperbound $1.00

THE BROWNIES: THEIR BOOK, *Palmer Cox*
The book that made the Brownies a household word. Generations of readers have enjoyed the antics, predicaments and adventures of these jovial sprites, who emerge from the forest at night to play or to come to the aid of a deserving human. Delightful illustrations by the author decorate nearly every page. 24 short verse tales with 266 illustrations. 155pp. 6⅝ x 9¼.
21265-3 Paperbound $1.50

THE PRINCIPLES OF PSYCHOLOGY,
William James
The full long-course, unabridged, of one of the great classics of Western literature and science. Wonderfully lucid descriptions of human mental activity, the stream of thought, consciousness, time perception, memory, imagination, emotions, reason, abnormal phenomena, and similar topics. Original contributions are integrated with the work of such men as Berkeley, Binet, Mills, Darwin, Hume, Kant, Royce, Schopenhauer, Spinoza, Locke, Descartes, Galton, Wundt, Lotze, Herbart, Fechner, and scores of others. All contrasting interpretations of mental phenomena are examined in detail—introspective analysis, philosophical interpretation, and experimental research. "A classic," *Journal of Consulting Psychology.* "The main lines are as valid as ever," *Psychoanalytical Quarterly.* "Standard reading . . . a classic of interpretation," *Psychiatric Quarterly.* 94 illustrations. 1408pp. 5⅜ x 8.
20381-6, 20382-4 Two volume set, paperbound $6.00

VISUAL ILLUSIONS: THEIR CAUSES, CHARACTERISTICS AND APPLICATIONS,
M. Luckiesh
"Seeing is deceiving," asserts the author of this introduction to virtually every type of optical illusion known. The text both describes and explains the principles involved in color illusions, figure-ground, distance illusions, etc. 100 photographs, drawings and diagrams prove how easy it is to fool the sense: circles that aren't round, parallel lines that seem to bend, stationary figures that seem to move as you stare at them — illustration after illustration strains our credulity at what we see. Fascinating book from many points of view, from applications for artists, in camouflage, etc. to the psychology of vision. New introduction by William Ittleson, Dept. of Psychology, Queens College. Index. Bibliography. xxi + 252pp. 5⅜ x 8½. 21530-X Paperbound $1.75

FADS AND FALLACIES IN THE NAME OF SCIENCE,
Martin Gardner
This is the standard account of various cults, quack systems, and delusions which have masqueraded as science: hollow earth fanatics. Reich and orgone sex energy, dianetics, Atlantis, multiple moons, Forteanism, flying saucers, medical fallacies like iridiagnosis, zone therapy, etc. A new chapter has been added on Bridey Murphy, psionics, and other recent manifestations in this field. This is a fair, reasoned appraisal of eccentric theory which provides excellent inoculation against cleverly masked nonsense. "Should be read by everyone, scientist and non-scientist alike," R. T. Birge, Prof. Emeritus of Physics, Univ. of California; Former President, American Physical Society. Index. x + 365pp. 5⅜ x 8. 20394-8 Paperbound $2.00

ILLUSIONS AND DELUSIONS OF THE SUPERNATURAL AND THE OCCULT,
D. H. Rawcliffe
Holds up to rational examination hundreds of persistent delusions including crystal gazing, automatic writing, table turning, mediumistic trances, mental healing, stigmata, lycanthropy, live burial, the Indian Rope Trick, spiritualism, dowsing, telepathy, clairvoyance, ghosts, ESP, etc. The author explains and exposes the mental and physical deceptions involved, making this not only an exposé of supernatural phenomena, but a valuable exposition of characteristic types of abnormal psychology. Originally titled "The Psychology of the Occult." 14 illustrations. Index. 551pp. 5⅜ x 8. 20503-7 Paperbound $3.50

FAIRY TALE COLLECTIONS, *edited by Andrew Lang*
Andrew Lang's fairy tale collections make up the richest shelf-full of traditional children's stories anywhere available. Lang supervised the translation of stories from all over the world—familiar European tales collected by Grimm, animal stories from Negro Africa, myths of primitive Australia, stories from Russia, Hungary, Iceland, Japan, and many other countries. Lang's selection of translations are unusually high; many authorities consider that the most familiar tales find their best versions in these volumes. All collections are richly decorated and illustrated by H. J. Ford and other artists.

THE BLUE FAIRY BOOK. 37 stories. 138 illustrations. ix + 390pp. 5⅜ x 8½.
21437-0 Paperbound $1.95

THE GREEN FAIRY BOOK. 42 stories. 100 illustrations. xiii + 366pp. 5⅜ x 8½.
21439-7 Paperbound $2.00

THE BROWN FAIRY BOOK. 32 stories. 50 illustrations, 8 in color. xii + 350pp. 5⅜ x 8½.
21438-9 Paperbound $1.95

THE BEST TALES OF HOFFMANN, *edited by E. F. Bleiler*
10 stories by E. T. A. Hoffmann, one of the greatest of all writers of fantasy. The tales include "The Golden Flower Pot," "Automata," "A New Year's Eve Adventure," "Nutcracker and the King of Mice," "Sand-Man," and others. Vigorous characterizations of highly eccentric personalities, remarkably imaginative situations, and intensely fast pacing has made these tales popular all over the world for 150 years. Editor's introduction. 7 drawings by Hoffmann. xxxiii + 419pp. 5⅜ x 8½.
21793-0 Paperbound $2.25

GHOST AND HORROR STORIES OF AMBROSE BIERCE, *edited by E. F. Bleiler*
Morbid, eerie, horrifying tales of possessed poets, shabby aristocrats, revived corpses, and haunted malefactors. Widely acknowledged as the best of their kind between Poe and the moderns, reflecting their author's inner torment and bitter view of life. Includes "Damned Thing," "The Middle Toe of the Right Foot," "The Eyes of the Panther," "Visions of the Night," "Moxon's Master," and over a dozen others. Editor's introduction. xxii + 199pp. 5⅜ x 8½.
20767-6 Paperbound $1.50

THREE GOTHIC NOVELS, *edited by E. F. Bleiler*
Originators of the still popular Gothic novel form, influential in ushering in early 19th-century Romanticism. Horace Walpole's *Castle of Otranto*, William Beckford's *Vathek*, John Polidori's *The Vampyre*, and a *Fragment* by Lord Byron are enjoyable as exciting reading or as documents in the history of English literature. Editor's introduction. xi + 291pp. 5⅜ x 8½.
21232-7 Paperbound $2.00

BEST GHOST STORIES OF LEFANU, *edited by E. F. Bleiler*
Though admired by such critics as V. S. Pritchett, Charles Dickens and Henry James, ghost stories by the Irish novelist Joseph Sheridan LeFanu have never become as widely known as his detective fiction. About half of the 16 stories in this collection have never before been available in America. Collection includes "Carmilla" (perhaps the best vampire story ever written), "The Haunted Baronet," "The Fortunes of Sir Robert Ardagh," and the classic "Green Tea." Editor's introduction. 7 contemporary illustrations. Portrait of LeFanu. xii + 467pp. 5⅜ x 8.
20415-4 Paperbound $2.50

EASY-TO-DO ENTERTAINMENTS AND DIVERSIONS WITH COINS, CARDS, STRING, PAPER AND MATCHES, *R. M. Abraham*
Over 300 tricks, games and puzzles will provide young readers with absorbing fun. Sections on card games; paper-folding; tricks with coins, matches and pieces of string; games for the agile; toy-making from common household objects; mathematical recreations; and 50 miscellaneous pastimes. Anyone in charge of groups of youngsters, including hard-pressed parents, and in need of suggestions on how to keep children sensibly amused and quietly content will find this book indispensable. Clear, simple text, copious number of delightful line drawings and illustrative diagrams. Originally titled "Winter Nights' Entertainments." Introduction by Lord Baden Powell. 329 illustrations. v + 186pp. 5⅜ x 8½. 20921-0 Paperbound $1.25

AN INTRODUCTION TO CHESS MOVES AND TACTICS SIMPLY EXPLAINED, *Leonard Barden*
Beginner's introduction to the royal game. Names, possible moves of the pieces, definitions of essential terms, how games are won, etc. explained in 30-odd pages. With this background you'll be able to sit right down and play. Balance of book teaches strategy — openings, middle game, typical endgame play, and suggestions for improving your game. A sample game is fully analyzed. True middle-level introduction, teaching you all the essentials without oversimplifying or losing you in a maze of detail. 58 figures. 102pp. 5⅜ x 8½. 21210-6 Paperbound $1.25

LASKER'S MANUAL OF CHESS, *Dr. Emanuel Lasker*
Probably the greatest chess player of modern times, Dr. Emanuel Lasker held the world championship 28 years, independent of passing schools or fashions. This unmatched study of the game, chiefly for intermediate to skilled players, analyzes basic methods, combinations, position play, the aesthetics of chess, dozens of different openings, etc., with constant reference to great modern games. Contains a brilliant exposition of Steinitz's important theories. Introduction by Fred Reinfeld. Tables of Lasker's tournament record. 3 indices. 308 diagrams. 1 photograph. xxx + 349pp. 5⅜ x 8. 20640-8 Paperbound $2.50

COMBINATIONS: THE HEART OF CHESS, *Irving Chernev*
Step-by-step from simple combinations to complex, this book, by a well-known chess writer, shows you the intricacies of pins, counter-pins, knight forks, and smothered mates. Other chapters show alternate lines of play to those taken in actual championship games; boomerang combinations; classic examples of brilliant combination play by Nimzovich, Rubinstein, Tarrasch, Botvinnik, Alekhine and Capablanca. Index. 356 diagrams. ix + 245pp. 5⅜ x 8½. 21744-2 Paperbound $2.00

HOW TO SOLVE CHESS PROBLEMS, *K. S. Howard*
Full of practical suggestions for the fan or the beginner — who knows only the moves of the chessmen. Contains preliminary section and 58 two-move, 46 three-move, and 8 four-move problems composed by 27 outstanding American problem creators in the last 30 years. Explanation of all terms and exhaustive index. "Just what is wanted for the student," Brian Harley. 112 problems, solutions. vi + 171pp. 5⅜ x 8. 20748-X Paperbound $1.50

SOCIAL THOUGHT FROM LORE TO SCIENCE,
H. E. Barnes and H. Becker
An immense survey of sociological thought and ways of viewing, studying, planning, and reforming society from earliest times to the present. Includes thought on society of preliterate peoples, ancient non-Western cultures, and every great movement in Europe, America, and modern Japan. Analyzes hundreds of great thinkers: Plato, Augustine, Bodin, Vico, Montesquieu, Herder, Comte, Marx, etc. Weighs the contributions of utopians, sophists, fascists and communists; economists, jurists, philosophers, ecclesiastics, and every 19th and 20th century school of scientific sociology, anthropology, and social psychology throughout the world. Combines topical, chronological, and regional approaches, treating the evolution of social thought as a process rather than as a series of mere topics. "Impressive accuracy, competence, and discrimination . . . easily the best single survey," *Nation*. Thoroughly revised, with new material up to 1960. 2 indexes. Over 2200 bibliographical notes. Three volume set. Total of 1586pp. 5⅜ x 8.
20901-6, 20902-4, 20903-2 Three volume set, paperbound $10.50

A HISTORY OF HISTORICAL WRITING, *Harry Elmer Barnes*
Virtually the only adequate survey of the whole course of historical writing in a single volume. Surveys developments from the beginnings of historiography in the ancient Near East and the Classical World, up through the Cold War. Covers major historians in detail, shows interrelationship with cultural background, makes clear individual contributions, evaluates and estimates importance; also enormously rich upon minor authors and thinkers who are usually passed over. Packed with scholarship and learning, clear, easily written. Indispensable to every student of history. Revised and enlarged up to 1961. Index and bibliography. xv + 442pp. 5⅜ x 8½.
20104-X Paperbound $3.00

JOHANN SEBASTIAN BACH, *Philipp Spitta*
The complete and unabridged text of the definitive study of Bach. Written some 70 years ago, it is still unsurpassed for its coverage of nearly all aspects of Bach's life and work. There could hardly be a finer non-technical introduction to Bach's music than the detailed, lucid analyses which Spitta provides for hundreds of individual pieces. 26 solid pages are devoted to the B minor mass, for example, and 30 pages to the glorious St. Matthew Passion. This monumental set also includes a major analysis of the music of the 18th century: Buxtehude, Pachelbel, etc. "Unchallenged as the last word on one of the supreme geniuses of music," John Barkham, *Saturday Review Syndicate*. Total of 1819pp. Heavy cloth binding. 5⅜ x 8.
22278-0, 22279-9 Two volume set, clothbound $15.00

BEETHOVEN AND HIS NINE SYMPHONIES, *George Grove*
In this modern middle-level classic of musicology Grove not only analyzes all nine of Beethoven's symphonies very thoroughly in terms of their musical structure, but also discusses the circumstances under which they were written, Beethoven's stylistic development, and much other background material. This is an extremely rich book, yet very easily followed; it is highly recommended to anyone seriously interested in music. Over 250 musical passages. Index. viii + 407pp. 5⅜ x 8.
20334-4 Paperbound $2.50

THE TIME STREAM
John Taine
Acknowledged by many as the best SF writer of the 1920's, Taine (under the name Eric Temple Bell) was also a Professor of Mathematics of considerable renown. Reprinted here are *The Time Stream*, generally considered Taine's best, *The Greatest Game*, a biological-fiction novel, and *The Purple Sapphire*, involving a supercivilization of the past. Taine's stories tie fantastic narratives to frameworks of original and logical scientific concepts. Speculation is often profound on such questions as the nature of time, concept of entropy, cyclical universes, etc. 4 contemporary illustrations. v + 532pp. 5⅜ x 8⅜.
21180-0 Paperbound $3.00

SEVEN SCIENCE FICTION NOVELS,
H. G. Wells
Full unabridged texts of 7 science-fiction novels of the master. Ranging from biology, physics, chemistry, astronomy, to sociology and other studies, Mr. Wells extrapolates whole worlds of strange and intriguing character. "One will have to go far to match this for entertainment, excitement, and sheer pleasure . . ."*New York Times*. Contents: The Time Machine, The Island of Dr. Moreau, The First Men in the Moon, The Invisible Man, The War of the Worlds, The Food of the Gods, In The Days of the Comet. 1015pp. 5⅜ x 8.
20264-X Clothbound $5.00

28 SCIENCE FICTION STORIES OF H. G. WELLS.
Two full, unabridged novels, *Men Like Gods* and *Star Begotten,* plus 26 short stories by the master science-fiction writer of all time! Stories of space, time, invention, exploration, futuristic adventure. Partial contents: *The Country of the Blind, In the Abyss, The Crystal Egg, The Man Who Could Work Miracles, A Story of Days to Come, The Empire of the Ants, The Magic Shop, The Valley of the Spiders, A Story of the Stone Age, Under the Knife, Sea Raiders,* etc. An indispensable collection for the library of anyone interested in science fiction adventure. 928pp. 5⅜ x 8.
20265-8 Clothbound $5.00

THREE MARTIAN NOVELS,
Edgar Rice Burroughs
Complete, unabridged reprinting, in one volume, of Thuvia, Maid of Mars; Chessmen of Mars; The Master Mind of Mars. Hours of science-fiction adventure by a modern master storyteller. Reset in large clear type for easy reading. 16 illustrations by J. Allen St. John. vi + 490pp. 5⅜ x 8½.
20039-6.Paperbound $2.50

AN INTELLECTUAL AND CULTURAL HISTORY OF THE WESTERN WORLD,
Harry Elmer Barnes
Monumental 3-volume survey of intellectual development of Europe from primitive cultures to the present day. Every significant product of human intellect traced through history: art, literature, mathematics, physical sciences, medicine, music, technology, social sciences, religions, jurisprudence, education, etc. Presentation is lucid and specific, analyzing in detail specific discoveries, theories, literary works, and so on. Revised (1965) by recognized scholars in specialized fields under the direction of Prof. Barnes. Revised bibliography. Indexes. 24 illustrations. Total of xxix + 1318pp.
21275-0, 21276-9, 21277-7 Three volume set, paperbound $7.75

HEAR ME TALKIN' TO YA, *edited by Nat Shapiro and Nat Hentoff*
In their own words, Louis Armstrong, King Oliver, Fletcher Henderson, Bunk Johnson, Bix Beiderbecke, Billy Holiday, Fats Waller, Jelly Roll Morton, Duke Ellington, and many others comment on the origins of jazz in New Orleans and its growth in Chicago's South Side, Kansas City's jam sessions, Depression Harlem, and the modernism of the West Coast schools. Taken from taped conversations, letters, magazine articles, other first-hand sources. Editors' introduction. xvi + 429pp. 5⅜ x 8½.   21726-4 Paperbound $2.50

THE JOURNAL OF HENRY D. THOREAU
A 25-year record by the great American observer and critic, as complete a record of a great man's inner life as is anywhere available. Thoreau's Journals served him as raw material for his formal pieces, as a place where he could develop his ideas, as an outlet for his interests in wild life and plants, in writing as an art, in classics of literature, Walt Whitman and other contemporaries, in politics, slavery, individual's relation to the State, etc. The Journals present a portrait of a remarkable man, and are an observant social history. Unabridged republication of 1906 edition, Bradford Torrey and Francis H. Allen, editors. Illustrations. Total of 1888pp. 8⅜ x 12¼.
            20312-3, 20313-1 Two volume set, clothbound $30.00

A SHAKESPEARIAN GRAMMAR, *E. A. Abbott*
Basic reference to Shakespeare and his contemporaries, explaining through thousands of quotations from Shakespeare, Jonson, Beaumont and Fletcher, North's *Plutarch* and other sources the grammatical usage differing from the modern. First published in 1870 and written by a scholar who spent much of his life isolating principles of Elizabethan language, the book is unlikely ever to be superseded. Indexes. xxiv + 511pp. 5⅜ x 8½. 21582-2 Paperbound $3.00

FOLK-LORE OF SHAKESPEARE, *T. F. Thistelton Dyer*
Classic study, drawing from Shakespeare a large body of references to supernatural beliefs, terminology of falconry and hunting, games and sports, good luck charms, marriage customs, folk medicines, superstitions about plants, animals, birds, argot of the underworld, sexual slang of London, proverbs, drinking customs, weather lore, and much else. From full compilation comes a mirror of the 17th-century popular mind. Index. ix + 526pp. 5⅜ x 8½.
            21614-4 Paperbound $3.25

THE NEW VARIORUM SHAKESPEARE, *edited by H. H. Furness*
By far the richest editions of the plays ever produced in any country or language. Each volume contains complete text (usually First Folio) of the play, all variants in Quarto and other Folio texts, editorial changes by every major editor to Furness's own time (1900), footnotes to obscure references or language, extensive quotes from literature of Shakespearian criticism, essays on plot sources (often reprinting sources in full), and much more.

HAMLET, *edited by H. H. Furness*
Total of xxvi + 905pp. 5⅜ x 8½.
            21004-9, 21005-7 Two volume set, paperbound $5.50

TWELFTH NIGHT, *edited by H. H. Furness*
Index. xxii + 434pp. 5⅜ x 8½.   21189-4 Paperbound $2.75

LA BOHEME BY GIACOMO PUCCINI,
translated and introduced by Ellen H. Bleiler
Complete handbook for the operagoer, with everything needed for full enjoyment except the musical score itself. Complete Italian libretto, with new, modern English line-by-line translation—the only libretto printing all repeats; biography of Puccini; the librettists; background to the opera, Murger's La Boheme, etc.; circumstances of composition and performances; plot summary; and pictorial section of 73 illustrations showing Puccini, famous singers and performances, etc. Large clear type for easy reading. 124pp. 5⅜ x 8½.

20404-9 Paperbound $1.50

ANTONIO STRADIVARI: HIS LIFE AND WORK (1644-1737),
W. Henry Hill, Arthur F. Hill, and Alfred E. Hill
Still the only book that really delves into life and art of the incomparable Italian craftsman, maker of the finest musical instruments in the world today. The authors, expert violin-makers themselves, discuss Stradivari's ancestry, his construction and finishing techniques, distinguished characteristics of many of his instruments and their locations. Included, too, is story of introduction of his instruments into France, England, first revelation of their supreme merit, and information on his labels, number of instruments made, prices, mystery of ingredients of his varnish, tone of pre-1684 Stradivari violin and changes between 1684 and 1690. An extremely interesting, informative account for all music lovers, from craftsman to concert-goer. Republication of original (1902) edition. New introduction by Sydney Beck, Head of Rare Book and Manuscript Collections, Music Division, New York Public Library. Analytical index by Rembert Wurlitzer. Appendixes. 68 illustrations. 30 full-page plates. 4 in color. xxvi + 315pp. 5⅜ x 8½.   20425-1 Paperbound $3.00

MUSICAL AUTOGRAPHS FROM MONTEVERDI TO HINDEMITH,
Emanuel Winternitz
For beauty, for intrinsic interest, for perspective on the composer's personality, for subtleties of phrasing, shading, emphasis indicated in the autograph but suppressed in the printed score, the mss. of musical composition are fascinating documents which repay close study in many different ways. This 2-volume work reprints facsimiles of mss. by virtually every major composer, and many minor figures—196 examples in all. A full text points out what can be learned from mss., analyzes each sample. Index. Bibliography. 18 figures. 196 plates. Total of 170pp. of text. 7⅞ x 10¾.

21312-9, 21313-7 Two volume set, paperbound $5.00

J. S. BACH,
Albert Schweitzer
One of the few great full-length studies of Bach's life and work, and the study upon which Schweitzer's renown as a musicologist rests. On first appearance (1911), revolutionized Bach performance. The only writer on Bach to be musicologist, performing musician, and student of history, theology and philosophy, Schweitzer contributes particularly full sections on history of German Protestant church music, theories on motivic pictorial representations in vocal music, and practical suggestions for performance. Translated by Ernest Newman. Indexes. 5 illustrations. 650 musical examples. Total of xix + 928pp. 5⅜ x 8½.   21631-4, 21632-2 Two volume set, paperbound $5.00

THE METHODS OF ETHICS, *Henry Sidgwick*
Propounding no organized system of its own, study subjects every major methodological approach to ethics to rigorous, objective analysis. Study discusses and relates ethical thought of Plato, Aristotle, Bentham, Clarke, Butler, Hobbes, Hume, Mill, Spencer, Kant, and dozens of others. Sidgwick retains conclusions from each system which follow from ethical premises, rejecting the faulty. Considered by many in the field to be among the most important treatises on ethical philosophy. Appendix. Index. xlvii + 528pp. 5⅜ x 8½.
21608-X Paperbound $3.00

TEUTONIC MYTHOLOGY, *Jakob Grimm*
A milestone in Western culture; the work which established on a modern basis the study of history of religions and comparative religions. 4-volume work assembles and interprets everything available on religious and folkloristic beliefs of Germanic people (including Scandinavians, Anglo-Saxons, etc.). Assembling material from such sources as Tacitus, surviving Old Norse and Icelandic texts, archeological remains, folktales, surviving superstitions, comparative traditions, linguistic analysis, etc. Grimm explores pagan deities, heroes, folklore of nature, religious practices, and every other area of pagan German belief. To this day, the unrivaled, definitive, exhaustive study. Translated by J. S. Stallybrass from 4th (1883) German edition. Indexes. Total of lxxvii + 1887pp. 5⅜ x 8½.
21602-0, 21603-9, 21604-7, 21605-5 Four volume set, paperbound $12.00

THE I CHING, *translated by James Legge*
Called "The Book of Changes" in English, this is one of the Five Classics edited by Confucius, basic and central to Chinese thought. Explains perhaps the most complex system of divination known, founded on the theory that all things happening at any one time have characteristic features which can be isolated and related. Significant in Oriental studies, in history of religions and philosophy, and also to Jungian psychoanalysis and other areas of modern European thought. Index. Appendixes. 6 plates. xxi + 448pp. 5⅜ x 8½.
21062-6 Paperbound $2.75

HISTORY OF ANCIENT PHILOSOPHY, *W. Windelband*
One of the clearest, most accurate comprehensive surveys of Greek and Roman philosophy. Discusses ancient philosophy in general, intellectual life in Greece in the 7th and 6th centuries B.C., Thales, Anaximander, Anaximenes, Heraclitus, the Eleatics, Empedocles, Anaxagoras, Leucippus, the Pythagoreans, the Sophists, Socrates, Democritus (20 pages), Plato (50 pages), Aristotle (70 pages), the Peripatetics, Stoics, Epicureans, Sceptics, Neo-platonists, Christian Apologists, etc. 2nd German edition translated by H. E. Cushman. xv + 393pp. 5⅜ x 8.
20357-3 Paperbound $3.00

THE PALACE OF PLEASURE, *William Painter*
Elizabethan versions of Italian and French novels from *The Decameron,* Cinthio, Straparola, Queen Margaret of Navarre, and other continental sources — the very work that provided Shakespeare and dozens of his contemporaries with many of their plots and sub-plots and, therefore, justly considered one of the most influential books in all English literature. It is also a book that any reader will still enjoy. Total of cviii + 1,224pp.
21691-8, 21692-6, 21693-4 Three volume set, paperbound $8.25

THE WONDERFUL WIZARD OF OZ, *L. F. Baum*
All the original W. W. Denslow illustrations in full color—as much a part of
"The Wizard" as Tenniel's drawings are of "Alice in Wonderland." "The
Wizard" is still America's best-loved fairy tale, in which, as the author expresses
it, "The wonderment and joy are retained and the heartaches and nightmares
left out." Now today's young readers can enjoy every word and wonderful picture of the original book. New introduction by Martin Gardner. A Baum
bibliography. 23 full-page color plates. viii + 268pp. 5⅜ x 8.
20691-2 Paperbound $1.95

THE MARVELOUS LAND OF OZ, *L. F. Baum*
This is the equally enchanting sequel to the "Wizard," continuing the adventures of the Scarecrow and the Tin Woodman. The hero this time is a little
boy named Tip, and all the delightful Oz magic is still present. This is the
Oz book with the Animated Saw-Horse, the Woggle-Bug, and Jack Pumpkin-
head. All the original John R. Neill illustrations, 10 in full color. 287pp.
5⅜ x 8.
20692-0 Paperbound $1.75

ALICE'S ADVENTURES UNDER GROUND, *Lewis Carroll*
The original *Alice in Wonderland*, hand-lettered and illustrated by Carroll
himself, and originally presented as a Christmas gift to a child-friend. Adults
as well as children will enjoy this charming volume, reproduced faithfully
in this Dover edition. While the story is essentially the same, there are slight
changes, and Carroll's spritely drawings present an intriguing alternative to
the famous Tenniel illustrations. One of the most popular books in Dover's
catalogue. Introduction by Martin Gardner. 38 illustrations. 128pp. 5⅜ x 8½.
21482-6 Paperbound $1.00

THE NURSERY "ALICE," *Lewis Carroll*
While most of us consider *Alice in Wonderland* a story for children of all
ages, Carroll himself felt it was beyond younger children. He therefore provided this simplified version, illustrated with the famous Tenniel drawings
enlarged and colored in delicate tints, for children aged "from Nought to
Five." Dover's edition of this now rare classic is a faithful copy of the 1889
printing, including 20 illustrations by Tenniel, and front and back covers
reproduced in full color. Introduction by Martin Gardner. xxiii + 67pp.
6⅛ x 9¼.
21610-1 Paperbound $1.75

THE STORY OF KING ARTHUR AND HIS KNIGHTS, *Howard Pyle*
A fast-paced, exciting retelling of the best known Arthurian legends for young
readers by one of America's best story tellers and illustrators. The sword
Excalibur, wooing of Guinevere, Merlin and his downfall, adventures of Sir
Pellias and Gawaine, and others. The pen and ink illustrations are vividly
imagined and wonderfully drawn. 41 illustrations. xviii + 313pp. 6⅛ x 9¼.
21445-1 Paperbound $2.00

Prices subject to change without notice.

Available at your book dealer or write for free catalogue to Dept. Adsci,
Dover Publications, Inc., 180 Varick St., N.Y., N.Y. 10014. Dover publishes more
than 150 books each year on science, elementary and advanced mathematics,
biology, music, art, literary history, social sciences and other areas.

PB-36057
Lot 5-19T (3 map)
Return with BK.